This book treats turbulence from the point of view of dynamical systems. The exposition centres around a number of important simplified models for turbulent behaviour in systems ranging from fluid motion (classical turbulence) to chemical reactions and interfaces in disordered systems.

In recent decades, turbulence has evolved into a very active field of theoretical physics. The origin of this development is the approach to turbulence from the point of view of deterministic dynamical systems, and in this book it is shown how concepts developed for low-dimensional chaotic systems, can be applied to turbulent states. Thus, the modern theory of fractals and multifractals now plays a major role in turbulence research, and turbulent states are being studied as important dynamical states of matter occurring also in systems outside the realm of hydrodynamics, i.e. chemical reactions or front propagation. The presentation relies heavily on simplified models of turbulent behaviour, notably shell models, coupled map lattices, amplitude equations and interface models, and the focus is primarily on fundamental concepts such as the differences between large and small systems, the nature of correlations and the origin of fractals and of scaling behaviour.

This book will be of use to graduate students and researchers who are interested in turbulence.

Cambridge Nonlinear Science Series 7

Dynamical Systems Approach to Turbulence

TITLES IN PRINT IN THIS SERIES

Dynamical Systems Approach to Turbulence

Tomas Bohr, Mogens H. Jensen,
The Niels Bohr Institute, University of Copenhagen

Giovanni Paladin and **Angelo Vulpiani**
University of L'Aquila   University of Rome 'La Sapienza'

CAMBRIDGE
UNIVERSITY PRESS

CAMBRIDGE UNIVERSITY PRESS
Cambridge, New York, Melbourne, Madrid, Cape Town, Singapore, São Paulo

Cambridge University Press
The Edinburgh Building, Cambridge CB2 2RU, UK

Published in the United States of America by Cambridge University Press, New York

www.cambridge.org
Information on this title: www.cambridge.org/9780521475143

First published 1998
This digitally printed first paperback version 2005

A catalogue record for this publication is available from the British Library

Library of Congress Cataloguing in Publication data
Dynamical systems approach to turbulence / Tomas Bohr . . . [et al.].
    p.   cm. – (Cambridge nonlinear science series)
Includes bibliographical references and index.
ISBN 0 521 47514 7
1. Turbulence–Mathematical models.   Bohr, Tomas, 1953–   .
II. Series.
TA357.T87D88   1998
620.1´064–dc21   97-10933   CIP

ISBN-13 978-0-521-47514-3 hardback
ISBN-10 0-521-47514-7 hardback

ISBN-13 978-0-521-01794-7 paperback
ISBN-10 0-521-01794-7 paperback

Contents

To the memory of Giovanni

During the first two weeks of June 1996, the four authors of this book were together in Copenhagen to complete the work. Two weeks later, on 29 June, Giovanni Paladin tragically died in a mountain climbing accident on Gran Sasso near L'Aquila in Italy. This was a terrible shock, and we, and many others, feel the incomprehensible loss of a dear colleague and close friend.

Giovanni was born in 1958 in Trieste. He received his education at the University of Rome and wrote his Master's thesis on the subject of 'Dynamical critical phenomena' under the supervision of Luca Peliti (1981). After the thesis he became interested in the theory of dynamical systems and chaotic phenomena, which led to the well-known work on multifractals in collaboration with Roberto Benzi, Giorgio Parisi and Angelo Vulpiani. He continued to develop new ideas within this field now in a very close collaboration with Angelo Vulpiani. From 1982 they co-authored more than 60 papers, among which is a widely cited review on multifractals and a book on products of random matrices. After his Ph.D. at the University of Rome 'La Sapienza' (1987), Giovanni started his 'travelling years'. First he went for a year to the Ecole Normale Superieure, Paris (1987–88) and visited the University of Chicago (1988); then, finally, he spent a year at The Niels Bohr Institute and Nordita in Copenhagen (1989–90). During this time he established links with many groups and individuals, and he kept returning to these places, where he was a treasured guest. While in Copenhagen, Giovanni and Angelo (who remained in close contact) started a collaboration on shell models for turbulence with Mogens Jensen, and this work forms an important part of our book. In 1990 Giovanni became Assistant Professor at the University of L'Aquila, and then in 1992 Associate Professor, again at the University of L'Aquila.

Giovanni was a very gifted and creative scientist. He mastered the techniques of statistical mechanics and dynamical systems to perfection and was able to draw analogies between the various subjects he worked on in a very elegant and productive way. Beyond his technical contributions in research he was also a very good teacher, with a keen interest in the education of young scientists. Six students prepared their Master's thesis under his supervision and two graduate students were working for the Ph.D. In addition Giovanni took great interest in scientific popularization, writing some contributions for encyclopaedias and taking part in many conferences for students and high-school teachers.

He was an extremely sweet and gentle person. Wherever he went, he made close friends immediately and he maintained personal contacts very carefully. One could always speak to Giovanni about anything, and he would listen and answer in his characteristic gentle and original way, always completely honest and deeply absorbed in science, literature, art and music (in particular, Mozart).

A funny aspect of Giovanni was his systematic absent-mindedness. He was basically able to lose anything: keys, books, papers, files, documents, money and so on. On the other hand he was lucky enough to find almost all the lost objects again. Among his friends and collaborators there was a sort of unwritten rule: *it was strictly forbidden to leave the only copy of any important thing with Giovanni.*

Giovanni loved the mountains and went to ski or climb as often as he could – almost every week. There, too, he had a large group of close friends, with whom he shared many adventures. One of his last outings was a long tour climbing and skiing down the volcanoes of northern Patagonia (October 1995). The practice of mountaineering added a new dimension to the purely intellectual side of his life, making it richer and more diverse.

All the many friends, colleagues and students of Giovanni miss him sorely.

We wish to dedicate this book to the memory of Giovanni Paladin.

Angelo, Mogens and Tomas

Preface

During the last few decades the theory of dynamical systems has experienced extremely rapid progress. In the 1980s we became attracted to this field by the enthusiasm and new insights generated by the works of Lorenz, Chirikov, Hénon and Heiles, Ruelle and Takens, Feigenbaum, and Libchaber, to cite just some of the most famous. It was surprising that concepts from low-dimensional dynamical systems, even seemingly abstract mathematical devices such as iterated maps, could be used to describe systems as complicated as unsteady fluids. For fluids under severe constraints, i.e. the experiments on Rayleigh–Bénard convection in small cells, the success was undisputed. For the understanding of turbulence, however, the success has been more limited. Turbulence, which implies spatial as well as temporal disorder, cannot be reduced to a low-dimensional system, and thus a large part of the theory of dynamical systems, in particular regarding bifurcation structures and symbolic dynamics, is basically inapplicable.

The aim of this book is to show that there are dynamical systems that are much simpler than the Navier–Stokes equations but that can still have turbulent states and for which many concepts developed in the theory of dynamical systems can be successfully applied. In this connection we advocate a broader use of the word 'turbulence', to be made precise in the first chapter, which emphasizes the common properties of a wide range of natural phenomena. Even for the case of fully developed hydrodynamical turbulence, which contains an extreme range of relevant length scales, it is possible, by using a limited number of ordinary differential equations (the so-called shell models), to reproduce a surprising variety of relevant features.

This book reflects to a large extent our own scientific interests during the last decade or so. These interests were strongly influenced by many outstanding colleagues. In particular we would like to express our thanks for guidance, inspiration and encouragement to P. Bak, G. Grinstein, L. P. Kadanoff, A. Libchaber, G. Parisi, I. Procaccia and D. Rand.

We are also deeply grateful to P. Alstrøm, E. Aurell, R. Benzi, L. Biferale, G. Boffetta, E. Bosch, O. B. Christensen, C. Conrado, A. Crisanti, P. Cvitanović, M. Falcioni, Y. He, J. M. Houlrik, G. Huber, C. Jayaprakash, J. Krug, R. Lima, R. Livi, J. Lundbek Hansen, V. L'Vov, D. Mukamel, E. Ott, A. W. Pedersen, A. Pikovsky, A. Provenzale, S. Ruffo, K. Sneppen, S. Vaienti, M. Vergassola, W. van de Water, I. Webman, R. Zeitak and Y.-C. Zhang, who participated in obtaining many of the results discussed in this book.

Finally we thank E. Aurell, M. Falcioni, J. Krug, Y. Pomeau, and N. Schörghofer for valuable comments on the first draft of the manuscript.

Introduction

The traditional description of turbulence (as summarized in the monograph by Monin and Yaglom [1971, 1975]) employs statistical methods, truncation schemes in the form of approximate closure theories and phenomenological models (e.g. Kolmogorov's theories of 1941 and 1962). The complementary point of view guiding our description of turbulence is to regard the Navier–Stokes equations, or other partial differential equations describing turbulent systems, as a deterministic dynamical system and to regard the turbulence as a manifestation of deterministic chaos.

In the case of fully developed turbulence the direct simulation of the Navier–Stokes equations is prohibitively difficult owing to the large range of relevant length scales. It is thus important to study simplified models, and a large part of this book is devoted to the introduction and investigation of such models. We shall give an introduction to the dynamical systems approach to turbulence and show the applicability of methods borrowed from dynamical systems to a wide class of dynamical states in spatially extended systems, for which we shall use the general term turbulence.

It is important to note that the dynamical models employed to describe turbulent states are not low-dimensional. In flows with high Reynolds numbers or in chaotic systems of large spatial extent, the number of relevant degrees of freedom is very large, and our primary interest is to explore properties that are well defined in the 'thermodynamic limit', where the system size (or Reynolds number) becomes very large.

Some of the main concepts and characteristics of this approach to turbulence are

1. The use of quantifiers of chaotic dynamics, such as Lyapunov exponents, entropies and dimensions. These concepts provide a very precise determination of e.g. the onset of turbulence and an understanding of the interplay between temporal chaos and spatial scales. In addition they provide information about the propagation of information and disturbances.

2. The use of the geometrical description of (multi)fractal objects, i.e. concepts borrowed from the thermodynamical formalism for dynamical systems. These concepts provide a framework for understanding subtle statistical properties, such as intermittency, and allow detailed comparison with experimental data.

3. Direct simulation of simplified dynamical models of turbulence, such as shell models, coupled maps or amplitude equations. Here it is important that we view these systems as non-linear dynamical systems in their full complexity, for which the more traditional study of stationary and periodic states and their stability is only the first step.

4. The fact that we deal with systems with a large number of degrees of freedom means that many concepts from statistical mechanics, and in particular from the theory of critical behaviour, become relevant. In fact, our goal is to understand the 'thermodynamic limit' of deterministically chaotic systems – as opposed to noisy systems (i.e. Langevin equations).

The layout of the book is as follows:

In **Chapter 1** we introduce various concepts and models to be described in detail later. After reading that chapter, the reader will know what we mean by turbulence and what the rest of the book is about. This should allow the reader to pick out which parts to read next.

In **Chapter 2** we go through the phenomenology of fully developed hydrodynamic turbulence, centred around the ideas of Kolmogorov. We introduce e.g. structure functions, scale invariance and the multifractal description of turbulence, and this should give a good background for the more detailed expositions in chapters 3, 6, 8 and 9.

Chapter 3 is mainly concerned with shell models for fully developed turbulence, which are introduced and treated in detail. The idea is to capture basic ingredients, such as conservation laws and the energy cascade in 3D turbulence, by a chaotic dynamical system with a reasonable number of equations. By studying a system of around 20 coupled differential equations one can obtain results on issues such as the scaling exponents of the structure functions, intermittency corrections and the probability distribution of velocity gradients, in agreement with experiments.

In **Chapter 4** coupled map lattices are introduced and selected topics are discussed in more detail. One main aim is to show that these models are helpful in understanding the interplay between chaos and turbulence, i.e. what happens when a chaotic system becomes larger, and to understand the influence of conservation laws and symmetry breaking. Spatio-temporal intermittency is treated in detail as an example of a deterministic 'phase transition'. and examples are shown of how to model aspects of turbulent Rayleigh–Bénard convection.

Chapter 5 is concerned with turbulence in amplitude equations, i.e. equations derived by expansions around an instability. It is centred around the complex Ginzburg–Landau equation, which models a rich variety of physical, chemical and biological systems in which a coherent periodic state exists. The main issues are the interplay between periodic states, spirals, turbulent states and so-called vortex–glass states, disordered states with many spirals. The chapter ends with a discussion of the Kuramoto–Sivashinsky equation and generalizations thereof.

In **Chapter 6** we discuss predictability in systems with many degrees of freedom. In chaotic systems the distance between two initially close trajectories diverges exponentially, which implies that prediction is feasible only up to a predictability time inversely proportional to the maximum Lyapunov exponent. This simple scenario fails in realistic situations, where many characteristic times are involved, especially if non-infinitesimal perturbations are applied. A relevant example is weather forecasting, where we focus on the prediction of the large-scale motion.

Chapter 7 deals with the dynamics of interfaces and surfaces. Here we are mostly dealing with noise-driven dynamical systems, which generate rough, scale-invariant fronts modelling e.g. the motion of a viscous fluid in a porous medium. Other fronts (e.g. flame fronts) can be modelled by deterministic equations, notably the Kuramoto–Sivashinsky equation, and one important issue is in what sense the deterministic and the noise-driven systems are alike. A large part of the chapter is devoted to 'extremal' models with non-local interactions, relevant for strongly pinned systems. The relation to self-organized criticality and to directed percolation is discussed in detail.

Lagrangian chaos is the subject of **Chapter 8**. The motion of a fluid particle is described by a low-dimensional dynamical system and can be chaotic even in the absence of (Eulerian) chaos in the velocity field. This fact is of great importance to mixing, transport and diffusion in fluids. The use of techniques of dynamical systems allows one to determine e.g. the scaling range of the Batchelor law for passive scalar fluctuations at small scales or the connection between variations of the effective Lyapunov exponent and the strong spatial fluctuations of the magnetic field in the dynamo problem.

This leads naturally to the problem of chaotic diffusion which is treated in **Chapter 9**. The main issue is: what does the structure of the velocity field tell us about the diffusion of fluid particles, e.g. whether it is anomalous and, if it is normal, how to compute the diffusion coefficient. In the last part we show that velocity fields generated by models from chapters 5 and 7 can give rise to anomalous diffusion.

Concepts from the theory of dynamical systems are used throughout the book, and we assume that the reader has some familiarity with these notions. For completeness we do, however, review some of the basic features, such as the calculation of Lyapunov exponents and the theory of the Hopf bifurcation in the **Appendices**. These also contain introductions to the theory of convective instabilities, linear front propagation, multifractality and directed percolation.

The aim of this book is to show that concepts and techniques developed in the context of chaotic dynamical systems play a key role in the understanding of turbulent states in spatially extended systems. We hope that we have managed to convey the richness and ubiquity of such turbulent states as well as the basic features which bind them together.

Chapter 1

Turbulence and dynamical systems

1.1   **What do we mean by turbulence?**

One of the main goals in the development of the theory of chaotic dynamical systems has been to make progress in the understanding of turbulence. The attempts to relate turbulence to chaotic motion got strong impetus from the celebrated paper by Ruelle and Takens [1971]. The idea was that, although a hydrodynamical system has a very large number of degrees of freedom, technically speaking infinitely many, most of them will be inactive at the onset of turbulence, leaving only a few interacting active modes, which nevertheless can generate a complex and unpredictable evolution.

The term *deterministic chaos* was thus introduced to indicate a strong sensitivity to initial conditions, that is, the exponential separation of nearby trajectories in phase space. In dissipative systems, when the temporal evolution is bounded in a limited region of the phase space, a small volume should fold, after an initial stretching due to the strong sensitivity on the initial state. Chaos arises from the competitive effect of repeated stretching and folding. Such a process produces very complex and irregular structures in phase space. The asymptotic motion evolves on a foliated structure called a strange attractor, usually with non-integer Hausdorff dimension. In other words, strange attractors are often fractals. The value of the Hausdorff dimension D_F provides some very important physical information since the evolution of the system can, in principle, be described by at most $2[D_F] + 1$ coupled ordinary differential equations, where [ ] denotes the integer part [Sauer et al. 1991].

The first and natural measure of the degree of chaoticity is the rate of the exponential increase of a small uncertainty in the initial state of the system. This is the maximum Lyapunov exponent and when it is positive, one generally says that the system is chaotic. That is the definition adopted in this book, although there are other possibilities, see, for example, Eckmann and Ruelle [1985].

In the field of deterministic chaos, one can, roughly speaking, distinguish two main classes of problems:

(a) the study of a dynamical system for which the evolution law is given;
(b) the extraction of qualitative and quantitative information from a time-signal obtained in an experiment.

The classical hydrodynamical experiments providing evidence of chaotic evolution are typically carried out in very small containers [Maurer and Libchaber 1979]. Although some systems prepared in a suitable way seem effectively to be ruled by low-dimensional dynamics, it is usually very hard to determine the appropriate reduced equations or even the correct dimensionality of phase space. Following the work of Packard et al. [1980] and Takens [1981], the so-called embedding techniques have been developed to solve this problem. Their use (e.g. via delay coordinates) allows for the determination of dimensions, Kolmogorov–Sinai entropy and Lyapunov exponents from a time series of only one variable such as the temperature at one point in a container.

In systems with a large phase space dimension, there are, however, severe limitations [Smith 1988, Eckmann and Ruelle 1992] in the use of embedding techniques for treatment and modelling of experimental data. In particular, for systems of large spatial extent it has proved very difficult to apply these methods. In large chaotic systems distant parts become uncorrelated. Thus, increasing the volume of the system, one adds more active degrees of freedom, and it is practically impossible to determine Lyapunov exponents or fractal dimensions from data taken at one point only. This does not imply that these quantities become ill-defined in large systems. Lyapunov exponents are, as always, defined in terms of the entire phase space, and in a large system one thus includes simultaneously the effects of many, spatially separated parts.

In large systems, just as in small ones, the existence of a positive Lyapunov exponent is the standard criterion for chaos. However, a chaotic system can be coherent (i.e. spatially ordered) or incoherent (spatially disordered). To distinguish, we shall use the word *turbulent* for systems that are chaotic and spatially incoherent. Since systems that are large enough will typically be incoherent we can simply define turbulence as *chaotic dynamics in a spatially extended system*.

Here the word *chaotic* means that the dynamics has positive Lyapunov exponents (at least one). The word *extended* has a slightly less precise meaning. The idea is that the system is so large that its size is unimportant. The size only determines overall (extensive) quantities, while changes in it do not affect the qualitative behaviour. This implies, for instance, that the number of positive Lyapunov exponents is large: it will scale with the system size, most likely in direct proportion to the volume. Attractor dimensions will scale in the same way. Thus, turbulence is *not* a low-dimensional chaotic phenomenon. Note that we shall always (at least implicitly) consider systems in the presence of arbitrarily small but finite noise, in order to get rid of very special, experimentally unobservable states.

Turbulent phenomena are, aside from the existence of many degrees of freedom, characterized by many time scales, somehow related to the different scales of motion. For instance, in fully developed hydrodynamical turbulence, which we discuss at length in this book, these times are the turn-over times of eddies of different length scales. In contrast to systems modelled in terms of random processes, such as Langevin equations, it is not possible to separate the degrees of freedom into two classes, corresponding to the slow and the fast modes. However, it is reasonable to expect that the degrees of freedom are likely to be organized in a hierarchical way so that the determination of the structure of such a hierarchy in the framework of the theory of dynamical systems could be a way to approach the turbulence puzzle.

Our definition of turbulence (as spatio-temporal chaos) suffers from a serious problem: it is very hard to decide whether an experimentally observed state is indeed turbulent. The determination of Lyapunov exponents requires very precise control of initial conditions well beyond present experimental capabilities. It is of great importance to find equivalent criteria for chaotic behaviour, for instance by means of correlation functions or the behaviour of advected tracer particles.

1.2   Examples of turbulent phenomena

We shall discuss the use of dynamical system techniques for the study of turbulence in a variety of different physical systems and mathematical models. In this section, we want to give some introductory examples, define paradigmatic models, which will be discussed in the following chapters, and show how turbulence occurs in typical systems.

1.2.1   Fluids

The motion of incompressible fluids is well known to produce turbulence. In typical open flow configurations, like pipe flows or shear flows, turbulence can be described by the Navier–Stokes equations discussed in detail in the next chapter. The control parameter for such flows is the Reynolds number

$$Re = \frac{VL}{v},$$

where V is a characteristic velocity, L is a characteristic length and v is the kinematic viscosity. At low Reynolds number the fluid motion is laminar. On the other hand at very large Re there is highly chaotic and irregular behaviour (fully developed turbulence). An idea of the increasing complexity is given in Fig. 1.1, which shows the flow around a cylinder. For moderate Re one has only temporal chaos and spatially coherent structures, while at very large Re the chaotic behaviour involves fluctuations on a wide spectrum of spatial and temporal scales.

The transition to turbulence can take place in a very non-uniform way, for example through the appearance of turbulent spots. Figure 1.2 shows a turbulent spot generated in an unstable boundary layer [Cantwell et al. 1978]. Inside the growing spot, the motion is fully turbulent; outside it is laminar and retains the characteristic boomerang shape during the growth. A similar phenomenon occurs in pipe flows. Here a 'slug' (first described by Reynolds [1883]) appears

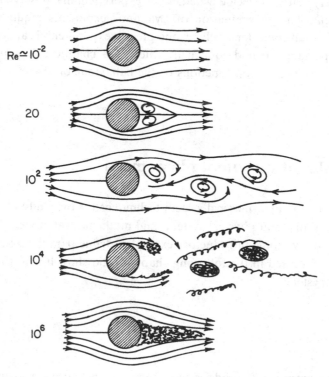

Figure 1.1 Schematic picture of a fluid around a cylinder at different Reynolds numbers.

and grows in size while being carried downstream. The growth velocity decreases at decreasing Reynolds number and vanishes at $Re \approx 2300$ [Tritton 1988]. We shall look at such localized turbulent states in chapters 4 and 5.

When temperature variations are important, we must extend the Navier–Stokes equations. A classical example is Rayleigh–Bénard convection, where a fluid in a closed container is heated from below. For a recent thorough review, see Cross and Hohenberg [1993]. Here the control parameter is the Rayleigh number

$$Ra = \frac{\alpha g \Delta T L^3}{\kappa \nu},$$

where α is the thermal expansion coefficient, κ is the heat diffusion coefficient, g is the gravitational acceleration, ΔT is the temperature difference from bottom to top and L is the size of the box. Figure 1.3(a) shows a turbulent Rayleigh–Bénard system, where 'plumes' shoot up from an unstable boundary layer [Sparrow et al. 1970]. The difference between low-dimensional chaos and turbulence is particularly clear in the experiments on a large box of helium carried out by Libchaber and co-workers [Heslot et al. 1987, Castaing et al. 1989]. In these experiments an enormous range of Rayleigh numbers was covered (up to 10^{15}). In Fig. 1.3(b) the Nusselt number N (the heat transfer through the system, normalized by the heat transfer due to pure conductance) is plotted as function of the Rayleigh number. At Rayleigh numbers of the order of 10^5 the spatial

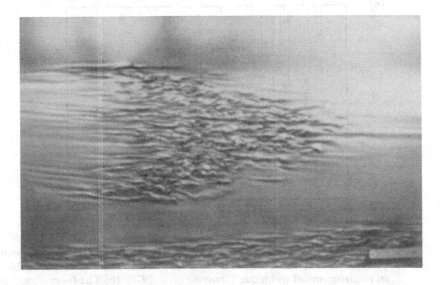

Figure 1.2 Turbulent 'Emmons' spot, as seen from above. On a flat plate, the transition from a laminar to a turbulent boundary layer takes place via the formation of localized spots of turbulent fluid, which move downstream while approximately maintaining their characteristic shape [Cantwell et al. 1978].

coherence in the container is lost and there is a transition from temporal chaos to fully developed turbulence. The further transition between 'weak' and 'strong' turbulence, seen in the slightly different power law for the Nusselt number, is probably a finite size effect – the size of the plumes becoming of the order of the system size; but within each of these two phases there is a large interval of system sizes (or equivalently, of Rayleigh numbers) where the system can be described as extended, since the qualitative behaviour remains unchanged under the changes of scale induced by the Rayleigh number. In systems where the coefficients defining the Rayleigh number are strongly temperature dependent one can find very different turbulent patterns. Figure 1.4 shows the intensity of light transmitted through a thin layer of carbon dioxide enclosed between

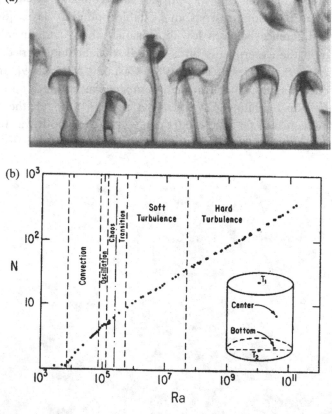

Figure 1.3 (a) Turbulence in the Rayleigh–Bénard convection is generated by 'plumes' that rise from the heated bottom surface. These are made visible by an electrochemical technique [Sparrow et al. 1970]. (b) The (normalized) heat transfer N through a convecting helium cell as a function of the Rayleigh number. The inset shows a sketch of the experimental cell, whose height and diameter are 8.7 cm. Coherence is lost at $Ra \approx 10^5$ and the system enters a turbulent state [Heslot et al. 1987].

two glass plates and heated from below [Morris et al. 1993]. The state shown here is strongly time dependent and looks turbulent. Very similar patterns have been obtained from a numerical solution of a truncation of the Navier–Stokes equations coupled to the temperature field in the so-called Boussinesq approximation [Decker et al. 1994]. For this system one might envisage a simplified dynamics in which the basic variables, instead of being the entire field configuration, are taken as the defects in the roll pattern, i.e. grain boundaries, dislocations and spiral centres.

1.2.2 Chemical turbulence

Chemical reactions provide beautiful examples of nonlinear dynamics. The non-linearities arise from the interactions between different species or within a given species [Nicolis and Prigogine 1977]. The Belousov–Zhabotinsky reaction [Zaikin and Zhabotinsky 1970] is an oscillatory reaction and will, without any external forcing, show time dependence in the concentration, in the form of limit cycles or even strange attractors [Roux et al. 1983]. These phenomena have been carefully studied in stirred reactors, where the concentrations are almost uniform over the entire volume. In the absence of stirring, inhomogeneities appear and play a crucial role. This can lead to stationary patterns (Turing structures) or non-stationary, disordered patterns that might qualify as turbulent. An example from Ouyang and Swinney [1991] is shown in Fig. 1.5 and is

Figure 1.4 'Spiral defect chaos' showing rolls, spirals and grain boundaries in a thin, convecting layer of CO_2 [Morris et al. 1993].

strongly reminiscent of a melted crystal. Very similar patterns can be observed in parametrically excited surface waves [Gollub 1991]. In oscillatory and excitable chemical systems one sees the formation of spiral waves. This happens naturally in systems that can perform a coherent cycle – such as lasers, growth of microorganisms and surface catalyses, to name a few. As long as inhomogeneities

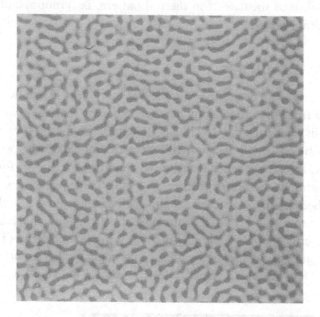

Figure 1.5 Turbulent state for a chlorite–iodide–malonic acid reaction in an acidic aqueous solution inside a gel [Ouyang and Swinney 1991].

Figure 1.6 Transient spiral state in the Belousov–Zhabotinsky reaction with ruthenium as catalyst. Courtesy of Flemming Jensen.

are suppressed by stirring, the coherent state can remain stable, but without stirring, the local variations in concentration give rise to variations in the local frequency of oscillations, which can trigger spiral waves. Figure 1.6 shows the characteristic spiral waves which develop in such a system. Close to the onset of a periodic oscillation (global Hopf bifurcation) such behaviour can be described by the complex Ginzburg–Landau amplitude equation

$$\partial_t A = \mu A - (1 + i\alpha)|A|^2 A + (1 + i\beta)\nabla^2 A \tag{1.1}$$

for a complex order parameter A, which represents the local strength of the (temporal) Fourier component corresponding to the frequency of the oscillating state [Cross and Hohenberg 1993]. We shall discuss the turbulence in this model in chapter 5.

1.2.3 Flame fronts

Combustion is another field with a rich variety of turbulent states. A good example [Markstein 1951] is shown in Fig. 1.7. Here the burning fronts of various hydrocarbon–air–nitrogen mixtures are shown. The gas is lead into a glass tube at Reynolds numbers of about 500 such that the flame front is almost steady. A theoretical model, known as the Kuramoto–Sivashinsky equation [Nepomnyashi 1974, Kuramoto 1984, Sivashinsky 1977], has turned out to be one of the simplest (perhaps the simplest) partial differential equation (PDE) that generates turbulent states. It has the form of a nonlinear reaction diffusion equation, but the linear term is unstable, with the instability centred at a particular length scale corresponding to the cellular flame patterns. If we denote by h the height of the flame front, the Kuramoto–Sivashinsky equation is

$$\partial_t h = -\nabla^2 h - \nabla^4 h + (\nabla h)^2. \tag{1.2}$$

This equation (in fact the equation for ∇h) occurs also as an approximation for the free surface of a fluid falling down a vertical plane [Nepomnyashchy 1974, Sivashinsky and Michelson 1980]. Further it appears generically as the phase equation for an unstable system, for example in the complex Ginzburg–Landau equation [Kuramoto 1984].

On the local scale, the Kuramoto–Sivashinsky equation is characterized by fronts which exhibit cellular structures determined roughly by the most unstable mode. On large scales, it is a propotype of a PDE which generates structures of all sizes possessing scale invariance both in space and time. If the dynamical evolution of the Kuramoto–Sivashinsky equation is initiated in a more or less 'flat' state, then, as time progresses, modes of larger and larger wavelength will be excited and one would say that the front (or the interface) 'roughens'.

It is interesting to remark that even in some PDEs where the irregular motion is determined by additive random noise (i.e. Langevin-like equations) instead of deterministic chaos, there are fronts which roughen. It is thus an important point to compare the scaling indices of the Kuramoto-Sivashinsky equation to those of other PDEs. We shall return to these issues in chapters 4, 5 and 7.

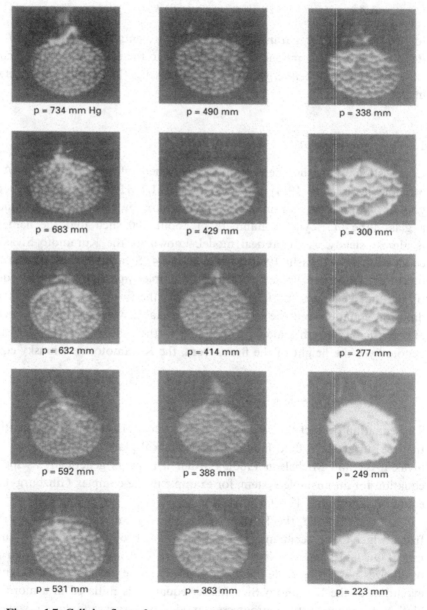

Figure 1.7 Cellular flame fronts generated in n-butane–air–nitrogen mixtures at various pressures. The diameter of the tube is 15.25 cm [Markstein 1951].

1.3 Why a dynamical system approach?

The evolution laws for the phenomena discussed in the previous section are given by partial differential equations, such as the Navier–Stokes equations for incompressible fluids. It is well known that partial differential equations can be approximated by a set of ordinary differential equations (ODEs), for example by the Galerkin truncation method. If one wants to get an accurate description it is necessary to take into account a very large number of equations. For instance, in fully developed hydrodynamical turbulence this number grows as a power of the Reynolds number, i.e. as $Re^{9/4}$. However, if the interest is focused only on the scaling laws and on the basic mechanism of the energy cascade at small scales, it is possible to reproduce the main features by a limited number of ODEs (the so-called shell models), extensively discussed in chapter 3. In the same way, one can catch some of basic features of thermal convection in terms of maps with asymmetric couplings which are able to model plumes and spots; see section 4.3.

The possibility of a dynamical system approach allows one to capture the fundamental physical mechanisms, such as the energy cascade in 3D turbulence. In particular, one can build a bridge between the traditional description in statistical terms (mainly structure functions, probability distribution functions, power spectra) and the dynamical behaviour in phase space.

1.4 Examples of dynamical systems for turbulence

We briefly discuss some examples of dynamical systems whose study is relevant in the description of turbulence in fluids and more generally in extended systems.

1.4.1 Shell models

In modelling turbulence on a computer, one has always to resort to some kind of discretization where the PDEs are transformed into a usually very large system of ordinary differential equations. The simulation of a turbulent flow is at the limit of the present capabilities of computers and, more importantly, a direct simulation does not necessarily provide any simple understanding of the phenomena.

An alternative and fruitful approach has been introduced by the works of Obukhov [1971], Gledzer [1973] and Desnyansky and Novikov [1974a,b]. The key idea is to mimic the Navier–Stokes equations by a dynamical system with N variables $u_1, u_2, \ldots, u_N$ each representing the typical magnitude of the velocity field on a certain length scale. The Fourier space is divided into N shells. Each shell k_n consists of the set of wave vectors $\mathbf{k}$ such that $k_0 2^n < |\mathbf{k}| < k_0 2^{n+1}$. The variable u_n is the velocity difference over a length $\sim k_n^{-1}$, so that there is one degree of freedom per shell. However, models with a large number of degrees of freedom per shell have also been introduced and analysed [Grossmann and Lohse 1992, Aurell et al. 1994a]. The most studied shell model has been introduced by Ohkitani and Yamada [1988] who used a set of complex variables $\mathbf{u} = (u_1, u_2, \ldots, u_N)$. The ODEs of the corresponding dynamical system are

$$\left(\frac{\mathrm{d}}{\mathrm{d}t} + vk_n^2 \right) u_n = \mathrm{i}[T_n(\mathbf{u})]^* + f_n, \tag{1.3}$$

where $T_n(\mathbf{u})$ is a bilinear expression which links neighbouring shells, e.g.

$$T_n(\mathbf{u}) = k_n u_{n+1} u_{n+2} - \frac{1}{2} k_{n-1} u_{n-1} u_{n+1} - \frac{1}{2} k_{n-2} u_{n-1} u_{n-2},$$

v is the viscosity and f_n the force acting on the nth shell.

This shell model has the same type of quadratic non-linearities as the 3D Navier–Stokes equations and the same symmetries. In the inviscid and unforced limit, $v = f_n = 0$, there are two quadratic invariants (energy and helicity) as well as conservation of the volume in the phase space. The asymptotic evolution given by (1.3) evolves on a strange attractor.

The number N of shells necessary to reproduce the behaviour observed at high Re is rather small since $N \sim \ln Re$. One thus has a dynamical system with a reasonably low number of degrees of freedom where standard methods and techniques of deterministic chaos can be used in order to relate the statistical description in terms of structure functions or intermittency of energy dissipation, and the dynamical properties, such as the spectrum of Lyapunov exponents, or the dimension of the strange attractor in phase space. Chapter 3 reviews various shell models proposed in the literature with a special emphasis on the comparison with experimental data and phenomenological results.

1.4.2 Coupled map lattices

One can go a step further and introduce discrete, coupled chaotic maps on a lattice (CMLs). This approach obviously does not give insight into the origin of turbulence: one simply takes for granted that small regions undergo chaotic motion. On the other hand, such models might give new insights about correlations among different parts and the relations among quantities like dimensions, correlation lengths and Lyapunov exponents. Coupled map lattices have become a useful tool in building up intuition and understanding about spatially extended systems; see Kaneko [1993].

As an example, we consider a two-dimensional, regular lattice indexed by (i, j), $i, j = 1, 2, \ldots, L$. Each site has attached a variable $u_{i,j}(n)$, where n is the discrete time and we assume periodic boundary conditions in i and j. In a simple diffusively CML [Kaneko 1984], the dynamics is given by

$$u_{i,j}(n+1) = (1-\epsilon)f(u_{i,j}(n)) + \frac{\epsilon}{4}\sum_{k,l} f(u_{k,l}(n)), \qquad (1.4)$$

where $\epsilon \in [0,1]$ and (k,l) are nearest neighbours of (i,j). The function $f(x)$ is some nonlinear map, such as the logistic map $f(x) = R\,x\,(1-x)$, that can sustain chaotic motion. It is well known that this map exhibits a period-doubling transition to chaos with accumulation point $R_c = 3.569\,9456\ldots$, and for $R_c < R < 4$ the attractor is either chaotic or periodic. A typical configuration, obtained by iterating (1.4) with $L = 30$ and random initial conditions, is shown in Fig. 1.8. This shows a snapshot of the system at a given time. The next timestep looks different – the system is both spatially and temporally disordered – but certain statistical features are preserved. Small R and large ϵ gives smooth spatial variations whereas large R and small ϵ makes the picture rugged with rapid spatial variations.

1.4.3 Cellular automata

In recent years a new class of dynamical systems has been studied intensively: the so-called cellular automata, where not only are space and time discrete, but

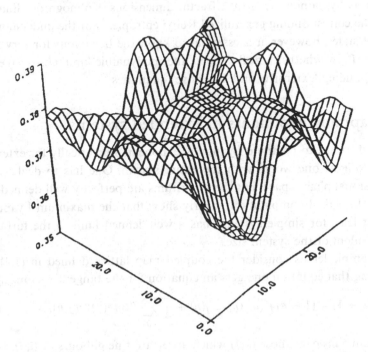

Figure 1.8 Snapshot of coupled map lattice. $R = 3.5732, \epsilon = 0.4$ and $L = 30$.

also the system variables themselves. Cellular automata have also been used to simulate the Navier–Stokes equations [Frisch et al. 1986], and to study various forms of chemical turbulence [Winfree et al. 1985, Oono and Kohmoto 1985].

For instance Oono and Kohmoto have described a form of chemical turbulence by the following rule applied to the field $A_j(t)$, which can take three possible integer values. Here j is the discrete one-dimensional space variable and t is the discrete time. First define the continuous variable

$$A'_j(t) = (1 - \alpha)A_j(t) + \frac{\alpha}{2}[A_{j+1}(t) + A_{j-1}(t)]$$ (1.5)

and then update $A_j(t)$ as

$$A_j(t + 1) = F(A'_j(t)),$$ (1.6)

where F takes on each of the three possible values in some range of its argument. This simple model leads to a surprising variety of different phases, e.g. displaying soliton-like excitations and turbulence.

1.5   Characterization of chaos in high dimensionality

For any finite-dimensional dynamical system it is possible to define in a simple way the quantities typically used for the characterization of low-dimensional chaos, such as Lyapunov exponents, fractal dimensions, Kolmogorov–Sinai entropy and the corresponding generalized Renyi entropies. For the understanding of turbulent states, however, it is essential to know the behaviour for very large system sizes, L, i.e. whether a well-defined 'thermodynamic limit' ($L \to \infty$) exists with corresponding 'extensive' and 'intensive' quantities.

1.5.1   Lyapunov exponents in extended systems

It is straightforward to compute Lyapunov exponents numerically for extended systems as long as one works with the entire system. One has to deal with a high-dimensional phase space, but the algorithms are perfectly well defined and functional. Numerical computations clearly show that the maximum Lyapunov exponent, at least for simple systems, has a well-defined limit in the turbulent state independent of the system size.

As an example let us consider the coupled map lattice defined in (1.4). By differentiating that equation one gets an equation for the tangent vectors $z_{i,j}(n)$,

$$z_{i,j}(n + 1) = (1 - \epsilon) f'(u_{i,j}(n)) z_{i,j}(n) + \frac{\epsilon}{4} \sum_{k,l} f'(u_{k,l}(n)) z_{k,l}(n),$$ (1.7)

where the sum runs over those (k, l) which are nearest neighbours of (i, j). For a d-dimensional lattice of linear size L, one can define Ld Lyapunov exponents in such a way that the sum of the first m ones characterizes the exponential growth or decay of an m-dimensional hyper-volume in phase space. Numerically the Lyapunov exponents are obtained by a simultaneous iteration of (1.4) and (1.7), as explained in detail in appendix C. The maximum Lyapunov exponent λ_1,

$$\lambda_1 = \lim_{n \to \infty} \frac{1}{n} \ln \frac{|\mathbf{z}(n)|}{|\mathbf{z}(0)|},$$

can be computed using the algorithm introduced by Benettin et al. [1980]. It does not depend on $\mathbf{z}(0)$, apart from non-generic choices. In principle λ_1 could depend on the initial condition $\mathbf{u}(0)$. However, it can be shown that almost all (in the sense of probability theory) initial conditions lead to the same value of λ_1, if the system has a unique 'natural' ergodic measure [Eckmann and Ruelle 1985]. Although there exist infinitely many different ergodic measures, experimental noise or numerical truncations are expected to select a unique 'natural' measure to which we refer in the following.

In a numerical computation, overflows can be avoided by a periodic rescaling of the tangent vector $\mathbf{z}(n)$. As a consequence of the linearity of the equation for $\mathbf{z}(n)$, such a trick does not introduce any approximation, as explained in appendix C.

The algorithm of Benettin et al. [1980] allows one to get a very precise determination of the Lyapunov exponents. One can make the time n as large as the computer can handle. It is worth stressing that there exists another, more intuitive, procedure for determining the maximum Lyapunov exponent, based directly on the separation of nearby trajectories. Assume again that we have obtained a non-transient turbulent state $u_{i,j}(0)$ at time $n = 0$. Now make a copy $u'_{i,j}(0)$ slightly different from $u_{i,j}(0)$, e.g. differing only in one lattice point, and look at the subsequent time evolution of the two replicas. In a large lattice one might expect that a small local change quickly gets ironed out, but in fact, if the system has positive Lyapunov exponents the two copies diverge exponentially in the sense that

$$|\mathbf{u}(n) - \mathbf{u}'(n)| = \sqrt{\sum_{i,j} [u_{i,j}(n) - u'_{i,j}(n)]^2} \sim e^{n\lambda}. \tag{1.8}$$

The application of this second procedure does not yield as accurate values for λ as the one based on (1.7), since (1.8) only applies as long as all differences are small, although it is interesting since it allows one to study the growth of an error in the nonlinear regime in the context of the loss of coherence in large systems and the predictability problem; see sections 4.2, 6.2 and 6.3.

1.5.2   Lyapunov spectra and dimension densities

The spectrum of Lyapunov exponents $\lambda_1 \geq \lambda_2 \cdots \geq \lambda_{L^d}$ can be determined by a generalization of the algorithms described in the previous section. To find the first m Lyapunov exponents one has to iterate m tangent vectors $z^{(1)}(0), z^{(2)}(0), ..., z^{(m)}(0)$. According to Benettin et al. [1980], one has

$$\lambda_1 + \ldots + \lambda_m = \lim_{n\to\infty} \frac{1}{n} \ln \left(\frac{|z^{(1)}(n) \times z^{(2)}(n) \times \ldots \times z^{(m)}(n)|}{|z^{(1)}(0) \times z^{(2)}(0) \times \ldots \times z^{(m)}(0)|} \right), \qquad (1.9)$$

where $\times$ indicates the external product so that $|z^{(1)}(n) \times z^{(2)}(n) \times \ldots \times z^{(m)}(n)|$ is the volume of an m-dimensional hyper-parallelepiped spanned by m tangent vectors at time n.

Also in this case, λ_i are independent of the choice of the initial tangent vectors and initial conditions if the system admits a natural probability measure. Appendix C describes the numerical procedure for the determination of the Lyapunov spectrum.

An example of a Lyapunov spectrum for the coupled map lattice (1.4) is shown in Fig. 1.9. In systems with many degrees of freedom, a central question is the existence of the 'thermodynamic' limit, $L \to \infty$, for the Lyapunov spectrum. In other words, one should understand whether λ_i versus $x = i/L^d$ converges to a well-defined density function $\Lambda(x)$ with $x \in [0, 1]$. Assuming the existence of such a limit function, the Kolmogorov–Sinai entropy H is proportional to the size of the system. Indeed, using the Pesin formula

$$H = \sum_i \lambda_i\, \theta(\lambda_i)\,, \qquad (1.10)$$

the entropy per degree of freedom is

$$h = \lim_{L\to\infty} \frac{H}{L^d} = \int_0^1 \Lambda(x)\, \theta(\Lambda)\, dx, \qquad (1.11)$$

where θ is the Heaviside function.

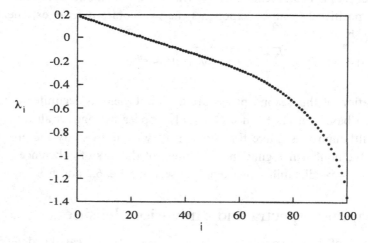

Figure 1.9 Lyapunov spectrum for (1.4), with $R = 3.75$, $\epsilon = 0.4$ and size 10×10.

For certain simple coupled map systems [Bunimovich and Sinai 1988, Gundlach and Rand 1993], this result can be proved, while in 3D hydrodynamic turbulence, it has been conjectured on the basis of heuristic arguments by Ruelle [1982]. Moreover, this behaviour was also found in some dissipative systems [Pomeau et al. 1984], symplectic systems [Livi et al. 1986, 1987, Eckmann and Wayne 1988, 1989] and in the product of random matrices [Paladin and Vulpiani 1986].

The existence of the Lyapunov density also implies that the number of positive Lyapunov exponents N_+, in a chaotic d-dimensional system, should scale as

$$N_+ \sim L^d \tag{1.12}$$

for large L.

The Kaplan–Yorke conjecture [Ott 1993],

$$D_\lambda = m + \frac{\sum_{i=1}^m \lambda_i}{|\lambda_{m+1}|}, \tag{1.13}$$

where m is the maximum integer such that $\sum_{i=1}^m \lambda_i$ is positive, relates the Lyapunov exponents to the information dimension D_I of the attractor; see appendix C. The dimension of the attractor D_λ in the phase space of an extended system will also be proportional to L^d, and thus the relevant intensive quantity is the dimension density

$$\delta_\lambda = D_\lambda / L^d. \tag{1.14}$$

The shape of the density $\Lambda(x)$, in particular the existence of a finite fraction of Lyapunov exponents close to zero, can be important in relation to the possibility of having a scaling law even in chaotic systems; see chapter 3.

Moreover, in a system with many degrees of freedom, it can happen that there are many different time scales such as the eddy turnover times in turbulent fluids. In this case, the predictability problem cannot be studied in terms of Lyapunov exponents, as in low-dimensional chaos. In fact, the growth of a non-infinitesimal perturbation is governed by a nonlinear mechanism which depends on the details of the system. It is thus possible to have a long predictability time on certain degrees of freedom that describe the physically relevant aspects (such as the large-scale motion in turbulence), even in the presence of a large maximum Lyapunov exponent [Boffetta et al. 1996]. One might observe a slow growth of a non-infinitesimal perturbation, quite similar to a power law instead of an exponential in time [Aurell et al. 1996]. These phenomena can be well understood only in terms of a dynamical system approach, as discussed in chapter 6.

1.5.3   Characterization of chaos in discrete models

In cellular automata, a priori, the concept of a Lyapunov exponent makes no
sense since it is impossible to look at the differential analogous to (1.7) when
the mapping is discrete. However, Bagnoli et al. [1992], using the concept of the
Boolean derivative and studying the local damage spreading, introduced a sort
of maximum Lyapunov exponent for 1D elementary cellular automata. A rather
natural characterization of chaos for cellular automata is the spreading velocity
of a perturbation [Wolfram 1986]. The reason is that one can regard the binary
string corresponding to the state of a one-dimensional cellular automaton as
the binary expression for a number, say between 0 and 1. In the models for
chemical turbulence such a definition would not seem to work, since the state
is supposed to be a coarse-grained description of the real state. Thus, Oono
and Kohmoto [1985] proposed to define turbulence in terms of the Shannon
entropy of the sequence of symbols for each site independently. If more than
half of these are positive, the state is considered turbulent. Oono and Kohmoto
actually approximated the entropy by that of an associated Markov process.
This approach was later refined by Oono and Yeung [1987], who introduced the
concept of 'weak' turbulence, based on a modification of the Shannon entropy.
It would seem of importance to study the relations between such methods and
the standard ones for continuous systems, since the discrete methods might be
applicable to experimental data.

1.5.4   The correlation length

The correlation length ξ of an extended system is usually defined as the inverse
decay rate of the correlation function. For instance, in system (1.4) one finds

$$\langle u_{i,j} \, u_{i',j'} \rangle - \langle u \rangle^2 \sim e^{-r/\xi}, \tag{1.15}$$

where r is the distance between the two lattice sites (i, j) and (i', j'). The lack
of coherence in large systems means that there exists a finite coherence length.
There is a simple, non-rigorous argument for this [Bohr et al. 1987], which we
shall give below. In chapter 4 we shall return to its consequences in more detail.
Consider a chaotic system where disturbances that are amplified can travel at
most with some finite speed, c, as is the case in systems with local interactions.
Now consider a small, local perturbation of size δ_0 at time $t = 0$ at point A – in
a system with even infinitesimal noise such perturbations will occur. Owing to
the positiveness of the Lyapunov exponent, the disturbance will increase as $e^{\lambda t}$
and it will spread with a velocity bounded by c. At point B a distance r away
from A no big changes thus occur before a time of the order of r/c at which
the perturbation has grown to $\delta_0 e^{\lambda r/c}$. If this quantity becomes large, say of the
order of one, one would expect A and B to become uncorrelated, and thus the

largest distance, ξ, over which coherence can be maintained should be bounded
as

$$\xi \leq \text{const } \frac{c}{\lambda}. \tag{1.16}$$

As we shall see in chapter 4, one can sometimes give a more accurate estimate
of c. In fact, it can depend on λ; but even so, (1.16) is valuable as an upper
bound.

The existence of a finite correlation length means that distant parts of the
system are effectively decoupled from one another. This implies that turbulent
motion in a large system cannot be described in terms of low-dimensional
strange attractors. Thus, the dimension of the strange attractor describing the
motion should grow linearly with the volume of the system and be infinite in
an infinite system.

In particular one could expect that the dimension density δ_λ scales as

$$\delta_\lambda \sim \frac{1}{\xi^d}. \tag{1.17}$$

The validity of this relation has been discussed by Egolf and Greenside [1994,
1995] and Bohr et al. [1994], for a particular example discussed in chapter 5.

1.5.5 Scaling invariance and chaos

One of the most striking features appearing in many extended systems is scaling
invariance. In 3D fully developed hydrodynamical turbulence the structure
function

$$S_2(r) = \langle (u(x+r) - u(x))^2 \rangle \tag{1.18}$$

for a component u of the velocity field grows as a characteristic power r^{ζ_2}
of the separation r at length scales varying in a suitable range (the so-called
inertial range). The typical temporal record and the spatial distribution of many
quantities exhibit self-similar properties.

Scaling laws are indeed typical of systems without a characteristic length,
such as a thermodynamical system at a critical point. At first glance, scaling
could thus seem inconsistent with the existence of a finite correlation length or
of a typical time and therefore with chaos since a positive maximum Lyapunov
exponent introduces a typical time $t_c \sim 1/\lambda_1$. On this basis, it is sometimes
argued that scaling appears only in situations of marginal chaos and that in
turbulent extended systems there should be a vanishing density of positive
Lyapunov exponents and a finite density of zero Lyapunov exponents. In fact,
chaotic extended systems can exhibit scaling laws in space and time. In section
3.3.3, we consider, for example, a dynamical model of turbulence where several
positive Lyapunov exponents coexist with scaling laws. At least on numerical

grounds, the positive part of the Lyapunov density seems to have a non-zero support.

Moreover, one can construct nonlinear field theories where, for the structure function, one has the scaling law $S_2(r) \sim r^{\zeta_2}$ and, at the same time, the gradient of u shows exponential decay $\langle \nabla u(x + r) \nabla u(x) \rangle \sim e^{-x/\lambda}$. Indeed, this is a generic feature of models with interface symmetry, which in many ways resembles Galilean invariance [Bohr et al. 1992]. The Kuramoto–Sivashinsky equation, or the coupled map systems with unbounded variables, studied in section 7.4.3, provide good examples where these properties can be studied in detail in the context of the dynamics of rough fronts. In the latter case *all* Lyapunov exponents can in fact be positive without disturbing the scaling behaviour.

In many cases, the possibility of building up the scaling in chaotic systems stems from the presence of many time scales related to the basic features of the system itself [Aurell et al. 1996] or to the anomalous fluctuations of the exponential rates of divergence between nearby trajectories [Crisanti et al. 1992]. These mechanisms supplement the scenario of 'systems at the edge of chaos' proposed for self-organized critical (SOC) systems by Bak et al. [1987]. Indeed, SOC models generally do not have a positive Lyapunov exponent. In fact, in some continuous models, the divergence rate between two nearby trajectories appears to be a power law [Chen et al. 1990] giving basis for speculations about the possibility of long-time forecasting in SOC. Actually, the most well known of these systems, the sand pile model [Bak et al. 1987], has a negative Lyapunov exponent [Caglioti and Loreto 1996] even if this fact does not imply that it is predictable since some randomness is present in its definition [Loreto et al. 1996].

In conclusion, scaling and chaos can coexist and their relation should be decided case by case by a careful analysis.

Chapter 2

Phenomenology of hydrodynamic turbulence

2.1   Turbulence as a statistical theory

The Navier–Stokes equations for the velocity field **v** of an incompressible fluid are

$$\partial_t \mathbf{v} + (\mathbf{v} \cdot \nabla)\mathbf{v} = -\frac{\nabla P}{\rho} + v\Delta \mathbf{v} + \mathbf{f}, \tag{2.1}$$

$$\nabla \cdot \mathbf{v} = 0,$$

where P is the pressure, ρ the density, $\mathbf{f}$ an external force and v the kinematic viscosity. These equations exhibit a transition from a laminar to a turbulent regime, with increasing values of a control parameter which measures the ratio between the nonlinear advective term and the linear viscous damping. This parameter is the Reynolds number

$$Re = V L/v,$$

where L and V are the characteristic length scale and velocity of the fluid. A direct numerical analysis of the Navier–Stokes equations at high Reynolds number is very difficult. Simple dynamical arguments (see section 2.3.5) show that the number of degrees of freedom, which should be taken into account for a good description of a turbulent flow, increases as $Re^{9/4}$. As a consequence, it is impossible to give a detailed description of the velocity field. At high Reynolds number, we can instead formulate a statistical theory of turbulence where the velocity field is considered to be a random variable $\mathbf{v}(\mathbf{x}, t)$ of the position $\mathbf{x}$ and the time t. For a nice discussion on the analogies between the statistical theory of turbulence and critical phenomena see Eyink and Goldenfeld [1994] and Frisch [1995]. We shall limit our consideration to the case of an incompressible homogeneous and isotropic fluid in a statistically steady stationary state, in such a way that spatial and temporal averages coincide with the averages taken over an *ensemble* of ideal copies of the physical system. As we are dealing with a system with many coupled degrees of freedom, it is reasonable to hope that a large class of initial probability distributions $p(\mathbf{v}, t = 0)$ of the velocity field evolve toward the same asymptote $p^*(\mathbf{v})$. In this framework the problem is to understand the statistical mechanics of turbulence by analytically determining its average properties.

2.1.1   Statistical mechanics of a perfect fluid

In a perfect fluid, i.e. $v = 0$, the evolution of the velocity field is given by the Euler equations, which conserve two quadratic functionals, the kinetic energy and the helicity:

$$E = \frac{1}{2} \langle v^2 \rangle, \qquad H = \frac{1}{2} \langle \mathbf{v} \cdot \boldsymbol{\omega} \rangle, \tag{2.2}$$

where $\boldsymbol{\omega} = \nabla \times \mathbf{v}$ is the vorticity. In this case, it is possible to construct a statistical mechanics as for a gas: by using the conservation laws and the conservation of the volume in phase space one obtains a gaussian distribution.

For simplicity, we do not take into account the helicity conservation. To be explicit, let us consider an incompressible inviscid fluid in a cube with periodic boundary conditions, so that the velocity field can be expanded in a Fourier series as

$$v_j(\mathbf{x}) = L^{-3/2} \sum_k e^{i\mathbf{k}\cdot\mathbf{x}} v_j(\mathbf{k}) \tag{2.3}$$

with $\mathbf{k} = 2\pi\mathbf{n}/L$ and $\mathbf{n} = (n_1, n_2, n_3)$ (n_i are integers). The variables $v_j(\mathbf{k})$ are not completely independent, since from the incompressibility condition and the fact that $\mathbf{v}(\mathbf{x})$ is real, it follows that

$$\mathbf{k} \cdot \mathbf{v}(\mathbf{k}) = 0 \quad \text{and} \quad \mathbf{v}(\mathbf{k}) = \mathbf{v}^*(-\mathbf{k}). \tag{2.4}$$

In any case, it is straightforward to introduce a new set of independent variables X_a, where now a labels the spatial component and the wave vector. By using an ultraviolet truncation, $\mathbf{v}(\mathbf{k}) = 0$ for $k > k_{\max}$, and by introducing (2.4) into the Euler equations one obtains a set of ordinary differential equations (ODEs) with the structure

$$\frac{dX_a}{dt} = \sum_{b,c} M_{abc} X_b X_c, \tag{2.5}$$

where $M_{abc} = M_{acb}$ and $M_{abc} + M_{bca} + M_{cab} = 0$ with $a = 1, \ldots, N \sim k_{max}^3$. We stress the fact that the ultraviolet truncation is necessary in order to avoid the infinite energy problems of the classical field theories.

It is easy to verify that (2.5) conserves the volume in the phase space as well as the energy $E = \frac{1}{2} \sum_a X_a^2$, i.e.

$$\sum_a \frac{\partial}{\partial X_a} \left(\frac{dX_a}{dt} \right) = 0 \quad \text{and} \quad \frac{dE}{dt} = 0. \tag{2.6}$$

These conservation laws are sufficient to construct the probability distribution of the variables $\{X_a\}$ [Kraichnan and Montgomery 1980]: using the ergodic hypothesis, one obtains the microcanonical probability measure

$$P_m(\{X_a\}) \sim \delta \left(\frac{1}{2} \sum_a X_a^2 - E \right). \tag{2.7}$$

It is well known that, in the limit $N \to \infty$, this is equivalent to the canonical measure

$$P_c(\{X_a\}) \sim \exp - \left(\frac{\beta}{2} \sum_a X_a^2 \right), \tag{2.8}$$

where the Lagrange multiplier β satisfies the relation

$$\langle X_a^2 \rangle = \frac{2E}{N} = \beta^{-1}. \tag{2.9}$$

In two dimensions, the helicity $H \equiv 0$, and there exists a second conserved quantity, the enstrophy

$$\Omega = \frac{1}{2} \int \omega^2 \, d^2x, \tag{2.10}$$

which is the mean square vorticity. In terms of the X variables, it can be written as

$$\Omega = \frac{1}{2} \sum k_a^2 X_a^2. \tag{2.11}$$

As a consequence, the microcanonical probability measure in 2D is

$$P_m(\{X_a\}) \sim \delta \left(\frac{1}{2} \sum_a X_a^2 - E \right) \delta \left(\frac{1}{2} \sum_a k_a^2 X_a^2 - \Omega \right) \tag{2.12}$$

and the corresponding canonical measure is

$$P_c(\{X_a\}) \sim \exp - \left(\frac{\beta_1}{2} \sum_a X_a^2 + \frac{\beta_2}{2} \sum_a k_a^2 X_a^2 \right), \tag{2.13}$$

where the Lagrange multipliers satisfy the relation

$$\langle X_a^2 \rangle = \frac{1}{\beta_1 + \beta_2 k_a^2}. \tag{2.14}$$

These results are, both in 2D and in 3D, well reproduced by numerical simulations [Kraichnan and Montgomery 1980].

2.1.2 Basic facts and ideas on fully developed turbulence

The limit $v \to 0$ (equivalent to $Re \to \infty$) is singular and cannot be interchanged with the limit $N \to \infty$. Therefore, the statistical mechanics of an inviscid fluid has a quite limited relevance to the behaviour of the Navier–Stokes equations at high Reynolds number. Recently some authors proposed the use of conservative statistical mechanics to justify some behaviours of real fluids, e.g Jupiter's red spot and the emergence of organized structures [Robert and Sommeria 1991,

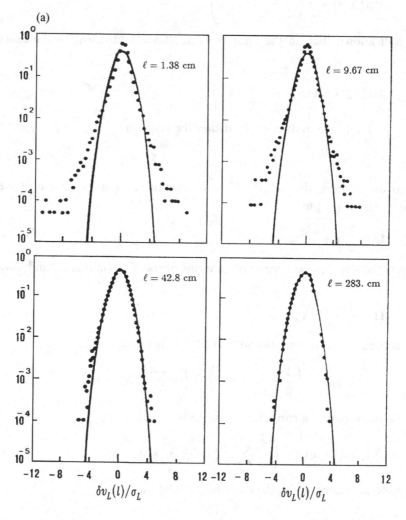

Figure 2.1 For caption see facing page.

Miller et al. 1992, Pasmanter 1994]. This approach seems rather interesting; however, its applicability is limited to some specific two-dimensional situations.

Actually, in a turbulent *viscid* fluid, the statistics of the velocity field at a fixed point is approximately gaussian. However, the one-point distribution is not sufficient to characterize the dynamics of the system. For instance, it does not allow one to distinguish the case where the field $v(x, t)$ is constant in space and given by a gaussian random variable $f(t)$ from the case where $v(x, t)$ and $v(x', t)$ are spatially uncorrelated random variables. In order to get information about the dynamics, we should consider the many-point statistics, starting with the two-point statistics. Usually, one studies the longitudinal velocity increments

$$\delta v_L(\ell) = v_1(x + r) - v_1(x),$$

where $r = (\ell, 0, 0)$. The probability distribution of $\delta v_L(\ell)$ is approximately gaussian only for large ℓ (of the order of the typical macroscopic length scale of

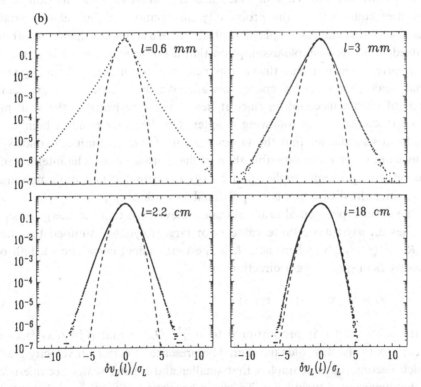

Figure 2.1 Log–linear plot of the experimental data for the probability distribution $\sigma_L P(\delta v_L(\ell))$ versus $\delta v_L(\ell)/\sigma_L$ with $\sigma_L^2 = \langle \delta v_L^2 \rangle$, at different length scales ℓ. (a) Experiments by Van Atta and Park [1972]; the full line indicates the standard gaussian. (b) Experiments by Herweijer and van de Water [1995]; the dashed line indicates the standard gaussian. Courtesy of W. van de Water, based on data described in Herweijer and van de Water [1995] and van de Water and Herweijer [1996].

the system), where the two fields are practically uncorrelated. With decreasing ℓ, the probability distribution of $\delta v_L(\ell)$ becomes more and more intermittent and skewed [Van Atta and Park 1972, Herweijer and van de Water 1995], as shown in Fig. 2.1.

For instance, hot anemometer measurements in grid turbulence at high Reynolds number show that the skewness

$$S(\ell) \equiv -\frac{\langle \delta v_L^3(\ell) \rangle}{\langle \delta v_L^2(\ell) \rangle^{3/2}} \tag{2.15}$$

and the kurtosis, or flatness,

$$K(\ell) \equiv \frac{\langle \delta v_L^4(\ell) \rangle}{\langle \delta v_L^2(\ell) \rangle^2} \tag{2.16}$$

have the values $K = 3$ and $S = 0$ at large $\ell \sim L$, indicating that the distribution at large scales is close to a gaussian. When $\ell \to 0$, they tend to $K \simeq 4$ and $S \simeq 0.4$ [Monin and Yaglom 1975] and the probability distribution of $\delta v_L(\ell)$ becomes equivalent to the probability distribution of the velocity gradients. These experimental results play a central role in a statistical theory of turbulence. Indeed, it is a reasonable assumption that a non-zero skewness is at the origin of a vortex stretching, and thus a nonlinear transfer of energy from large toward small scales, the so-called energy cascade. We shall discuss in some detail the basis of such a mechanism, since it lies at the very heart of the shell models to be discussed in the following chapter. On the other hand, a high value of the kurtosis indicates that the velocity gradients are intermittent; that is, large values are much more probable than in the gaussian case. The intermittency of the velocity gradients implies that energy is dissipated in a quite non-uniform way, since the dissipation is proportional to $v \langle (\partial v/\partial x)^2 \rangle$.

The other experimental result of major relevance is that the energy dissipation reaches an asymptotic finite value ϵ for large Reynolds number, here defined as $Re = \sqrt{\langle v^2 \rangle}\, L/v$, where here L is the longitudinal correlation length of the velocity field (say along a direction 1):

$$L = \int_0^\infty d\ell \; \langle v_1(\mathbf{x} + \mathbf{r}) v_1(\mathbf{x}) \rangle / \langle v_1^2 \rangle. \tag{2.17}$$

It turns out that ϵ is proportional to $\langle v^2 \rangle^{3/2}/L$, so that $\langle (\partial v_1/\partial x_1)^2 \rangle$ diverges linearly with the Reynolds number. The presence of very high velocity gradients which become singular implies that smaller and smaller scales become relevant in the dynamics of turbulence. To give a heuristic justification of the singularity of the velocity gradients when $Re \to \infty$, it is convenient to write the Navier–Stokes equations in terms of the vorticity $\omega = \text{rot } \mathbf{v}$ since one can easily show that $\epsilon = \frac{1}{2} v \langle \omega^2 \rangle$. One thus has

$$\frac{\partial \omega}{\partial t} + (\mathbf{v} \cdot \nabla) \omega = (\omega \cdot \nabla)\mathbf{v} + v\Delta\omega \tag{2.18}$$

and the equation for the enstrophy density $\Omega = \frac{1}{2}\langle\omega^2\rangle$ is

$$\frac{1}{2}\frac{\partial\langle\omega^2\rangle}{\partial t} = \sum_{i,j}\left\langle\omega_i\,\omega_j\frac{\partial v_i}{\partial x_j}\right\rangle - v\left\langle\left(\frac{\partial\omega_i}{\partial x_j}\right)^2\right\rangle. \tag{2.19}$$

The term $\langle\omega_i\,\omega_j\frac{\partial v_i}{\partial x_j}\rangle$ is responsible for the growth of the enstrophy up to a value proportional to $Re \sim v^{-1}$. For instance, consider a vorticity field directed along the x_1 axis, so that $\omega_i\,\omega_j\frac{\partial v_i}{\partial x_j} = \omega_1^2\frac{\partial v_1}{\partial x_1}$. This term produces an increase of $\langle\omega^2\rangle$ if there is a positive correlation between ω_1^2 and $\frac{\partial v_1}{\partial x_1}$. The term $\frac{\partial v_1}{\partial x_1}$ represents the stretching along x_1 of a small volume containing the vorticity ω_1. One thus expects that the stretching of a small volume of vorticity ω_1 along the same direction of the vorticity vector should produce a growth of ω^2, in order to ensure the conservation of the angular momentum. This rough argument illustrates the main mechanism of the energy transfer from large toward small scales.

We can also link the stretching of the vorticity lines to the experimental results for the skewness. After some lengthy algebraic manipulations [Monin and Yaglom 1975, Lesieur 1990], (2.19), for $v = 0$, can be rewritten as

$$\frac{\partial\langle\omega^2\rangle}{\partial t} = \frac{7}{6}\frac{S(0)}{\sqrt{15}}\langle\omega^2\rangle^{3/2}, \tag{2.20}$$

where $S(0) = \lim_{\ell\to 0}S(\ell) \simeq 0.4$. One thus sees that the term characterizing the vortex stretching is proportional to the skewness of the probability distribution of the velocity gradients. If we assume that $S(0)$ is constant in time, the equation can be easily integrated to give

$$\langle\omega^2(t)\rangle = \frac{\langle\omega^2(0)\rangle}{\left(1 - \frac{7}{12}\frac{S(0)}{\sqrt{15}}\sqrt{\langle\omega^2(0)\rangle}\,t\right)^2}. \tag{2.21}$$

The experimental result $S(0) > 0$, which stems from the positive correlation between $\omega_i\,\omega_j$ and $\frac{\partial v_i}{\partial x_j}$, is therefore essential for obtaining a solution which blows up in a finite time

$$t^* = \frac{12\sqrt{15}}{7S(0)}\langle\omega^2(0)\rangle^{-1/2}. \tag{2.22}$$

Let us stress that the previous arguments are not rigorous and, at present, it is not clear whether the solutions of the Euler equations explode in a finite time [Frisch 1995]. More importantly, the value of such a result (if true) is questionable for the Navier–Stokes equations at high Reynolds number, since the limit $v \to 0$ is singular. In fact, at $v = 0$ there is energy equipartition and no transfer through a cascade mechanism. However, it is reasonable to expect that the density of energy dissipation ϵ reaches an asymptotic value at $Re \to \infty$, as a consequence of the enstrophy divergence.

2.1.3   The closure problem

The statistical mechanics of turbulence cannot be formulated using the standard approach which was used above for the non-viscous fluid. An alternative is provided by the closure approach. In order to describe this let us formally write the Navier–Stokes equations as

$$\frac{\partial v}{\partial t} = v\,v + \nu \Delta v, \tag{2.23}$$

where $v\,v$ indicates the nonlinear terms. The lowest level of statistical description is given by the knowledge of the first moments of v, say $\langle v\,v \rangle$ (which indicates a generic two-point correlation of the velocity field), since $\langle v \rangle = 0$ for isotropic systems. We ignore the vectorial character of the velocity field to illustrate the problem in a simplified context.

The interested reader can find the detailed derivation of the closure equations in Leslie [1973], Monin and Yaglom [1975] and Orszag [1977]. The equation for $\langle v\,v \rangle$ is of the type

$$\frac{\partial \langle v\,v \rangle}{\partial t} = \langle v\,v\,v \rangle, \tag{2.24}$$

while for $\langle v\,v\,v \rangle$ one has

$$\frac{\partial \langle v\,v\,v \rangle}{\partial t} = \langle v\,v\,v\,v \rangle, \tag{2.25}$$

where we have not written down the viscous linear terms that are not relevant in this context. At any level, the number of equations is one less than the number of variables. This problem is typical of nonlinear equations. An analogous situation appears in the kinetic theory of gases described by the Bogolyubov–Born–Green–Kirkwood–Yvon hierarchy [Résibois and De Leener 1977]. In isotropic homogeneous turbulence, the average value of $\mathbf{v}(\mathbf{x})$ and of its Fourier transform $\mathbf{v}(\mathbf{k})$ vanish so that the simplest non-trivial quantity to be considered is the two-point correlation

$$\langle v_i(\mathbf{k}) v_j(-\mathbf{k}) \rangle = \frac{1}{4\pi k^2} \left(\delta_{ij} - \frac{k_i k_j}{k^2} \right) E(k). \tag{2.26}$$

Here $E(k)\,dk$ represents the contribution of the kinetic energy to the wave numbers in a shell between k and $k + dk$, so that the total energy of the system is $E = \int E(k)\,dk$. The first evolution equation of the hierarchy, corresponding to (2.24), is

$$\left(\frac{\partial}{\partial t} + 2\nu k^2 \right) E(k) = T(k) \tag{2.27}$$

with a three-point correlation term

$$T(k) = -\mathrm{Im} \left(\sum_{lm} 4\pi k^2 k_\ell \int_{\mathbf{q}+\mathbf{p}=\mathbf{k}} P_{mlm}(\mathbf{k}, \mathbf{q}) d^3\mathbf{p} \right) \tag{2.28}$$

and

$$\langle u_i(\mathbf{k})\, u_j(\mathbf{q})\, u_l(\mathbf{p}) \rangle = P_{ijk}(\mathbf{k}, \mathbf{q})\, \delta(\mathbf{k} + \mathbf{p} + \mathbf{q}). \qquad (2.29)$$

The hierarchy can be stopped at this level by a closure hypothesis on the four-point correlations. The simplest assumption [Millionshchikov 1941a,b] is that the fourth-order cumulants are vanishing, i.e.

$$\langle vvvv \rangle = \sum \langle vv \rangle \langle vv \rangle, \qquad (2.30)$$

where the sum is over the three possible combinations of $\langle vv \rangle$. The resulting set of integro-differential equations give unphysical results, as shown by Ogura [1963] who found, by a numerical integration, that $E(k)$ becomes negative in the range of wave numbers corresponding to the energy containing eddies (large length scales). This is because the so-called 'quasi-normal approximation' overestimates the values of the third-order moments, in contrast to what happens in real fluids, where they saturate. In fact, the fourth-order cumulants neglected by the quasi-normal approximation provide a damping mechanism. In order to take such an effect into account, without opening the equation hierarchy, it is convenient to introduce a linear damping term of the type $\mu \langle vvv \rangle$ in (2.24) with an arbitrary parameter, the damping rate $\mu(k)$. From dimensional counting one sees that $\mu(k) = c\sqrt{k^3 E(k)}$, where the constant c has to be fixed a posteriori. Even this eddy-damped quasi-normal approximation is not sufficient to prevent a negative energy spectrum. It is actually possible to obtain a non-negative energy spectrum, by a markovian assumption for the relaxation mechanisms due to nonlinear effects. This approximation is called the eddy-damped quasi-normal markovian (EDQNM) approximation and has been widely used in the statistical predictability theory of turbulence [Leith and Kraichnan 1972].

A similar closure approach is due to Heisenberg [1948a,b], who independently discovered the Kolmogorov scaling for the energy spectrum. Integrating (2.27) up to a wave number k, one finds that

$$\frac{\partial}{\partial t} \int_0^k dq\, E(q) = -2\nu \int_0^k E(q)\, q^2\, dq + \int_0^k T(q)\, dq. \qquad (2.31)$$

The term

$$\Pi(k) = -\int_0^k T(q)\, dq \qquad (2.32)$$

is the energy flux through the wave number k, that is, the energy transferred in a unit time from the wave numbers $q < k$ toward the ones with $q > k$. The Heisenberg approximation consists in modelling the nonlinear energy transfer by an effective dissipation due to the high wave numbers

$$\Pi(k) = 2\nu_T(k) \int_0^k E(q)\, q^2\, dq \qquad (2.33)$$

with the introduction of a turbulent viscosity $v_T(k)$ depending on $q > k$. In other words, small eddies are assumed to absorb energy from larger eddies. By a dimensional argument, a possible choice for the turbulent viscosity is

$$v_T(k) = A_H \int_k^\infty dq \sqrt{E(q)/q^3}, \tag{2.34}$$

where A_H is a non-dimensional constant. One then obtains an integro-differential equation for the energy spectrum $E(k, t)$. Some qualitative information on the form of the solution can be obtained for large wave numbers $k \gg L^{-1}$ (L is the typical length scale of the system). Indeed, almost all the energy of the system is contained in the large length scales around L, so that

$$\langle \epsilon \rangle = \frac{\partial}{\partial t} \int_0^\infty E(q)\, dq \approx \frac{\partial}{\partial t} \int_0^k E(q)\, dq. \tag{2.35}$$

The equation of the energy spectrum for $k \gg L^{-1}$ becomes

$$\langle \epsilon \rangle = 2 \left(v + A_H \int_k^\infty \sqrt{E(q)/q^3}\, dq \right) \int_0^k E(p)\, p^2\, dp. \tag{2.36}$$

After some algebraic manipulations, the analytic solution of this last equation behaves as

$$E(k) \sim \begin{cases} \langle \epsilon \rangle^{2/3} k^{-5/3} & L^{-1} \ll k < k_D, \\ k^{-7} & k > k_D, \end{cases} \tag{2.37}$$

where the dissipative wave number $k_D = (\langle \epsilon \rangle\, v^{-3})^{1/4}$ is the inverse of the Kolmogorov length scale $\ell_D = k_D^{-1}$ below which the molecular viscosity overwhelms the nonlinear transfer. The scaling at wave numbers in the inertial range $[L^{-1}, k_D]$ is the famous 5/3 Kolmogorov law, while the k^{-7} power law in the viscous range is an artefact of the approximation, since one expects exponential damping at very high wave numbers.

We can now discuss the problem of the singularities of the inviscid Euler equation. From the EDQNM approximation, one has the following scenario for the unforced Navier–Stokes equations [Lesieur 1990]. Starting with an energy spectrum $E(k, 0)$ concentrated only at large scales, i.e. $E(k, 0)$ significantly different from zero only for $k \sim L^{-1}$, one has a fast increase of the enstrophy $\int k^2 E(k, t)\, dk$ up to a time $t^* \sim \left(\int k^2 E(k, 0)\, dk \right)^{-1/2}$. Because of the presence of the dissipative term the singularity is suppressed, so that for $t > t^*$ one has $\epsilon = O(1)$ and the Kolmogorov spectrum $E(k) \sim k^{-5/3}$ is established. This scenario is supported by some numerical simulations; see Grauer and Sideris [1991] and Frisch [1995].

2.2   Scaling invariance in turbulence

Before discussing the Kolmogorov derivation of the 5/3 law in the framework of the energy cascade scenario, we must repeat that it is obtained only on the basis of phenomenological arguments and is not derived from the Navier–Stokes equations. It gives a deeper insight into the fundamental mechanisms of a turbulent fluid and is compatible with the scaling invariance exhibited by the Navier–Stokes equations when $Re \rightarrow \infty$. In fact, one expects that the statistical properties of turbulence at small scales are ruled by scaling laws which are universal, i.e. independent of boundary conditions, type of fluid or external force. It is also reasonable to expect that although the system is described by many degrees of freedom, only few parameters are relevant for the small-scale statistics. In that case, the exponents of the power laws can be estimated using the cascade picture. Following Richardson [1922], it is commonly believed that the energy is transferred from large to small length scales via a local cascade in the wave number k-space. The idea is that a force term, pumping energy at a constant rate, is able to create disturbances ('eddies') at large scales ($\approx L$). Because of the nonlinear terms of the Navier–Stokes equations, a disturbance creates disturbances on smaller (though comparable) scales and so on, up to the Kolmogorov length ℓ_D, where the viscous damping overwhelms the nonlinear transfer and energy is dissipated into heat by molecular friction. The typical velocity of an eddy of scale ℓ is $\delta v(\ell) = |\mathbf{v}(\mathbf{x} + \mathbf{r}) - \mathbf{v}(\mathbf{x})|$, with $|\mathbf{r}| = \ell$, where we have not considered the vectorial nature of the velocity field to simplify the notation.

The self-similar aspect of the cascade becomes transparent by noting that the Navier–Stokes equations are invariant under the scaling transformation $\ell \rightarrow \Lambda \ell$ when

$$v \rightarrow \Lambda^h v, \qquad v \rightarrow \Lambda^{1+h} v, \qquad t \rightarrow \Lambda^{1-h} t \qquad (2.38)$$

with an arbitrary exponent h.

The theory of Kolmogorov [1941], often called K41, is based on a hypothesis of global scaling invariance which has as its assumption that the energy transfer rate is uniform in space and is independent of the length scale ℓ in the inertial range $[\ell_D, L]$ and on the assumption that the mean rate of energy dissipation per unit mass ϵ has a finite non-zero value. This invariance is obviously broken in a real fluid, although it is reasonable to expect that, when $Re \rightarrow \infty$, it is restored in a statistical sense at small scales (the inertial range) and far from the boundaries.

Kolmogorov [1941] also derived one of the few exact results which can be obtained directly from the Navier–Stokes equations, showing that the longitudinal velocity increment satisfies the equation

$$\langle \delta v_L^3(\ell) \rangle - 6\nu \frac{d}{d\ell} \langle \delta v_L^2(\ell) \rangle = -\frac{4}{5} \langle \epsilon \rangle \ell$$

under the assumption of homogeneity, isotropy and finite density of energy dissipation. In the inertial range, the term containing the viscosity can be neglected so that

$$\langle \delta v_L^3(\ell) \rangle = -\frac{4}{5} \langle \epsilon \rangle \ell. \tag{2.39}$$

In the absence of fluctuations of δv, (2.39) implies that

$$\delta v_x(\ell) = |\mathbf{v}(\mathbf{x} + \mathbf{r}) - \mathbf{v}(\mathbf{x})| \sim \ell^h, \qquad |\mathbf{r}| = \ell, \tag{2.40}$$

with $h = 1/3$ at any point $\mathbf{x}$ of the fluid so that

$$E(k) \sim C_K \, \epsilon^{2/3} \, k^{-\gamma} \qquad \text{with } \gamma = 1 + 2h = 5/3. \tag{2.41}$$

In the K41 theory, the so-called Kolmogorov constant C_K is a universal number, whose value cannot be determined by dimensional arguments. From a fit of the data using (2.41), one estimates that $C_K = 1.7 \pm 0.2$. From (2.39) one also sees that the K41 theory predicts a skewness factor independent of ℓ:

$$S(\ell) = \frac{4}{5} C_K^{-3/2}, \qquad \ell_D \ll \ell \ll L.$$

However, there is clear experimental evidence that the power γ differs slightly from the K41 prediction ($\gamma \approx 1.70 > 5/3$) [Monin and Yaglom 1975]. In this case there is no reason to expect that C_K is universal.

The exponent h can be regarded as the Hölder exponent of the velocity and determines the strength of singularity of the gradients

$$\nabla v(\mathbf{x}) = \lim_{\ell \to 0} \frac{\delta v_x(\ell)}{\ell}, \tag{2.42}$$

in the limit of infinite Reynolds number, where $\ell_D \to 0$. The exponent h can be determined from the fact that $\langle \epsilon \rangle = O(1)$ and from the assumption of global scaling invariance. One thus obtains the Kolmogorov value $h = 1/3$. This result can also be obtained by the following argument of dimensional type. The energy density $\tilde{\epsilon}(\ell)$ transferred in unit time at a scale ℓ is given by the ratio between the density of kinetic energy proportional to $\delta v(\ell)^2$ and the lifetime of the eddy at the same length scale:

$$\tau(\ell) \sim \frac{\ell}{\delta v(\ell)},$$

and so

$$\tilde{\epsilon}(\ell) \sim \frac{\delta v^3(\ell)}{\ell}. \tag{2.43}$$

The dimensional relation (2.43) is assumed to be valid in all the approaches described in this chapter and implies a strong link between fluctuations of the velocity increments and those of the energy transfer rate $\tilde{\epsilon}$.

Under the transformation (2.38), one has

$$\tilde{\epsilon} \to \Lambda^{3h-1} \tilde{\epsilon}. \tag{2.44}$$

The value $h = 1/3$ thus guarantees that the energy transfer rate $\tilde{\epsilon}(\ell)$ is constant through the whole inertial range $[\ell_D, L]$.

The assumption of global scaling is not verified by the experimental measurements of the scaling exponents ζ_p of the structure functions:

$$\langle |\delta v_x(\ell)|^p \rangle \sim \ell^{\zeta_p}. \tag{2.45}$$

Global scaling invariance neglects fluctuations, and so the average operation in (2.45) can be ignored and $\zeta_p = p/3$. On the other hand, experimental data [Anselmet et al. 1984, van de Water and Herweijer 1996] show unambiguous deviations from such a prediction: ζ_p appears to be a nonlinear function of p. The breaking of the global scaling invariance is also evident when one looks at the density of energy dissipation:

$$\epsilon(\mathbf{x}) = \frac{1}{2} \nu \sum_{i,j} \left(\frac{\partial u_i}{\partial x_j} + \frac{\partial u_j}{\partial x_i} \right)^2. \tag{2.46}$$

Numerical simulations show that the kinetic energy is dissipated non-uniformly on stretched structures with a fractal appearance [Siggia 1981]. In other words, global scaling invariance seems not to be satisfied in turbulent fluids, even though there is good evidence of non-trivial scaling laws for the structure functions at large Reynolds number.

The same type of scaling argument has been pursued by Novikov and Steward [1964] and Mandelbrot [1974, 1982] with the additional hypothesis that in the limit of infinite Reynolds number the energy dissipation covers in a uniform way a fractal structure with fractal dimension $D_F < 3$. The resulting intermittency is rather strong since there are regions of full Lebesgue measure in the fluid where there is a tiny amount of energy dissipation and the velocity field is differentiable, i.e. $h = 1$. On the active regions covering the fractal, the velocity field is therefore more singular than in the Kolmogorov theory, since the average dissipation $\langle \epsilon \rangle$ must remain constant at increasing Re. Under the scaling transformation (2.38), the fraction of active fluid (from the dissipation point of view) scales as Λ^{3-D_F} so that the mean energy dissipation rate scales as

$$\langle \epsilon \rangle \to \Lambda^{3-D_F+3h-1} \langle \epsilon \rangle. \tag{2.47}$$

In order to have a finite, non-zero value of $\langle \epsilon \rangle$, the Hölder exponent on the fractal is

$$h = \frac{D_F - 2}{3}. \tag{2.48}$$

Let us stress that from a mathematical point of view, the active regions have vanishing Lebesgue measure as $Re \to \infty$, even if at finite Reynolds number the dissipation introduces a natural cut-off, the Kolmogorov length, which scales as $\ell_D \sim (Re)^{-1/(1+h)}$. In the fractal approach the structure functions obey the scaling law

$$\langle |\delta v(\ell)|^p \rangle \sim \ell^{3-D_F} \ell^{hp} = \ell^{\zeta_p}$$

with

$$\zeta_p = p\frac{(D_F - 2)}{3} + 3 - D_F. \tag{2.49}$$

This result is still in disagreement with the experimental data, which indicate that ζ_p is nonlinear in p: the hypothesis of uniform dissipation on a fractal is not satisfactory, since the active regions themselves seem to be intermittent.

2.3   Multifractal description of fully developed turbulence

In order to take the experimental result for the scaling exponent ζ_p into account, Parisi suggested that fully developed turbulence possesses only a local scaling invariance, that is, the Hoelder exponent h can vary at different points of the fluid [Parisi and Frisch 1985, Benzi et al. 1984]. This weaker ansatz allows one to describe the intermittent nature of energy dissipation in terms of the multifractal approach widely used in describing the scaling properties of ergodic probability measures on strange attractors of dynamical systems with few degrees of freedom. Local scaling invariance implies detailed predictions of many different aspects of turbulence which can be verified in numerical and experimental tests. In this sense, the validity of the multifractal approach has an a posteriori justification.

2.3.1   Scaling of the structure functions

The assumption of statistical scaling invariance at high Reynolds number implies that there exists an inertial range of lengths where the velocity increments scale as

$$\delta v_{\mathbf{x}}(\ell) \sim v_0 \ell^h. \tag{2.50}$$

Here $v_0 = |V_0|$ is the absolute value of the characteristic velocity V_0 on the typical macroscopic length L. For the sake of simplicity, we assume $L = 1$. Spatial variations of h lead to a strong spatial intermittency in the magnitude of the gradients and thus in the energy dissipation.

The main assumption of the multifractal description is that $h < 1$ on a fractal set F which can be regarded as a superposition of subsets $\Omega(h)$ each consisting of points x such that $\delta v_x \sim \ell^z$ with $z \in [h, h + dh]$. The complement of F is covered by regions which have a non-zero volume in the 3D space where the velocity field can be linearized, i.e. $h \geq 1$, so that the gradients remain small at high Re.

One can satisfy local scaling invariance together with the existence of scaling exponents for all the moments of $\delta v(\ell)$ by assuming that the set F is a multifractal [Parisi and Frisch 1985, Paladin and Vulpiani 1987a]. By this term one means that each subset $\Omega(h)$ is itself a fractal with fractal dimension $D(h)$. The probability of picking up a singularity exponent h scales in the inertial range as the fraction of the coarse-grained measure of the volume of $\Omega(h)$ over the total volume, that is,

$$P_\ell(h)\,dh = \ell^{3-D(h)}\rho(h)dh, \tag{2.51}$$

where $\rho(h)$ is a smooth function of h independent of ℓ. From (2.51) one has

$$\langle \delta v^p(\ell) \rangle \sim \int \ell^{hp+3-D(h)}\,dh,$$

and a simple saddle point estimate gives

$$\zeta_p = \min_h\,[h\,p + 3 - D(h)] = h^*\,p + 3 - D(h^*) \tag{2.52}$$

with

$$h^*(p) = \frac{d\zeta_p}{dp}.$$

The Kolmogorov theory corresponds to the case of only one singularity $h = 1/3$ with $D(h = 1/3) = 3$.

The log–normal theory introduced by Kolmogorov [1962] and Obukhov [1962], often referred to as K62, can be regarded as a particular limit of the multifractal approach. Essentially the K62 theory assumes that the variable

$$\tilde{\epsilon}_x(\ell) = \frac{1}{\ell^3}\int_{\Lambda_x(\ell)} \epsilon(\mathbf{y})\,d\mathbf{y}, \tag{2.53}$$

where $\Lambda_x(\ell)$ is a cube of edge ℓ centred on $\mathbf{x}$, is distributed according to a log–normal distribution fully characterized by two parameters,

$$\langle \epsilon \rangle = \langle \tilde{\epsilon}_x(\ell) \rangle,$$

and μ, given by

$$\langle (\ln \tilde{\epsilon}_x(\ell) - \langle \ln \tilde{\epsilon}_x(\ell) \rangle)^2 \rangle = \mu \ln(L/\ell).$$

This corresponds to a parabolic shape for $D(h)$ in (2.52) and to the structure function exponents

$$\zeta_p = \frac{p}{3} + \frac{\mu}{3}p(3 - p). \tag{2.54}$$

The log–normal distribution is a good approximation for a wide class of phenomena (see appendices C and E) even if it fails to describe the moments of large order [Orszag 1970]. The ζ_p in (2.54) is a parabola with a maximum at $p^* = 3(2 + \mu)/2\mu$. It follows that the derivative $d\zeta_p/dp$ is not bounded, so that the corresponding singularity $h^*(p) \to -\infty$ as $p \to \infty$. This contradicts a rigorous result which guarantees that for an incompressible fluid ζ_p is a non-decreasing function of p so that h cannot be negative [Frisch 1995]

The computation of $D(h)$, or equivalently of ζ_p, from the Navier–Stokes equations is the goal of the multifractal approach, although it is a very difficult task which is still far from being solved. This chapter and the next one describe the first steps:

(1) the phenomenological approach using a multiplicative random process with parameters obtained by a data fit, the so-called random beta model;

(2) the study of simplified dynamical systems, the shell models, describing the energy cascade.

2.3.2   Multiplicative models for intermittency

Multiplicative models of the energy cascade (the so-called absolute and weighted curdling) have been introduced by Novikov and Steward [1964] and Mandelbrot [1974, 1982]. Among the fractal models, the beta model of Frisch et al. [1978], which is a reformulation of the absolute curdling, has become very popular. This model describes the energy cascade in real space, looking at eddies of size $\ell_n = r^{-n}\ell_0$ with $r > 1$. ℓ_0 is the length scale at which the energy is injected and the arbitrary value of the ratio is $r = 2$. At the nth step of the cascade, a mother eddy of size ℓ_n splits into daughter eddies of size ℓ_{n+1}. At each step the daughter eddies cover only a fraction β ($0 < \beta < 1$) of the mother volume. In this way, one has a multiplicative process which builds up a Cantor set on which the energy dissipation concentrates in the limit of infinite Reynolds number corresponding to $n \to \infty$. After n steps only a fraction β^n of the fluid is active. From the definition of fractal dimension the volume of the active regions scales as ℓ^{-D_F}, so that $\beta^n \sim \ell_n^{3-D_F}$ and the fractal dimension is

$$D_F = 3 + \log_2\beta. \tag{2.55}$$

The value $\beta = 1$ corresponds to the Kolmogorov picture, where the energy dissipation is distributed over the whole fluid.

The multifractal description as proposed by Parisi makes no use of the concept of cascade. However, one can introduce a multifractal multiplicative model by assuming that β is a random variable instead of being a constant. In fact, the random beta model [Benzi et al. 1984] was the first multifractal model able to fit the scaling exponents of the structure functions. In that model, at a scale ℓ_n

there are N_n mother eddies. The daughter eddies of size ℓ_{n+1} cover a fraction $\beta_{n+1}(k)$ of the mother k, with $k = 1, \ldots, N_n$.

The velocity increment at a scale ℓ_n is given by the typical velocity $v_n(k)$ of the corresponding eddy whose lifetime is $\tau_n \sim \ell_n/v_n$. The energy transfer rate from mother to daughter is therefore

$$\epsilon_n \sim \frac{v_n^3}{\ell_n}. \tag{2.56}$$

As a consequence of the fact that the energy transfer is constant throughout the cascade, one has

$$\frac{v_n^3(k)}{\ell_n} = \beta_{n+1}(k) \frac{v_{n+1}^3(k)}{\ell_{n+1}}. \tag{2.57}$$

After n cascade steps, an eddy is individuated by a particular history of fragmentations (a sequence of random variables $\beta_1, \ldots, \beta_n$), so that it covers a fraction $\prod_i \beta_i$ of the fluid volume, and

$$v_n \sim v_0 \ell_n^{1/3} \prod_{i=1}^{n} \beta_i^{-1/3}. \tag{2.58}$$

Let us now assume that the β_i's are independent, identically distributed random variables. In this case the structure functions $\langle v_n^p \rangle$ scale with exponents

$$\zeta_p = \frac{p}{3} - \log_2 \overline{\beta^{1-p/3}}, \tag{2.59}$$

where the overbar indicates the average over the probability distribution of β. If β is a constant, i.e. $\beta_n(k) = 2^{D_F-3}$ for all n and k, one recovers the result of the beta model (2.49). Phenomenological arguments suggest restricting the choice of the β probability distribution to a dichotomic distribution with a free parameter x:

$$P(\beta) = x\,\delta(\beta - 1) + (1 - x)\,\delta(\beta - 2^{3h_{\min}-1}). \tag{2.60}$$

The two limiting cases are $x = 1$, corresponding to the K41 theory, and $x = 0$, corresponding to the usual beta model, where at the end of the cascade the energy dissipation concentrates on a fractal structure with dimension

$$D_F = 2 + 3h_{\min}. \tag{2.61}$$

One may conjecture that the main mechanism of the eddy fragmentation (beyond the Kolmogorov-like one) is the creation of two-dimensional vorticity sheets, so that $h_{\min} = 0$. Using the probability distribution (2.60) with $h_{\min} = 0$, the exponents (2.59) become

$$\zeta_p = p/3 - \log_2(x + (1 - x)\,2^{p/3-1}). \tag{2.62}$$

The multifractal spectrum is given by the Legendre transform of (2.62),

$$D(h) - 3 = (3h - 1)\left[1 + \log_2\left(\frac{1 - 3h}{1 - x}\right)\right] + 3h\,\log_2\left(\frac{x}{3h}\right),$$
$$h \in [0, 1/3]. \tag{2.63}$$

A fit of the experimental data of Anselmet et al. [1984] for the ζ_p exponents (see Fig. 2.2) gives the value $x = 7/8$.

She and Levèque [1993] have proposed a different random multiplicative model with a hierarchy of fluctuation structures associated with vortex filaments. The resulting formula for the exponents of the structure functions,

$$\zeta_p = \frac{p}{9} + 2\left(1 - \left(\frac{2}{3}\right)^{p/3}\right), \tag{2.64}$$

allows one to get an extremely good fit of the experimental data. This corresponds to assuming that the energy transfer rate has a log–Poisson distribution [Dubrulle 1994] and can be obtained from the random beta model by taking in (2.60) the simultaneous limits $x \to 1$ and $n(1 - x) = C$ when the cascade step $n \to \infty$. In other words the probability x of having a space-filling eddy fragmentation of Kolmogorov type depends on the length scale and tends to unity for the lowest eddy sizes considered. This kind of approach, also related to infinitely divisible laws [Novikov 1994], is interesting, although one cannot hope to reach a deep understanding of the intermittency phenomenon only by multiplicative processes.

Let us stress that all the random multiplicative processes of this section construct a probability measure proportional to the density of energy dissipation. The scaling of the velocity increments is then obtained by the dimensional relation (2.43) identifying the transfer rate $\tilde{\epsilon}(\ell)$ at a scale ℓ with the coarse grained energy dissipation density; see (2.53). However, it is possible to introduce a more sophisticated process which directly constructs a multiaffine velocity field [Benzi et al. 1993b], as discussed in appendix F.

Up to this point, all the multiplicative models have been a phenomenological way to describe the data for the structure functions, taking into account the invariance under rescaling of the Navier–Stokes equations. However, the multifractal approach is also able to give new predictions which can be verified by experiments and numerical simulations. This is the most important test of its physical relevance and heuristic power. In the following, we describe some consequences of multifractality: the shape of the probability distribution function of the velocity gradients [Benzi et al. 1991], the appearance of pseudo-algebraic laws (multiscaling) in an intermediate dissipative range of scales [Frisch and Vergassola 1991] and the number of degrees of freedom of a turbulent flow [Paladin and Vulpiani 1987b].

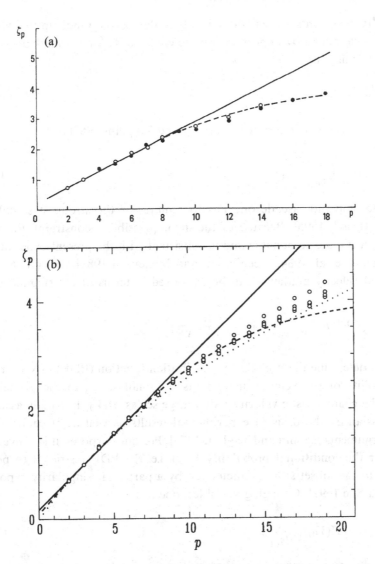

Figure 2.2 (a) The structure function exponents ζ_p, plotted vs p. The dots and circles correspond to the data of Anselmet et al. [1984]. The curve indicates (2.62) with $x = 7/8$ and the line indicates (2.49) with $D_F = 2.83$ [Benzi et al. 1984]. (b) The exponents ζ_p versus p for the longitudional structure functions measured in jet- and grid-turbulence. The lines represent the prediction of several models for ζ_p. Full line: Kolmogorov theory; dashed line: prediction by She and Levèque [1994], equation (2.64); dotted line: asymptotic prediction, for large p, given by a geometric constraint of the multifractal model, $\zeta_p \simeq 1 + h_{min}/p$, with $h_{min} = 0.16$ [Herweijer and van de Water 1995, van de Water and Herweijer 1996].

2.3.3   Probability distribution function of the velocity gradients

The velocity increments are self-similar only in the inertial range up to a viscous cut-off ℓ_D, determined by imposing that the effective Reynolds number on scale ℓ is equal to unity, i.e.

$$\frac{\delta v(\ell_D)\ell_D}{\nu} \sim 1,$$

so that ℓ_D is itself a function of h [Paladin and Vulpiani 1987b],

$$\ell_D(h) \sim \left(\frac{\nu}{v_0}\right)^{\frac{1}{1+h}}. \tag{2.65}$$

In order to stop the cascade one has to require for the strongest singularity $h_{min} > -1$ [Frisch 1995]. Because of the incompressibility constraint, the value $h_{min} = 0$ seems more reasonable and is consistent with the present experimental data [Anselmet et al. 1984, Sreenivasan and Meneveau 1988]. Using (2.65), the longitudinal velocity gradients can be expressed in terms of the singularities h as

$$|s| \approx \frac{\delta v(\ell_D)}{\ell_D} = v_0 \ell_D^{h-1} = v_0^{\frac{2}{1+h}} \nu^{\frac{h-1}{h+1}}. \tag{2.66}$$

In order to determine the probability distribution function (PDF) of s, we relate the probability of the velocity increments on small scales to the probability $\Pi(V_0)$ of the characteristic velocity V_0 on large scales. $\Pi(V_0)$ is usually assumed to be gaussian on the basis of experimental results as well as of central limit theorem arguments [Monin and Yaglom 1975]. For our purposes, it is convenient to consider the conditional probability $P_h(s)$, i.e. the PDF restricted to points belonging to the subset $\Omega(h)$ characterized by a particular singularity exponent [Frisch and She 1991]. Changing variables, one has

$$P_h(s) = \Pi(V_0) \left|\frac{dV_0}{ds}\right|, \tag{2.67}$$

and expressing V_0 in terms of s via (2.66), it follows that

$$P_h(s) \sim \left(\frac{\nu}{|s|}\right)^{\frac{1-h}{2}} \exp\left(-\frac{\nu^{1-h}|s|^{1+h}}{2\langle V_0^2\rangle}\right). \tag{2.68}$$

The prediction of the K41 theory is obtained by neglecting intermittency effects, i.e. by assuming that $h = 1/3$ uniformly in the fluid. One thus finds that

$$P(s) \sim \left(\frac{\nu}{|s|}\right)^{1/3} \exp\left(-C\,\nu^{2/3}\,|s|^{4/3}\right), \tag{2.69}$$

with $C = (2\langle V_0^2 \rangle)^{-1}$. Let us recall that in a fractal picture, the velocity gradients are very small in the non-active zones where $h \geq 1$. Therefore, we must take into account the presence of an additional delta function in the PDF:

$$P(s) = \overline{P}_F(s) + \gamma\,\delta(s), \tag{2.70}$$

where $\overline{P}_F$ is the PDF restricted to the active zones covering the fractal set F and γ is a normalization factor, such that $\int P(s)\,ds = 1$. For instance, $\overline{P}_F(s) = P_{h_F}$ in the beta model where the singularity value on F is constant and equal to $h_F = (D_F - 2)/3$.

Let us now derive the form of the PDF in the multifractal approach [Benzi et al. 1991]. As a matter of fact, when there is a hierarchy of singularities, the probability of observing a gradient value s related to a given singularity h is $P_h(s)\,P_{\ell_D}(h)$, where $P_\ell(h)$ is given by (2.51). It follows that the conditional probability is given by a weighted integral over the singularities:

$$\overline{P}_F(s) = \int dh\, P_h(s)\, P_{\ell_D}(h)$$

$$\sim \int dh\, \rho(h) \left(\frac{v}{|s|}\right)^{2 - \frac{h + D(h)}{2}} \exp\left(-\frac{v^{1-h}\,|s|^{1+h}}{2\langle V_0^2 \rangle}\right). \tag{2.71}$$

An analytic estimate of the integral is not easy, since one cannot apply a saddle point method as for the analogous integral giving the structure functions. This is due to the fact that the gradients scale as $|s| \sim v^{\frac{h-1}{h+1}}$. Even for strong singularities $h \approx 0$, one has $|s| \sim v^{-1}$, and we therefore expect that $|s|/v$ should not be very large.

Instead of inserting the function $D(h)$ given by (2.63) with $x = 7/8$ into (2.71), one can obtain a simpler formula for the PDF, by a direct estimate. The results are of course equivalent, although the latter is more transparent. From (2.71), the probability distribution of the velocity increments is

$$P(v_n) = \int \Pi(V_0)\,dV_0 \int \delta\left(v_n - v_0\,\ell_n^{1/3} \prod_{i=1}^{n} \beta_i^{-1/3}\right) \prod_{i=1}^{n} \beta_i\,\mu(\beta_i)\,d\beta_i, \tag{2.72}$$

where $\mu(\beta)$ is the probability density of the β_i's. Since the β_i's are identically distributed according to the dichotomic distribution (2.60), the integral can be reduced to the sum

$$P(v_n) \sim \sum_{K=0}^{n} \binom{n}{K} x^{n-K}\,(1-x)^K\,2^{4K/3}\,\ell_n^{-1/3} \exp(-C\,2^{2K/3}\,\ell_n^{-2/3}\,v_n^2), \tag{2.73}$$

where $C = (2\langle V_0^2 \rangle)^{-1}$ and $\ell_n = 2^{-n}$.

It is important to stress that both the K41 theory and the beta model predict a PDF with stretched exponentials of the form $\exp(-c\,|s|^t)$, with $t > 1$. In a log–linear plot of the PDF, such a form implies that the curve should be convex, which is in contradiction to the qualitative features of the experimental and numerical data [Van Atta and Park 1972, Castaing et al. 1990, Vincent and Meneguzzi 1991]. They seem, however, consistent with an effective stretched exponent $t < 1$. Figure 2.3 shows the log–linear plot of the probability distribution $P(v_n)$ of the velocity increments v_n. One observes the change of the PDF from a gaussian form at large scales to an exponential-like form at small scales. The multifractal prediction thus seems consistent with the available data for the PDF of the velocity gradients with an effective stretched exponent $t < 1$.

The PDF of the gradients is obtained by (2.71) computed at the step N which corresponds to the viscous cut-off (2.65), $v_N \ell_N / v = 1$, namely

$$\ell_N^2 \equiv 2^{-2N} \sim \frac{v}{s}.\qquad(2.74)$$

Therefore, the cascade stops at $N = \ln(s/v)/(2\ln 2)$, and the conditional PDF is

$$\overline{P}_F(s) = \sum_{K=0}^{N} \binom{N}{K} x^{N-K} (1-x)^K \left(\frac{v}{|s|}\right)^{(1+2k)/3} \exp[-C \, v^{(2+k)/3} \, |s|^{(4-k)/3}],$$

$$(2.75)$$

where $k = K / N$. The K41 prediction (2.69) corresponds to considering only the term $K = 0$, with $x = 1$. When $N(s)$ is not large, the main contribution to the sum is given by the first K terms. With increasing $|s|$, the PDF becomes sensitive to higher K-terms, i.e. to stronger singularities. A direct inspection of (2.73) shows that the largest contributions to the sum are given by K around a $K^*(s)$ which exhibits a very weak dependence on s, probably logarithmic. The absence of any dominant contribution implies that one never has a pure exponential (or

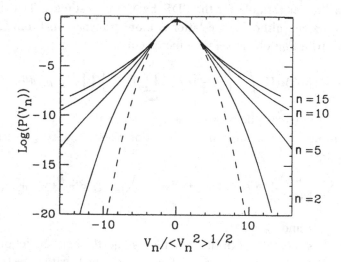

Figure 2.3 Log–linear plot of the normalized probability distribution $P(v_n)$ versus v_n/σ_n, where $\sigma_n^2 = \langle v_n^2 \rangle$ with $x = 7/8$, $\langle V_0^2 \rangle = 1$, $n = 2, 5, 10, 15$ (full lines). The dashed line indicates the standard gaussian [Benzi et al. 1991].

a pure stretched exponential) form of the PDF. The maximal value of $N(s)$ is limited by the Reynolds number of the experiment: typical values are $N = 14$–16. In Fig. 2.4, we show the comparison between the numerical data of Vincent and Meneguzzi [1991] and (2.75), with the same parameters $h_{min} = 0$ and $x = 7/8$, used by the fit [Benzi et al. 1984] of the structure function exponents ζ_p.

2.3.4 Multiscaling

We have seen that the cut-off scale ℓ_D varies with h and the Reynolds number according to (2.65). This implies a new form of universality for the energy spectrum $E(k)$ in an intermediate dissipation range [Frisch and Vergassola 1991, Jensen et al. 1991b].

In the integral of (2.52), the variable h varies between h_{min} and $h(\ell)$, which is given by the relation $\ell \sim v^{\frac{1}{1+h(\ell)}}$. If $h(\ell)$ is larger then the value h_{SP} for which the quantity $2h + 3 - D(h)$ has a minimum, i.e. $\ell > \eta_K = \ell_D(h_{SP}) \sim Re^{-3/4}$, one has

$$\langle \delta v(\ell)^2 \rangle \sim \ell^{\zeta_2}, \qquad \ell > \eta_K. \tag{2.76}$$

η_K is the length scale where the intermediate dissipation range starts; for $\eta_{min} < \ell < \eta_K$, the leading contribution to $\langle \delta v(\ell)^2 \rangle$ is given by the upper limit $h(\ell)$:

$$\langle \delta v(\ell)^2 \rangle \sim \ell^{2h(\ell)+3-D(h(\ell))}, \qquad \eta_{min} < \ell < \eta_K. \tag{2.77}$$

This intermediate range extends up to η_{min} which is the viscous cut-off related to the strongest singularity h_{min}. Only below this scale are all the singularities turned off and one enters into the dissipation range where an exponential decay is expected on the basis of analyticity arguments. From an experimental point of view the most interesting quantity is the energy spectrum

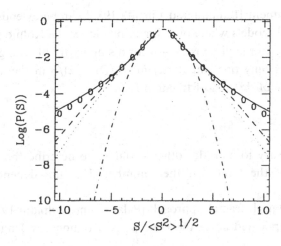

Figure 2.4 Log–linear plot of the PDF of the gradients $\overline{P}_F(s)$ versus s/σ, where $\sigma^2 = \langle s^2 \rangle$, $\langle V_0^2 \rangle = 1$ and $v = 10^{-3}$. The data of Vincent and Meneguzzi [1991] are indicated by circles; the full line is the multifractal prediction (2.75) with $x = 7/8$; the K41 prediction and the result of the unifractal beta model with $D_F = 2.83$ ($h \simeq 0.27$) are respectively the dotted and the dashed lines [Benzi et al. 1991].

$E(k) \sim k^{-1} \langle |\delta v(\ell = 1/k)|^2 \rangle$. A simple calculation shows that one can find a universal function both in the inertial range $\ell > \eta_K$ and in the intermediate dissipation range $\eta_K > \ell > \eta_{\min}$ by introducing the multiscaling transformation [Kadanoff et al. 1989, Wu et al. 1990]

$$F(\theta) = \frac{\ln E(k)}{\ln(1/\nu)} \quad \text{and} \quad \theta = \frac{\ln k}{\ln(1/\nu)}. \tag{2.78}$$

From (2.76) and (2.77), one obtains

$$F(\theta) = -(1 + \zeta_2) \theta \quad \text{for} \quad \theta < \frac{1}{1 + h_K}, \tag{2.79}$$

$$F(\theta) = -2 - 2\theta + \theta D \left(h = -1 + \frac{1}{\theta} \right) \quad \text{for} \quad \frac{1}{1 + h_K} < \theta < \frac{1}{1 + h_{\min}}. \tag{2.80}$$

The prediction of a universal scaling of multifractal type for the energy spectrum can be tested experimentally. Gagne and Castaing [1991] have analysed a wide sample of data with Reynolds numbers ranging from 10^3 to 10^6 obtaining a good agreement with a scaling of multifractal type.

2.3.5 Number of degrees of freedom of turbulence

In the framework of the K41 theory, dimensional arguments allow one to estimate the number of degrees of freedom of turbulence. In order to get a satisfactory description of a turbulent flow, one needs a resolution of the length scales given by the Kolmogorov length $\ell_D \sim (\nu^3/\epsilon)^{1/4}$. Therefore, the number of grid points per unit volume is

$$N \sim \left(\frac{L}{\ell_D} \right)^3 \sim Re^{9/4}. \tag{2.81}$$

This classical Landau argument [Landau and Lifshitz 1987] can be extended to the fractal and multifractal models where one should introduce the Kolmogorov length (2.65) corresponding to the viscous cut-off for a singularity of strength h. In the beta model, there is only one singularity $h_F = (D_F - 2)/3$ in the active regions so that the number of degrees of freedom is

$$N \sim \left(\frac{L}{\ell_D(h_F)} \right)^{D_F} \sim Re^{\frac{3D_F}{1+D_F}}. \tag{2.82}$$

Obviously, it is also necessary to consider other variables to describe the non-active 'laminar' regions of the fluid, but their number does not depend on Re.

In the framework of the multifractal approach [Paladin and Vulpiani 1987b], the situation is more complicated since the concept of Kolmogorov length is not well defined. There is a whole spectrum of singularities, each one of them stopped at a different cut-off given by (2.65). Since the number of eddies of size ℓ characterized by a singularity h scales as $\ell^{-D(h)}$, the number of degrees of freedom is given by an integral over the possible cut-offs,

$$N \sim \int dh \left(\frac{L}{\ell_D(h)} \right)^{D(h)} \sim Re^{\delta}, \tag{2.83}$$

where δ is obtained by a saddle point estimate of (2.83),

$$\delta = \max_h \frac{D(h)}{1+h}. \tag{2.84}$$

Using the multifractal function $D(h)$ given by (2.63) with $x = 7/8$ one has $\delta \approx 2.2$, which is quite close to the K41 value $9/4 = 2.25$. Although this result is important from a theoretical point of view, it has no practical relevance because it is very hard to locate the moving active regions of the fluid during a numerical simulation. As one usually works with a fixed grid or with pseudo-spectral methods, the relevant resolution scale is the cut-off $\ell_{\min}$ related to the strongest singularity,

$$\ell_{\min} \sim Re^{-\frac{1}{1+h_{\min}}}, \tag{2.85}$$

so that the effective number of equations necessary to obtain an accurate description is

$$N_{\text{eff}} \sim \left(\frac{L}{\ell_{\min}} \right)^3 \sim Re^{\frac{3}{1+h_{\min}}}. \tag{2.86}$$

This number is much larger than the number of degrees of freedom since if $h_{\min} = 0$, one has $N_{\text{eff}} \sim Re^3$ while $N \sim Re^{2.2}$.

2.4   Two-dimensional turbulence

The phenomenology of two-dimensional turbulence is rather different from the three-dimensional one. The main reason stems from the fact that in the 2D Euler equations the vorticity is a material invariant, i.e. conserved along stream lines. Instead of (2.18) one has

$$\frac{\partial \omega}{\partial t} + (\mathbf{v} \cdot \nabla) \omega = v \Delta \omega. \tag{2.87}$$

For $v = 0$, (2.87) becomes $D_t \omega = 0$, where D_t is the lagrangian time derivative. It follows that there is an infinite number of integrals of motion, since the global integral of any functional of vorticity is conserved. However, in truncated Euler equations all these invariants disappear with the exception of energy and enstrophy [Kraichnan and Montgomery 1980].

From (2.87) one finds that the energy dissipation is $O(v)$, so that it is impossible to have a forward energy cascade as in 3D. On the other hand, one can repeat the Kolmogorov dimensional arguments of section 2.2 for enstrophy instead of energy.

An important difference with the 3D situation is that in 2D, molecular viscosity cannot remove energy efficiently. Since enstrophy contains two more derivatives than energy, it is dissipated at large scales. As a consequence, one expects that energy is constrained to flow towards large scales. This implies that in numerical investigations one has to add to the right-hand side of the Navier–Stokes equation an infrared viscous term $-v'(-\Delta)^{-\gamma}\mathbf{v}$ in order to stop such an inverse energy flow. The exponent γ is rather arbitrary.

The first approach, due to Kraichnan [1967] and Batchelor [1969], proceeds along the lines of the Kolmogorov cascade picture for 3D. In the forward transport range, a direct cascade of enstrophy implies, by dimensional arguments, that the energy spectrum scales as $E(k) \sim k^{-3}$. Similarly, in the backward transport range, an inverse cascade of energy gives the scaling law $E(k) \sim k^{-\frac{5}{3}}$.

The evidence for the existence of cascade processes in 2D is not very clear. In the forward range, an energy spectrum steeper than k^{-3} implies that the most important interactions are not local in wave number space. A scaling k^{-3} is marginal, and the dimensional predictions based on a cascade process with local interactions in k-space are questionable [Rose and Sulem 1978]. In fact, most numerical investigations of the 2D Navier–Stokes equations report spectra which are significantly steeper than k^{-3} [Basdevant and Couder 1986, Benzi et al. 1986]. The true state of the 2D Navier–Stokes equations in the inverse cascade range is even more controversial. Some simulations report a $k^{-\frac{5}{3}}$ spectrum in agreement with the dimensional predictions [Frisch and Sulem 1984, Smith and Yakhot 1993]. However, the latest investigation over much longer time scales finds instead a k^{-3} spectrum extending to scales about one order of magnitude larger than the forcing [Borue 1994].

Finally, let us mention the fact that the scaling laws for the velocity increment $\delta v(\ell)$ are trivial. It is easy to show [Paladin and Vulpiani 1987a] that, because of the regularity properties of the Navier–Stokes equation in 2D [Rose and Sulem 1978], one has

$$\langle |\delta v(\ell)|^p \rangle \sim \ell^p.$$

The interesting structure functions are those of the vorticity increments

$$\langle |\omega(x + \ell) - \omega(x)|^p \rangle \sim \ell^{x_p},$$

which are expected to scale with anomalous exponents x_p [Benzi et al. 1990].

In any case, scaling laws are not able to characterize the statistical properties in a complete way. Indeed, a typical vorticity field of a two-dimensional turbulent flow is characterized by the presence of long-lived coherent vortices immersed in a low-energy background turbulent field, see, for example, Benzi et al. [1986], [1987], Babiano et al. [1987]. While the latter is reasonably well-described by a classic Batchelor–Kraichnan statistical approach, the coherent vortices behave as individual entities which cannot be studied with standard statistical approaches.

Chapter 3

Reduced models for hydrodynamic turbulence

3.1   Dynamical systems as models of the energy cascade

A direct numerical simulation of the Navier–Stokes equations in the turbulent regime, i.e. at high *Re*, is difficult since the number of degrees of freedom which is necessary to describe the flow increases as a power of the Reynolds number (see section 2.3.5). However, these degrees of freedom are probably organized in a hierarchical way, so that simplified dynamical systems could be useful for the understanding of the scaling invariance. Borrowing some ideas introduced by the renormalization group approach for second-order phase transitions, one can also argue that the statistical properties of intermittency, at least in isotropic homogeneous turbulence, might be quite independent of the detailed dynamics of the Navier–Stokes equations. If that is true, it would imply that dynamical systems sharing the same 'symmetries' of the Navier–Stokes equations should be characterized by the same intermittency effects. Unfortunately, it is neither clear whether we already know all the 'symmetries' of the Navier–Stokes equations nor whether universality arguments, à la renormalization group, can be assumed in a theory of turbulence. In order to improve our present knowledge, we need to study and hopefully solve simplified models with the same 'phenomenological' properties as the Navier–Stokes equations. It is thus useful to analyse chaotic dynamical systems which model the energy cascade, instead of the complete Navier–Stokes equations, using an approach to the intermittency problem proposed by many authors, such as Gledzer [1973], Siggia [1978], Grappin et al. [1986] and others. These dynamical systems are closer to the original problem

48

than random multiplicative processes and exhibit a large variety of interesting properties. In this chapter, we mainly discuss the so-called GOY model, a shell model introduced by Ohkitani and Yamada [1989] which has one complex variable per shell and can be regarded as a generalization of the Gledzer [1973] model, where the variables are real. In these models, there exist two quadratic invariants (energy and helicity) and volumes in phase space are conserved, in the absence of viscosity and forcing.

The basic idea of shell models is to consider a discrete set of wave vectors, 'shells', in Fourier k-space, and construct a set of ordinary differential equations, taking into account only a few, typically one or two, variables per shell. The form of the coupling terms among the various shells is chosen in accordance with the main symmetries of the Navier–Stokes equations. Standard shell models have a relatively small number of degrees of freedom, so that they can be analysed as a dynamical system. The set of ordinary differential equations are derived under the assumption that the most relevant mechanism for the behaviour of the velocity field, $\mathbf{v}$, is given by a cascade transfer from large to small scales. In the huge literature that exists today on shell models, one finds interesting results about the properties of static solutions for the Desnyansky and Novikov model as well as numerical and analytical studies of the GOY model.

3.2 A brief overview on shell models

Shell models for the energy cascade were proposed in the early 1970s and mainly developed by the Russian school [Obukhov 1971, Desnyansky and Novikov 1974a,b and Gledzer 1973]. At the beginning, the idea was to find a particular closure scheme which is able to reproduce the Kolmogorov spectrum in terms of an attractive fixed point of appropriate differential equations for the velocity field averaged over shells in Fourier space. The velocity variable u_n is given by the mean energy of the nth wave number shell

$$u_n(t) = \sqrt{\int_{k_n}^{k_{n+1}} 2\,E(q,t)\,\mathrm{d}q} \tag{3.1}$$

and can also be regarded as the velocity increment $|v(x) - v(x+\ell)|$ on an eddy of scale $\ell \sim k_n^{-1}$. A shell of radius

$$k_n = r^n k_0 \qquad \text{with } r > 1 \tag{3.2}$$

contains the wave numbers with modulus k such that $k_n < k < k_{n+1}$ and the standard, although arbitrary, choice for the ratio between neighbouring shells is $r = 2$. The wave numbers k_n are thus equidistant on a logarithmic scale.

The main criteria for building up the evolution equations are the following:

(a) the linear term for u_n is given by $-vk_n^2 u_n$;

(b) the nonlinear terms for u_n are quadratic combinations of the form $k_n u_{n'} u_{n''}$;

(c) in the absence of forcing and damping, the energy $\frac{1}{2} \sum_n |u_n|^2$ is conserved;

(d) the interactions among shells are local in k-space (i.e. n' and n'' are close to n).

These criteria directly stem from an analysis of the Navier–Stokes equations, apart from point (d), which, at this level, is like a sort of closure approximation.

On the other hand, a more ambitious goal is to introduce shell models, which can be regarded as truncations of the Navier–Stokes equations. In this case the variable u_n represents the Fourier transform of the velocity field rather than an average value, and the dynamics of the corresponding shell model should be chaotic. Moreover, one has to impose at least the further constraint of the conservation of volume in phase space in the absence of forcing and damping. In this context, point (d) is still a reasonable assumption, since one expects that only local interactions are relevant for the energy transfer. In fact, the locality of the energy cascade in k-space is supported by recent calculations using diagrammatic techniques [L'vov and Procaccia 1995]. The main qualitative difference with the Navier–Stokes equations is that k is a scalar and the spatial structures, such as vortices or filaments, are lost. This is a weak point of shell models as 'realistic' models of turbulence, and whether they are reasonable can be decided only a posteriori by the results. One hopes that the coherent structures widely observed in turbulent fluids do not affect the nature of scaling, at least on a qualitative level. In other words, one is looking for a chaotic dynamical system with a reasonable number of degrees of freedom (typically less than one hundred) that reproduces the phenomenology of turbulence *in vitro*. The advantage is the possibility of studying the strange attractor in phase space, with the standard tools of analysis of deterministic chaos such as fractal dimensions, Lyapunov spectrum, dynamical intermittency, multifractality. This might allow one to make the bridge between ergodic properties of the dynamical evolution on the strange attractors in an infinite-dimensional space and local scaling invariance in the real 3D space, which is at the very basis of the comprehension of turbulence from a statistical mechanics point of view.

Before starting with a discussion of the different shell models, we briefly describe a class of models for the turbulent cascade, introduced by Eggers [1992] and Beck [1994], that follow a similar approach, although with a discrete time stochastic dynamics. These models are in some sense a bridge between the phenomenological random β model and the shell models.

Eggers [1992] introduced a stochastic model of turbulence which conserves energy and ensures equipartition of energy in the inviscid limit. The model exhibits anomalous scaling laws whose exponents can be calculated analytically.

In Beck [1994], the length scales are still separated on shells, $\ell_k = 2^{-k}$ and instead of having a constant driving force at large scales, as in the GOY model, the system at large scales is driven by a simple deterministic, chaotic map $T(x)$, which is taken as the fully developed logistic map $T(x) = 1 - 2x^2$. It appears to be important that the map is not close to the onset of chaos in order to get results which are close to experimental data. Expressed in the velocity variable $u_k(n)$ at the kth shell and at time τn, the Beck model takes the form

$$x(n+1) = T[x(n)],$$
$$u_1(n+1) = \lambda_1 u_1(n) + x(n),$$
$$u_k(n+1) = \lambda_k u_k(n) + c\xi_{k-1}(n)(1 - \lambda_{k-1})u_{k-1}(n),$$

where $\xi_k(n)$ are independent random variables uniformly distributed in $[0,1]$, c has the role of β^{-1} in the β model, and $\lambda_k = b^k$, with $b < 1$, are damping factors which take into account the viscous effects. The random term in the evolution equation of u_k mimics the momentum loss of the $(k-1)$th level. In this model, the PDF for the velocity differences is in very good agreement with experimental data and also the structure function exponents compare nicely with experiments.

3.2.1 The model of Desnyansky and Novikov

Desnyansky and Novikov [1974b] introduced a model where the energy is conserved and the interactions are only among nearest neighbour shells. These constraints lead to the following evolution equations for the real variables u_n:

$$\left(\frac{d}{dt} + \nu k_n^2\right) u_n = k_n \left(u_{n-1}^2 - 2u_n u_{n+1} - 2^{1/3} C \left(u_{n-1}u_n - 2u_{n+1}^2\right)\right) + f_n \quad (3.3)$$

with $n = 0, 1, \ldots, \infty$, boundary condition $u_{-1} = 0$ and an external forcing independent of time on the first shell, $f_n = f \delta_{n,0}$. Note that when $\nu = f = 0$, the volume in phase space is not conserved in contrast with the case of the Navier–Stokes equations. The parameter C is not fixed and determines the type of asymptotic scaling of the variables u_n. Bell and Nelkin [1977] made an accurate analysis of the model at varying C. Independently of the C-value, in the limit of an infinite number of shells, the equations have an attractive fixed point corresponding to a spectrum

$$E(k_n) \sim k_n^{-5/3} F(k_n/k_D), \quad (3.4)$$

where $k_D = (\epsilon/v^3)^{1/4}$ is the inverse of the Kolmogorov length, and the energy dissipation is

$$\epsilon = \sum_n f_n u_n = f u_0. \tag{3.5}$$

For $C < 1$, one recovers the Kolmogorov 5/3 law with

$$\lim_{x \to 0} F(x) = F_0 \neq 0, \tag{3.6}$$

while for $C > 1$, the function $F(x)$ is non-analytic in x, namely for $x \to 0$

$$F(x) \sim x^{-\xi}, \qquad \xi = 2\log_2 C. \tag{3.7}$$

In the limit $v \to 0$, the corresponding spectrum scales as

$$E(k) \sim k^{-(5/3+\xi)} \qquad (C > 1) \tag{3.8}$$

and the energy dissipation vanishes as

$$\epsilon \sim v^{9\xi/(8+3\xi)}. \tag{3.9}$$

In other words, the energy is not dissipated on small scales and should cascade towards large scales. This in contrast with the 3D phenomenology of turbulence where the energy dissipation per unit mass ϵ is observed to saturate to a finite non-zero value as $Re \to \infty$.

Let us stress again that models governed by a stable fixed point cannot reproduce intermittency and have a global scaling invariance with trivial power laws for the structure functions. Here, for instance, one has $\zeta_p = p/3$ for $C < 1$ and $\zeta_p = (\xi - 1)/2 + p/3$ for $C > 1$.

3.2.2   The model of Gledzer, Ohkitani and Yamada (the GOY model)

In order to obtain a model of the intermittency of energy dissipation, it is necessary to consider a shell model which exhibits a chaotic dynamics instead of having a fixed point as an asymptotic solution. In this case, the fluctuations of the chaoticity degree in general generate a multifractal probability measure on the strange attractor of the system in a high-dimensional phase space. Such a behaviour is expected to induce a breaking of the global scaling invariance and corrections to the Kolmogorov laws for the structure functions. Here we discuss the GOY model where the variables u_n are complex, the interactions are among nearest and next nearest neighbour shells, and there is conservation of volume in phase space, beyond energy conservation. Under these constraints, the evolution equations have the form

$$\left(\frac{\mathrm{d}}{\mathrm{d}t} + v k_n^2 \right) u_n = \mathrm{i} (a_n k_n u_{n+1}^* u_{n+2}^* + b_n k_{n-1} u_{n-1}^* u_{n+1}^* + c_n k_{n-2} u_{n-1}^* u_{n-2}^*)$$
$$+ f \delta_{n,4}, \tag{3.10}$$

with $n = 1, \ldots, N$, $k_n = r^n k_0$ ($r = 2$), and boundary conditions

$$b_1 = b_N = c_1 = c_2 = a_{N-1} = a_N = 0, \tag{3.11}$$

f is an external forcing, usually on the fourth mode.

In a numerical integration the number of shells N has to be chosen in such a way that the inertial range ends at a shell $n_D \approx N$ (typically $n_D = N - 3$). The coefficients of the nonlinear terms must follow the relation

$$a_n + b_{n+1} + c_{n+2} = 0 \tag{3.12}$$

in order to satisfy the conservation of $\sum_n |u_n|^2$ (energy) when $f = v = 0$. Moreover, they are defined modulo a multiplicative factor (related to a time rescaling), so that one can fix $a_n = 1$. In some respects, this model can be regarded as a crude truncation of the Navier–Stokes equations rather than a set of closure equations, since the main symmetries of the Navier–Stokes equations are respected. The constraints (3.12) still leave a free parameter δ so that one can set

$$a_n = 1, \qquad b_n = -\delta, \qquad c_n = -(1-\delta), \tag{3.13}$$

and equation (3.10) becomes

$$\left(\frac{d}{dt} + vk_n^2 \right) u_n = i k_n \left(u_{n+1}^* u_{n+2}^* - \frac{\delta}{r} u_{n-1}^* u_{n+1}^* - \frac{1-\delta}{r^2} u_{n-1}^* u_{n-2}^* \right) + f \delta_{n,4},$$

with $r = k_{n+1}/k_n = 2$. The parameter δ plays an important role in both the static and the dynamical properties of the model since it is related to a second quadratic invariant of the equations. Indeed, the nonlinear terms of the shell model conserve the quadratic quantities

$$Q = \sum k_n^\alpha |u_n|^2, \tag{3.14}$$

with $z = r^\alpha = 2^\alpha$ satisfying the equation

$$1 - \delta z - (1-\delta)z^2 = 0, \tag{3.15}$$

with roots $z = 1$ and $z = \delta - 1$ corresponding to

$$\alpha = 0 \quad \text{and} \quad \alpha = -\ln_r(\delta - 1). \tag{3.16}$$

The first solution corresponds to energy ($Q = E$), while the second one corresponds either to an enstrophy-like invariant

$$\Omega_\alpha = \sum_n k_n^\alpha |u_n|^2, \qquad \delta > 1, \tag{3.17}$$

or to a helicity-like invariant

$$H_\alpha = \sum_n (-1)^n k_n^\alpha |u_n|^2, \qquad \delta < 1, \tag{3.18}$$

where the power $\alpha = -\ln_r |1 - \delta|$. Kadanoff [Kadanoff et al. 1995] was the first to remark that one has to choose $\alpha = 1$ in order to construct an invariant with the correct dimensions to play the role of the 'true' helicity $H = \int \omega(\mathbf{x}) \cdot \mathbf{v}(\mathbf{x}) d\mathbf{x} = \sum \mathbf{k} \times \mathbf{v}(\mathbf{k}) \cdot \mathbf{v}(\mathbf{k})$ in the shell model. In fact, from a 'realistic' point of view, there are only two relevant δ values. The first value is $\delta = 1 + r^{-2} = 5/4$, giving $\alpha = 2$, so that the second quadratic invariant is the enstrophy

$$\Omega = \sum k_n^2 |u_n|^2, \tag{3.19}$$

as in two-dimensional turbulence. The second value is $\delta = 1 - r^{-1} = 1/2$, giving $\alpha = 1$, so that the second quadratic invariant is the 'helicity'

$$H = \sum_n (-1)^n k_n |u_n|^2, \tag{3.20}$$

as in three-dimensional turbulence.

The value $\delta = 1$ is thus the threshold between two classes of GOY models: a family of 2D-like models with conservation of the generalized enstrophy Ω_α and a family of three-dimensional type with conservation of generalized helicity H_α. Both these families have been studied at varying δ. In the former case a direct cascade of generalized enstrophy Ω_α is expected for $\alpha \leq 2$ [Yamada and Ohkitani 1988a], in contrast with the direct cascade of energy in the latter case [Biferale et al. 1995a]. However, a study of the truly two-dimensional shell model corresponding to $\alpha = 2$ [Aurell et al. 1994b] has revealed that in this case there is no direct enstrophy cascade and the dynamical behaviour can be explained in terms of a formal statistical equilibrium very similar to the approach to two-dimensional inviscid hydrodynamics of Onsager, Hopf and Lee (see section 2.1). In this model, there exist a forward flux of enstrophy and a backward flux of energy that are due to mean diffusive drifts from a source to two sinks, as we shall discuss in section 3.7. It can be shown that formal statistical equilibrium is present for $1 < \delta \leq 5/4$, while for $5/4 < \delta < 2$ there is a direct cascade of generalized enstrophy [Ditlevsen and Mogensen 1996].

It is important to stress that the scaling law of Kolmogorov ($u_n \sim k_n^{-1/3}$) for all δ values is a fixed point of the inviscid unforced equations when $N \to \infty$ and the infrared (small n) boundary conditions (3.11) are neglected. In section 3.3.1 we show that the Kolmogorov scaling remains a fixed point of the shell model with forcing and finite viscosity, and plays a key role in the dynamics. Actually, the existence of a second quadratic invariant and thus of a second fixed point seems a necessary, although not sufficient, condition for observing multifractal scaling laws. It is an intriguing and open problem to understand whether helicity conservation is also relevant for intermittency in the Navier–Stokes equations.

Let us mention that recently the GOY model has been studied by varying both the parameter δ and the ratio between neighbour shells $r = k_{n+1}/k_n$ [Schörghofer et al. 1995]. It is also possible to introduce GOY models with two free coefficients, without imposing $a_n + b_{n+1} + c_{n+2} = 0$, so that the two quadratic invariants are the generalized helicity (3.20) and the generalized energy $E_y = \sum_n |u_n|^2 k_n^{-y}$ with $y > 0$, instead of energy [Ditlevsen 1996, Ditlevsen and Mogensen 1996].

Another interesting generalization of the GOY model has been proposed by Biferale and Kerr [1995] and Benzi et al. [1996]. They noted that the structure of the second quadratic invariant in the GOY model is not fully consistent with the helicity in the Navier–Stokes, even if they have the same dimensions and are not positive-definite. Indeed, the equations of the GOY model have an asymmetry between odd and even shells which does not have any counterpart in physical flows. To overcome this problem one has to introduce two dynamical variables per shell, one transporting positive helicity u_n^+ and the other negative helicity u_n^-. In order to choose the nonlinear interaction term in the evolution equations, it is possible to look at the triad interactions in the Fourier-helicity decomposition of the Navier–Stokes equations in Fourier space [Waleffe 1992], where the two independent components of the velocity field at each wave number correspond to two pure helical waves. The resulting equations are

$$\left(\frac{d}{dt} + \nu k_n^2\right) u_n^+ = i k_n \left(u_{n+1}^- u_{n+2}^+ + b\, u_{n-1}^- u_{n+1}^+ + c\, u_{n-1}^- u_{n-2}^+\right)^* + f^+ \delta_{n,4},$$

$$\left(\frac{d}{dt} + \nu k_n^2\right) u_n^- = i k_n \left(u_{n+1}^+ u_{n+2}^- + b\, u_{n-1}^+ u_{n+1}^- + c\, u_{n-1}^+ u_{n-2}^-\right)^* + f^- \delta_{n,4},$$

where the coefficients b and c are given by two algebraic equations obtained by enforcing respectively the conservation of the energy

$$E = \sum_n |u_n^+|^2 + |u_n^-|^2$$

and of the helicity

$$\Omega = \sum_n k_n \left(|u_n^+|^2 - |u_n^-|^2\right).$$

The first equation is $1 + rb + r^2 c = 0$, while for the second there are the four possibilities $1 - r^2 b + r^4 c = 0$ (model 1), $1 - r^2 b - r^4 c = 0$ (model 2), $1 + r^2 b - r^4 c = 0$ (model 3), $1 + r^2 b + r^4 c = 0$ (model 4). Two of these models (1 and 4) coincide with the 3D and 2D versions of the GOY model. On the other hand, model 2 and model 3 exhibit different properties. In particular, model 3 has an intermittent dynamical behaviour and scaling exponents ζ_p that are more stable under changes of the free parameters than in the GOY model.

The fact that the wave numbers k_n are equidistant in a logarithmic scale allows one to simulate flows with very high Reynolds numbers. The shell number N is usually chosen by imposing the condition that the inertial range is as large as possible, say $n_D = N - \tilde{n}$ with $\tilde{n} = 3 - 5$, where $\tilde{n}$ is the shell number in the dissipative range. Therefore, once the viscosity is fixed, there is an optimal N

and when $v \to 0$, $N - \tilde{n} \sim -\ln v$ since the viscous cut-off $K_D = k_0\, 2^{n_D}$ diverges with the power of the viscosity (in the K41 theory $K_D \sim \langle \epsilon \rangle^{1/4}\, v^{-3/4}$).

3.2.3   Hierarchical shell models

The shell models discussed up to now cannot describe spatial structures, such as vortices or filaments, since they consider only one variable for each shell. To take the geometrical structure into account, it is necessary to introduce a more realistic description of the turbulent flow. The basic idea, based on the existence of scaling laws, is to consider for the shell n a set of variables $\mathbf{u}_{n,j}$, where $j = 1, \ldots, N_n = L^n$, and L is an integer larger than 1. Each variable $\mathbf{u}_{n,j}$ interacts only with the 'mother' $\mathbf{u}_{n-1,j_M}$, 'grandmother' $\mathbf{u}_{n-2,j_{GM}}$, the L 'daughters' on the shell $n+1$ and the L^2 'granddaughters' on the shell $n+2$. See Fig. 3.1 for the hierarchical structure of the couplings.

Therefore, the evolution equation has the form

$$\frac{\mathrm{d}\, u_{n,j}^{(a)}}{\mathrm{d}\, t} = \sum_{n_1, n_2, j_1, j_2, b, c} C^{(b,c)}(n_1, n_2, j_1, j_2) u_{n_1, j_2}^{(b)}\, u_{n_2, j_2}^{(c)} - v k_n^2 u_{n,j}^{(a)} + F_{n,j}^{(a)}, \qquad (3.21)$$

where a, b, c take values $1, 2, 3$, the coefficients $C^{(b,c)}(n_1, n_2, j_1, j_2)$ are $O(k_n)$ if n_1 and n_2 are in the range $[n-2, n+2]$ and zero in the other cases, and $F_{n,j}^{(a)}$ is a forcing term. The introduction of the variables $\mathbf{u}_{n,j}$ is basically equivalent to expanding the velocity field $\mathbf{u}(\mathbf{x}, t)$ in momentum space in terms of the Weierstrass–Fourier series

$$\mathbf{v}(\mathbf{x}, t) = \sum_{n,j} \mathbf{u}_{n,j}(t) e^{i k_{n,j} \cdot \mathbf{x}}, \qquad (3.22)$$

where $|\mathbf{k}_{n,j}| = 2^n$, $j = 1, \ldots, L^n$.

For 3D turbulence, hierarchical models have been widely studied by Grossmann and co-workers [Eggers and Grossmann 1991, Grossmann and Lohse

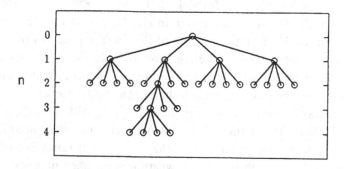

Figure 3.1 Hierarchical tree with $L = 4$.

1992]. For 2D, Aurell et al. [1994a] have introduced a similar model where, instead of the Weierstrass–Fourier series, a wavelet expansion is used:

$$\mathbf{v}(\mathbf{x}, t) = \sum_{n,j} \mathbf{u}_{n,j}(t) \psi_{n,j}(\mathbf{x}), \qquad (3.23)$$

where $\psi_{n,j}(\mathbf{x})$ is a wavelet function with characteristic length k_n^{-1}.

These sophisticated models, however, have a very large number of degrees of freedom and their numerical simulation is almost as difficult as the standard numerical simulations of the Navier–Stokes equations, e.g. by pseudo-spectral methods, although they allow one to reach a much higher Reynolds number. We shall not enter in a detailed discussion of the hierarchical models and refer the reader to the literature [Eggers and Grossmann 1991, Grossmann and Lohse 1992, Aurell et al. 1994a].

3.2.4 Continuum limit of the shell models

It is interesting to define an appropriate continuous limit of the GOY model. The most direct way is to consider the ratio between the shells as $r \to 1$. One thus obtains the partial differential equation for the complex function $u(k, t)$ [Parisi 1990]:

$$\frac{\partial u^*}{\partial t} = -\mathrm{i} \left(k u^2 + 3 k^2 u \frac{\partial u}{\partial k} \right) + F(k) - \nu k^2 u^*. \qquad (3.24)$$

If we neglect the forcing $F(k)$ and the viscosity, the energy $E(t) = \int |u(k, t)|^2 \, \mathrm{d}k/k$ is a conserved quantity and the, apparently unusual, $\mathrm{d}k/k$ integration step comes from the original logarithmically equispaced shell structure. It is also worth stressing that in x-space, (3.24) becomes a non-local integro-differential equation. The discretization of (3.24) with k equispaced on a logarithmic scale leads back to finite difference equations very similar to those of the GOY model if energy conservation is enforced. However, in this straightforward continuum limit, the second quadratic invariant (3.18) disappears and all GOY models describing three-dimensional behaviour (i.e. with control parameter $\delta \in [0, 1]$) lead to the same partial differential equation (PDE). It is therefore doubtful that equation (3.24) has the same qualitative behaviour as (3.10) with $\delta = 1/2$, since the existence of a second quadratic invariant seems to be necessary to obtain a chaotic time evolution and multifractal scaling laws. It is an open question to find a one-dimensional PDE that is able to model the energy cascade process with chaotic dynamics.

It is, however, possible to find two coupled PDEs which satisfy both energy and helicity conservation by taking the continuum limit of the Biferale–Kerr models defined at the end of section 3.2.1. In fact, models 1, 2 and 3 lead to the same two coupled PDEs for u^+ and u^-, namely

$$\partial_t u^+(k) = ik \left(4ku^- \partial_k u^+ + 2ku^+ \partial_k u^- + (2+\alpha)u^+ u^- - \alpha u^- u^- \right)^*$$
$$-\nu k^2 u^+ + f(k),$$

and the equation for u^- is obtained by changing all helicity indices.

Taking the continuum limit is an irreversible process: trying to come back to a logarithmically equispaced k space, one does not recover the original equations. However, two conserved quantities survive (in the unforced and inviscid limit) corresponding to the continuum analogous to energy and generalized helicity:

$$E = \int \frac{dk}{k}(|u^+|^2 + |u^-|^2), \qquad H_\alpha = \int \frac{dk}{k}k^\alpha \left(|u^+|^2 - |u^-|^2 \right).$$

The most interesting difference between the continuous expression and the analogue for the GOY model is that now also helicity conservation is well defined. It is possible that this set of PDEs has a much richer dynamics than the corresponding PDE (3.24) obtained from the GOY model.

3.3 Dynamical properties of the GOY models

The numerical simulations of the GOY model are performed by using standard algorithms for the integration of ordinary differential equations such as the second-order Adams–Bashforth scheme

$$u_n(t + \Delta t) = e^{-\nu k_n^2 \Delta t} u_n(t) + \frac{1 - e^{-\nu k_n^2 \Delta t}}{\nu k_n^2} \left(\frac{3}{2}g_n(t) - \frac{1}{2}g_n(t - \Delta t) \right), \quad (3.25)$$

where $g_n(t)$ stands for the right hand side of (3.10). This 'slaved scheme' is particularly convenient since it takes into account the fast damping of the high-wave number modes.

3.3.1 Fixed points and scaling

In correspondence to the two quadratic invariants, the GOY model has two types of static solutions (fixed points of the equations) characterized by the same scaling exponents:

(1) The Kolmogorov-like solution

$$u_n^{K41} = k_n^{-1/3} g_1(n),$$

with $g_1(n)$ being a periodic function of period three.

(2) The fluxless-like solution

$$u_n^{fl} = k_n^{-z} g_2(n),$$

where still $g_2(n)$ is a function of period three and $z = (-\log_2(\delta - 1) + 1)/3$.

As long as one is interested in scaling laws, the presence of superimposed periodic oscillations could seem disappointing. Indeed this means that there is a set of fixed points u_n^{K41} with the same modulus and different phases. Nevertheless, the existence in the phase space of such a K41-like manifold, instead of a single point, will turn out to be relevant for the dynamical properties of the model.

In order to focus only on the power law scaling it is useful to study the static behaviour of the ratios:

$$q_n = u_{n+3}/u_n. \qquad (3.26)$$

Note that the same set of observables has already been used to describe some exotic (chaotic) behaviours of the energy cascade in a different class of shell models [Biferale et al. 1994].

In terms of the q_n's, a static, unforced and inviscid solution of equations (3.10) can be generated by the iterations of the following one-dimensional complex rational linear transformation:

$$q_n = \frac{\delta}{2} + \frac{(1-\delta)}{4q_{n-1}}. \qquad (3.27)$$

The map (3.27) has two fixed points q^{K41}, q^{fl} corresponding to the two possible scaling behaviours for the u_n's:

(1) $q^{K41} = \dfrac{1}{2} \rightarrow u_n \sim u_n^{K41},$

(2) $q^{fl} = \dfrac{(\delta - 1)}{2} \rightarrow u_n \sim u_n^{fl}.$

The first fixed point is ultraviolet (UV) stable for $0 < \delta < 2$ and infrared (IR) stable for any other value of δ. For the second fixed point the stability properties are, of course, opposite. For UV (IR) stable we mean that the fixed point is asymptotically approached by starting from any initial condition and by iterating the ratio-map (3.27) forward (backward). From a physical point of view, forward (backward) iteration of the map (3.27) means a static cascade of fluctuations from large (small) scales to small (large) scales. In the GOY model, the UV stability is the relevant one, since it has been constructed to model cascade processes from large toward small lengths. As long as the main dynamical mechanism driving the time evolution of (3.10) is a forward cascade of energy (like in 3D turbulence), that is for $0 < \delta < 1$, we expect that the system spends a large fraction of its time near the K41-like fixed points.

Let us also stress the importance of the parameter δ from a dynamical point of view. To do this, we introduce the total flux of energy, Π_n, through the nth shell [Pisarenko et al. 1993]:

$$\Pi_n = \text{Im} \left[k_n u_n u_{n+1} \left(u_{n+2} + \frac{(1-\delta)}{2} u_{n-1} \right) \right], \qquad (3.28)$$

where we have written only the terms coming from the nonlinear transfer of energy. From (3.28), one expects that on increasing the value of δ from 0 to 1, there is a depletion of the forward transfer of energy (the coefficient in front of the coupling term for small scales goes to zero). Indeed, numerical integration of the GOY model for $0 < \delta < 1$ [Biferale et al. 1995a] has shown that the main dynamical effect is a forward transfer of energy.

On the other hand, by setting, for example, $\delta = 5/4$, the fluxless-like point $u_n \sim k_n^{-1}$ dominates the dynamics, although it is UV unstable. In fact, the GOY model with $\delta = 5/4$ was introduced to describe a direct enstrophy cascade in 2D turbulent flows. This is actually false, since the model for this δ value gives a state of statistical equilibrium; see section 3.7 for a discussion.

Expression (3.28) for the flux of energy also clarifies why the static solution u_n^{fl} is called 'fluxless'. Whenever two shells u_{n+2} and u_{n-1} get trapped by this static fixed point, the flux throughout the shell n is completely inhibited, i.e. $\Pi_n = 0$ (apart from viscous and forcing terms). As we will see in the following, the presence of dynamical barriers for the forward cascade of energy is probably the main cause of the intermittent nature of the dynamical evolution.

3.3.2   Transition from a stable fixed point to chaos

It is interesting to study the dynamical properties of the GOY model in the whole 'forward-energy cascade' range of parameters ($0 < \delta < 1$) where the two quadratic invariants are the energy and the quantity $\sum_n (-1)^n k_n^\alpha |u_n|^2$ ($\alpha = -\log_2 |1 - \delta|$). Often the GOY model is studied numerically and analytically only for $\delta = 1/2$, where the second invariant corresponds to helicity. In this case, the most striking result is that the scaling exponents of the structure functions ζ_p are a nonlinear function of p, indicating the presence of intermittency which can be described by the multifractal approach [Jensen et al. 1991a]. Moreover, the values of ζ_p (for $\delta = 1/2$) are very similar to those measured experimentally in real fluids [Anselmet et al. 1984, van de Water and Herweijer 1996].

It is an open problem to relate the multifractality of energy dissipation in 3D real space to the multifractality of the natural probability measure on the strange attractor embedded in $2N$ phase space. One is tempted to argue that the second invariant is quite relevant for the detailed multifractal properties of intermittency since it controls the backward flow of the energy in the cascade.

It is remarkable that for $0 < \delta \leq 0.3843\ldots$ the GOY model with viscosity and forcing different from zero has a stable fixed point satisfying a Kolmogorov-like scaling in the inertial range.

For example, in Fig. 3.2, one sees the ratios q_n at the fixed point, as obtained from a numerical integration with $\delta = 0.05$. The numerical solution exactly coincides with the result predicted by the 'forward' iteration of the ratio-map

(3.27) in the inertial range (from the forced shell to the beginning of the viscous range).

The scaling at the fixed point is not exactly Kolmogorov like ($q_n = 1/2$ for all n's) because of the damped oscillation due to the fact that the ratio-map (3.27) does not start exactly at its fixed point. The oscillations decrease with increasing δ, but for small δ they completely mask the presence of the Kolmogorov scaling unless one considers a much larger number of shells. This is, obviously, an effect due to the presence of the infrared boundary conditions (3.11) at small n's.

It is worth stressing that it is possible to follow the fixed point (with non-zero viscosity and forcing) even for values of δ where it is unstable. By looking at the eigenvalues of the stability matrix of the fixed point, one can show that there is a Hopf bifurcation at $\delta_c = 0.3843$, since a couple of complex conjugate eigenvalues have a real part which passes from negative to positive [Biferale et al. 1995a].

The critical value δ_c, where the fixed point becomes unstable, can be estimated by a simple heuristic argument. Indeed, if the K41 fixed point is stable, the second quadratic invariant scales as

$$H_\alpha = \sum_n (-1)^n k_n^\alpha |u_n|^2 \sim \sum_n (-1)^n k_n^{\alpha-2/3}$$

with $\alpha = -\log_2|1 - \delta|$. It is thus natural to expect that the second invariant be-

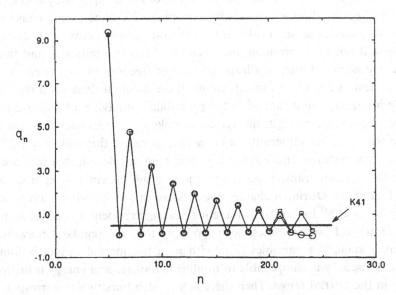

Figure 3.2 Values of the ratios q_n at the K41 fixed point of the GOY model equations with $\delta = 0.05$. Circles are the output of the numerical integration, while squares correspond to the values of the ratio map (3.27). The straight line corresponds to the exact K41 scaling $q_n = 1/2$ for all n's [Biferale et al. 1995a].

comes relevant and is able to intermittently reverse the forward energy transfer, when $\alpha - 2/3 > 0$, that is $\delta > \delta_c = 0.37\ldots$ Such an argument can be easily repeated for a shell thickness r other than 2, giving $\delta_c = 1 - r^{-2/3}$.

Above δ_c, a stable limit cycle appears with a period of $T_1 \approx 90$ natural time units (n.u.). This limit cycle loses stability at $\delta = 0.3953$, and for $0.3953 < \delta < 0.398$, the attracting set is a torus. The two periods of rotations are $T_1 \approx 90$ n.u. and $T_2 \approx 8$ n.u. At $\delta = 0.398$ there is a third transition to an aperiodic attractor with a positive maximum Lyapunov exponent. The transition to chaos thus seems well described by the Ruelle–Takens scenario. This fact does not mean, of course, that the transition to turbulence in real fluids follows such a scenario with increasing Reynolds number.

We have also to note that the transition from a fixed point to a chaotic attractor in the GOY model does not always occur via a Hopf bifurcation if one allows modifications in the forcing. For instance, with a forcing on the first shell instead of the fourth, instability can also occur via the passage of a single real eigenvalue through zero, in contrast with the Ruelle–Takens scenario. More interestingly, Kadanoff et al. [1997] have provided some numerical evidence that in the limit of an infinite number of shells (and $v \to 0$), an infinite number of eigenvalues becomes unstable simultaneously. In other words, they conjecture that the transition to chaos occurs via a phase transition scenario which is peculiar to high-dimensional systems.

In any case, at $\delta > 0.398$, the time evolution of the dissipative system (3.10) is chaotic and confined on a strange attractor in the $2N$-dimensional phase space. This result provides strong evidence that the interaction between shells plays a fundamental role in determining the strength of the intermittency, and that the correct symmetries still leave a large amount of freedom to the system.

Let us now add some comments about these different dynamical regimes. A possible interpretation is that when the probability of having a backward energy transfer is not large enough, the system is able to transfer energy in the most efficient way via a non-intermittent cascade. Above the threshold $\delta = 0.398$ for the transition to chaos, backward energy flows are so relevant that they are able to stop this type of transfer. As a consequence the system may charge energy on the first shells. During a charge, one observes a time-varying scaling, i.e. the velocity $|u_n| \sim k_n^{-h(t)}$ has an 'instantaneous' scaling exponent $h(t)$ which increases from $1/3$ toward larger values, corresponding to more regular behaviour. At a certain instant, the variables $|u_n|$ (with n in the inertial range) become so small that viscosity is comparable to nonlinear transfer and energy is dissipated directly in the inertial range. Then there is a sudden burst which corresponds to a discharge of the energy accumulated in the first modes. This is a completely different way of dissipating energy, which could be the origin of multifractality.

The charge–discharge scenario for intermittency has a counterpart in the Lyapunov analysis of the shell model, where only a few degrees of freedom seem to be relevant for the chaotic properties of the system. Although the Lyapunov dimension of the attractor (at least at $\delta = 0.5$) is proportional to the total number of shells of the GOY model, only a few Lyapunov exponents are positive and there is a large fraction of almost zero Lyapunov exponents. By an analysis of the Lyapunov eigenvectors, it can be shown that they correspond to marginal degrees of freedom which concentrate on the inertial range of wave numbers. There are only a few degrees of freedom which are chaotic in a very intermittent way. In fact, during the charge, the energy dissipation stays very low, and the effective maximum Lyapunov exponent is almost zero. When there is an energy burst, there is also a large chaoticity burst, i.e. a very large value of the effective maximum Lyapunov exponent, with a localization of the corresponding eigenvector on the dissipative wave numbers at the end of the inertial range (see section 3.4.2).

The existence of few active degrees of freedom, in a sea of marginal ones, suggests that, at least for the dynamics of some global observables, an appropriate one-dimensional map could capture the essence of the dynamics. To verify this idea, we can choose a variable which can be interpreted as the local singularity, or instantaneous scaling exponent of velocity, that is,

$$ h(t) = \frac{1}{3} \frac{1}{(N-13)} \sum_{n=7}^{N-7} \log_2 |u_n/u_{n+3}|. \tag{3.29} $$

The Kolmogorov scaling corresponds to $h = 1/3$, and a laminar signal has $h = 1$. The choice of the ratio $q_n = u_{n+3}/u_n$ is intended to minimize the effect of period-three oscillation proper to the fixed point structures.

We find numerically that for $\delta < 0.385$ the local singularity has the constant value $h = 1/3$ up to an error smaller than 10^{-2}, as expected. At $\delta = 0.396$, where the dynamics evolves on a torus, the scaling exponent $h(t)$ exhibits very small oscillations with two characteristic frequencies around $h = 1/3$. With increasing δ, the signal $h(t)$ becomes less and less regular, with a broadening of the probability distribution of h. The maximum scaling exponent is $h_{max} \approx 1$ in both cases, while the minimum one, h_{min}, decreases with δ. Note that a value $h(t) < 1/3$ corresponds to a velocity field more singular than that given by the Kolmogorov scaling. Such an instantaneous scaling exponent is realized during the fast energy burst due to the discharge, while during the charge the h-value slowly fluctuates in an almost regular way around $h \approx 1/3$ and eventually increases from $h \approx 1/3$ up to $h \approx 1$. We can thus hope to describe the most relevant features of the dynamics by looking at the one-dimensional map $h(t + \Delta t)$ versus $h(t)$ with an appropriate time delay Δt, which is shown in Fig. 3.3. Although it is quite noisy, it is somewhat similar to a map of the Pomeau–Manneville type. The channel close to the diagonal is due to the charge periods while the relaminarization corresponds to a fast energy burst (the discharge process) when a small $h(t + \Delta t)$ follows a rather large $h(t)$.

A further complication arises since we are dealing with a dynamical system with many degrees of freedom. Roughly speaking, the majority of them act as a noisy term which induces vertical (temporal) oscillation on the one-dimensional map. A picture close to the real mechanisms that are present in the model seems therefore to be a '1.5'-dimensional map. This will permit us to include, more accurately, the shell-time structure of the symmetries that govern the dynamics of the energy transfer. However, it is reasonable to expect that their statistical effect *on the mean quantities* is not very important, at least near the transition to chaos. Therefore, we have studied the two cases $\delta = 0.42$ (slightly above the transition) and $\delta = 0.5$ (the usual value for the shell model). One sees that the laminar channel of the 1D map becomes fatter at increasing δ, but the relaminarization mechanism is robust. As is well known, the dynamical behaviour of $h(t)$ may be affected by 'random' oscillations of the one-dimensional map $y = h(t + \Delta t)$ versus $x = h(t)$, close to the diagonal $x = y$. In particular, this mechanism may also be responsible for the broadening of the probability distribution of the instantaneous scaling exponent h with increasing δ. In practice, the presence of many marginal degrees of freedom is revealed by 'random' oscillations in the form of the one-dimensional map, without consequences for the qualitative picture.

It still remains an open problem whether the charge–discharge intermittency

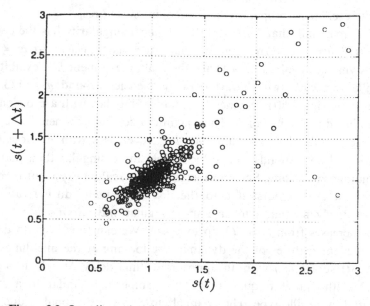

Figure 3.3 One dimensional map obtained by plotting the instantaneous scaling exponent $s(t + \Delta t) = 3\, h(t + \Delta t)$ versus $s(t)$ with $\Delta t = 0.6$ at $\delta = 0.42$ [Biferale et al. 1995a].

is also present in the Navier–Stokes equations as a consequence of the existence
of a second quadratic invariant.

3.3.3 The Lyapunov spectrum

It is widely accepted that the statistical properties of a turbulent flow can be
described by a dynamical evolution on a strange attractor in a high-dimensional
phase space. An important characterization of the ergodic properties of the
attractor is then given by the Lyapunov spectrum.

The GOY model with N shells has a chaotic attractor in a $2N$-dimensional
phase space for a large range of δ-values (in particular for $\delta = 1/2$ and
$\delta = 5/4$). Using standard techniques (see appendix C), one can thus compute
the $2N$ Lyapunov exponents $\lambda_1 \geq \lambda_2 \geq \ldots \geq \lambda_{2N}$.

However, when the number of shells $N \to \infty$ and the viscosity vanishes (with
$N \sim -\ln \nu$, so that the number of shells in the inertial range is proportional to
N), knowledge of the Lyapunov spectrum gives interesting physical information
only if there exists an asymptotic density function $\Lambda(x)$ with $x = i/D$, where D
is the dimension of the attractor and $i = 1, \ldots, 2N$ (see section 1.5.2).

The existence of this limit has been discussed by Ruelle [1982] in the Navier–
Stokes equations and numerically investigated in high-dimensional maps and
symplectic systems [Paladin and Vulpiani 1986, Livi et al. 1986, 1987, Eckmann
and Wayne 1988, 1989, Isola et al. 1990, Bohr et al. 1995], we shall discuss
this point again in section 4.3. The main question in this context is to find out
whether there is a finite density of positive Lyapunov exponents or whether it
vanishes for $\nu \to 0$, so that one has $\Lambda(x) = 0$ for $x \leq x_c$ ($x_c > 0$) and the system
becomes only marginally chaotic.

From the Lyapunov spectrum one can also estimate the fractal dimension D
of the strange attractor by the Kaplan and Yorke formula (1.13),

$$D = p + \frac{\sum_{j=1}^{p} \lambda_j}{|\lambda_{p+1}|},$$

where p is the maximum integer such that $\sum^p \lambda_j > 0$.

However, in a high-dimensional system, it is not sufficient to compute the
Lyapunov spectrum, but one also needs to determine the directions in phase
space where on average there is an expansion with rate λ_j. This allows one to
make a *Lyapunov–Fourier correspondence* between Lyapunov exponents λ_j and
a set of wave numbers.

To compute the Lyapunov spectrum in the GOY model, one considers the
linear variational equations

$$\frac{dz_i}{dt} = \sum_{j=1}^{2N} A_{ij} \cdot z_j, \qquad i = 1, \ldots, 2N \tag{3.30}$$

for the time evolution of an infinitesimal increment $z = \delta U$, where $A_{nj} \equiv \partial F_n / \partial U_j$ is the Jacobian matrix of (3.10), and $U \equiv (\text{Re}(u_1), \text{Im}(u_1), \ldots, \text{Re}(u_N), \text{Im}(u_N))$. The solution for the tangent vector z can thus be formally written as $z(t_2) = M(t_1, t_2) \cdot z(t_1)$, with $M = \exp \int_{t_1}^{t_2} A(\tau) d\tau$. The orthonormal Lyapunov basis is then given by the $2N$ eigenvectors f of the matrix $M^\dagger M$ in the limit $t \to \infty$, and depends on the starting point U in phase space. It is also possible to introduce [Orszag et al. 1987] a stability basis $e^{(i)}$ given by the eigenvectors of the matrix M. Note that a generic tangent vector $z(t)$ is projected by the evolution along $e^{(1)}$, i.e. $z(t) = c\, e^{(1)} \exp(\lambda_1 t)$, apart from small corrections $O(\exp - |\lambda_1 - \lambda_2| t)$.

Ohkitani and Yamada [1988] performed a detailed stability analysis for $\delta = 1/2$ and $\delta = 5/4$ and provided good numerical evidence that the density function $\Lambda(x)$ does exist in the limit $v \to 0$, as shown in Fig. 3.4.

Under this assumption, the Lyapunov dimension of the attractor D is related to the dissipative wave number k_D by $D \sim \log_2 k_D \sim -\ln v$. This means that the dimension of the attractor increases proportionally to the size of the inertial range since $k_n \sim 2^n$ and there are two degrees of freedom per shell. It is also interesting that the Kolmogorov entropy $H = \sum_{i=1}^{p} \lambda_i$ is proportional to the maximum Lyapunov exponent, indicating that the GOY model has a finite density of positive Lyapunov exponents. However, only a small fraction of the spectrum is formed by positive exponents, the dominating part consisting

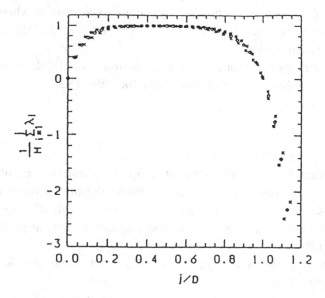

Figure 3.4 $\sum_{i=1}^{j} \lambda_i / H$ versus j/D, where H is the Kolmogorov entropy and D is the Kaplan–Yorke dimension, for the GOY model with $\delta = 1/2$, $v = 10^{-7}$, $10^{-8}, 10^{-9}$ [Ohkitani and Yamada 1988].

of slightly negative, almost vanishing Lyapunov exponents. A finite density of zero Lyapunov exponents has been obtained by Ruelle [1982] for the discrete spectrum of an operator that linearizes the Navier–Stokes equations in the (unifractal) beta model if the fractal dimension of the energy dissipation set is $D_F \leq 2.5$. This kind of 'weak chaos' [Chen et al. 1990] usually appears also in symplectic systems near integrability [Livi et al. 1987].

In contrast to the Lyapunov exponents, which give information about the relevant time scales, the corresponding Lyapunov tangent vectors can, as anticipated, resolve the interplay between various length scales during time evolution. Actually, the time average of the jth Lyapunov vector gives a measure of the localization on the various shells related to the instability of the jth Lyapunov exponent. Figure 3.5 shows the components of the time average of the Lyapunov eigenvector in the wave number basis k_i.

The main conclusions following from this figure are:

(a) Lyapunov eigenvectors related to the most negative Lyapunov exponents span the subspace of the dissipative modes following the end of the inertial scaling range. This strong *Lyapunov–Fourier correspondence* is

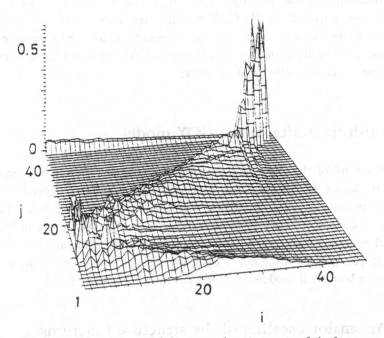

Figure 3.5 Time average of the squared components of the Lyapunov eigenvectors $\langle |f_i^{(j)}|^2 \rangle$ where $\mathbf{f}^{(j)}$ is the eigenvector corresponding to the Lyapunov exponent λ_j and the subscript i indicates its component on the ith shell, for the GOY model with $\delta = 1/2$. Note that the inertial range extends roughly between $i = 10$ and 40 and that the Lyapunov exponents λ_j with j between 10 and 40 are almost null [Ohkitani and Yamada 1989].

expected since the viscous damping is responsible of the strongest contraction rates, so that $\lambda_{2j} = \lambda_{2j-1} \approx -v k_j^2$ for $N \geq j \geq D$.

(b) More interestingly, a large proportion of the Lyapunov exponents are very close to zero. Their eigenvectors span a subspace in phase space corresponding to the inertial wave number shells. This weak *Lyapunov–Fourier correspondence* between eigendirections and inertial range suggests that power laws in turbulence are connected to the large number of marginal eigenvalues in the spectrum of $\mathbf{M}^\dagger \mathbf{M}$.

(c) There is a small fraction of positive Lyapunov exponents, probably a finite density, when $v \to 0$ (with $N \sim -\ln v$). However, there is no correspondence between positive Lyapunov exponents and small wave number shells in the 3D model. In fact, the time average of the first eigenvector $\mathbf{e}^{(1)}$ is not concentrated on the small wave numbers, but spreads in the whole inertial range [Jensen et al. 1991a].

These results indicate that only the positive Lyapunov exponents give rise to the intermittent corrections to power laws, as we shall discuss in section 3.4.2. In this picture, a vanishing density of positive Lyapunov exponents would be seen as the indication that the Kolmogorov scaling becomes valid as $Re \to \infty$. It is also worth noting that in the 2D GOY model, where there are no intermittency corrections to the classical value for the exponents of the structure functions (see section 3.7) the time average of $\mathbf{f}^{(1)}$ concentrates on small wave numbers and does not spread into the 'scaling' range.

3.4   Multifractality in the GOY model

In the framework of the GOY model, it is possible to analyse the link between anomalous scaling of the structure functions (the nonlinear dependence on p of the exponents ζ_p) and dynamical intermittency exhibited by the chaotic evolution. The relation between multifractality of energy dissipation in the 3D space and multifractality of the ergodic probability measure on the strange attractor in phase space is the major open problem of any theory of the scaling invariance in both shell models and fully developed turbulence.

3.4.1   Anomalous scaling of the structure functions

Numerical simulations of the model at $\delta = 1/2$ show that the exponents ζ_p are not linear in p, and can be fitted by the formula (2.62) given by the random β model with $x = 0.88$. The value of the free parameter x is very close to the one used to fit the experimental data of Anselmet et al. [1984] ($x = 7/8$). Moreover, the intermittency of the energy dissipation exhibited by the model is consistent with the multifractal approach described in chapter 2.

The exponents ζ_p can be expressed in terms of the generalized fractal dimensions D_q, related to the scaling of $\epsilon_{\mathbf{x}}(r)$, which is the coarse-grained energy dissipation given by the integral of the energy dissipation over a sphere of radius r around $\mathbf{x}$. The D_q's are defined via the relation [Halsey et al. 1986, Paladin and Vulpiani 1987a]

$$\langle \epsilon_{\mathbf{x}}(r)^q \rangle \sim r^{(q-1)(D_q-3)}. \tag{3.31}$$

As in the K41 theory, one usually assumes that the energy transfer rate is

$$\epsilon(r) \sim \frac{\delta v^3(\ell)}{r}. \tag{3.32}$$

In this way, the D_q are related to the multifractal spectrum of singularities of the velocity field:

$$D_q = \frac{\zeta_{3q} + 2q - 3}{q - 1}. \tag{3.33}$$

Shell models open up the possibility of testing the validity of the dimensional relation (3.32) and of the Taylor hypothesis [Taylor 1938]. This hypothesis plays a fundamental role in experiments since it makes it possible to infer the two-point statistics of an observable A, that is, the statistics of

$$\delta A(\mathbf{r}) = A(\mathbf{x} + \mathbf{r}, t) - A(\mathbf{x}, t)$$

from time measurements of $A(\mathbf{x}, t)$ at one point $\mathbf{x}$. For instance, suppose we have a flow in a channel with mean speed V, so that a fluid element at the point $\mathbf{x}$ will be at $\mathbf{x} + \mathbf{r}$ after a time delay $\tau = r/V$. Taylor then conjectured that the statistics of

$$\delta A(\mathbf{x}, \tau) = A(\mathbf{x}, t + \tau) - A(\mathbf{x}, t)$$

is the same as $\delta A(\mathbf{r}, t)$, for r much smaller than the typical large length of the system L (here the channel width). For homogeneous isotropic turbulence, this hypothesis can still be formulated by identifying V with the characteristic velocity of the energy containing eddies of length scale L, i.e. $V = \langle \delta V(L)^2 \rangle^{1/2}$. Such reasoning suggests studying the temporal signal of the energy dissipation

$$\epsilon(t) = v \sum_n k_n^2 |u_n(t)|^2$$

in order to construct a coarse-grained probability measure

$$p_i(\tau) = \int_{(i-1)\tau}^{i\tau} \epsilon(t) \, dt$$

and to determine the exponents $\overline{D}_q$ of the time scaling

$$\sum_i (p_i(\tau))^q \sim \tau^{(q-1)\overline{D}_q}.$$

If the Taylor hypothesis $\tau \sim r$ holds, we can estimate the generalized dimensions D_q defined in (3.31) by

$$D_q = \overline{D}_q + 2$$

since the time signal can be regarded as a generic intersection of a fractal structure with a line. In the GOY model this numerical determination of D_q provides clear evidence (see Fig. 3.6) that the dimensional arguments (3.32) and (3.33) are indeed correct.

Moreover, Fig. 3.7 shows that the GOY model has a PDF of the velocity increments which is in good agreement with the multifractal prediction (2.14), where $D(h)$ is obtained by the random β model fit of the structure function exponents of the shell model.

In section 3.5, we shall see that it is possible to obtain an independent analytic estimate of the exponents ζ_p by a closure theory approach [Benzi et al. 1993a]. The results are in good agreement with the values of the numerical simulations and provide a strong indication that the non-linearity of ζ_p versus p in the shell model is not an artefact of the finite number of shells considered.

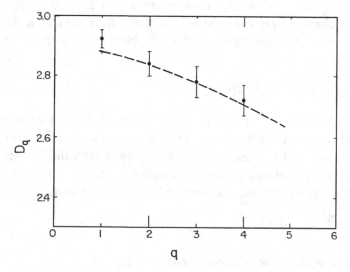

Figure 3.6 Generalized dimensions D_q (dots) obtained numerically compared against D_q (dashed line) obtained with equation (3.33) using ζ_p [Jensen et al. 1992].

3.4.2 Dynamical intermittency

In section 3.3.3, it was already conjectured that intermittency is due to the presence of a finite density of positive Lyapunov exponents. In fact, the average of the first eigenvector $\mathbf{e}^{(1)}$ is not concentrated on the small wave numbers, but spreads throughout the whole inertial range [Yamada and Ohkitani 1987, 1988a]. We want here to link the multifractal corrections with the time evolution of the effective maximum Lyapunov exponent and of its eigenvector. Therefore, one has to compute the response after a time τ to an infinitesimal perturbation, defining an effective maximum Lyapunov exponent as

$$\gamma_t(\tau) \equiv \frac{1}{\tau} \ln \left| \frac{\mathbf{z}(t+\tau)}{\mathbf{z}(t)} \right|.$$

The value of γ is an indication of the chaoticity of the system, at a given instant. The square modulus of the projection of the first eigenvector on the nth shell

$$p(k_n) \equiv \frac{|e_1(k_n)|^2}{\sum_j |e_1(k_j)|^2}$$

can be interpreted as the fraction of the largest instability localized over the shell k_n.

In the laminar phase the values of $p(k_n)$ for different n spread around the forced mode and over the whole inertial range while in the intermittent regime they are significatively different from zero only for few shells k_n around a dissipative shell k_D. The intermittent behaviour is thus produced by strong

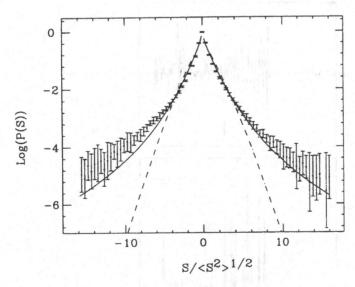

Figure 3.7 Log–linear plot of the PDF of the gradients $P(s)$. The full line is the multifractal prediction given by equation (2.14); the dashed line is the K41 prediction; dots with error bars are the numerical data [Biferale et al. 1992].

bursts of chaoticity along a direction corresponding to the dissipative shells followed by a contraction back to the laminar phase.

In order to give a quantitative description of the above scenario, we focus attention on one of the dissipation wave numbers $k_{15} = k_D$ in a numerical integration with $N = 19$ shells. The effective Lyapunov exponent γ, the energy dissipation ϵ estimated by $|u_{15}|^2$, and the chaoticity fraction on the shell k_D, estimated by $p_D \equiv p(k_{15})$, present a very strong temporal intermittency. The peaks of the three temporal sequences are strongly correlated; see Fig. 3.8.

The sequence shows that there is a correspondence between energy bursts and large deviations of γ and localization of the eigenvector on the dissipative wave numbers. Note the laminar period (very small ϵ) during the first 50 time units corresponding to almost vanishing γ and p_D. This indicates that strong chaoticity, that is, large values of the effective Lyapunov exponent, localizes on

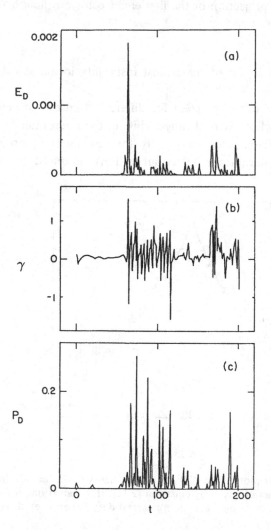

Figure 3.8 Temporal sequence (200 time units) of ϵ (a), the effective Lyapunov exponent γ (b), and p_D (c) for $N = 19$, $v = 10^{-6}$, $k_0 = 2^{-4}$ and $f = 0.005 \times (1 + i)$ [Jensen et al. 1991a].

dissipative wave numbers, corresponding to high values of the energy dissipation [Jensen et al. 1991a].

3.4.3 Construction of a 3D incompressible velocity field from the shell models

The shell models render it possible to generate an artificial turbulent velocity field $v(x, t)$ in the 3D real space, by a simple numerical algorithm. Such a field can be very useful for testing the presence of intermittent correction in the relative diffusion of passive contaminants without resorting to a direct integration of the Navier–Stokes equations.

Let us recall that the dynamical variables of shell models (e.g. u_n in the GOY model) can be considered as the coefficients of a Fourier expansion at wave vectors k in a shell of radius $|k| = k_n = k_0 2^n$. The choice of a discretization of the k-space via a geometrical progression of the k_n can appear somewhat strange for a standard Fourier transform. However, it is natural in the inertial range of fully developed turbulence where there are power laws for the velocity increments. For instance, to generate a one-dimensional self-affine field $f(x)$ whose increments scale with an exponent h, i.e. $|f(x + r) - f(x)| \sim r^h$, one can use the Mandelbrot–Weierstrass function

$$f(x) = \lim_{N \to \infty} \sum_{n=-N}^{N} 2^{-nh} \cos(2^n x + \phi_n),$$

where ϕ_n are random phases. The function $f(x)$ can be regarded as a Fourier decomposition with a geometrical discretization of the modes (see appendix F).

In order to build up a 3D incompressible velocity field, we should start by introducing a set of vectors k_n,

$$k_n = k_n e_n,$$

where $e_n = (e_n^{(1)}, e_n^{(2)}, e_n^{(3)})$ are vectors of unit norm. Each component $v_j(x, t)$, $j = 1, 2, 3$ of the field is thus obtained by a sort of inverse Fourier transform

$$v_j(x, t) = \sum_{n=1}^{N} C_n^{(j)} [u_n(t) e^{ik_n \cdot x} + \text{c. c.}].$$

The coefficients $C_n^{(j)}$ are random numbers $O(1)$. A careful choice is crucial to satisfy the incompressibility constraint. A moment of reflection shows that they should satisfy the condition

$$\sum_{j=1}^{3} C_n^{(j)} e_n^{(j)} = 0 \quad \forall n$$

to have div $v = 0$.

However, there is a subtle point here. The field of an intermittent shell model (such as the GOY model at $\delta = 1/2$) exhibits a multiaffine scaling after one performs both a spatial and a temporal average. However, at any fixed time, there is self-affine scaling when a spatial average is performed. This is because the intermittency in the GOY shell model is only a mirror of the dynamical intermittency since there is no spatial structure.

Such an inconvenience occurs even when using an algorithm for the construction of a multiaffine field that directly generalizes the standard procedure for obtaining the self-affine Mandelbrot–Weierstrass function (see appendix F and Benzi et al. [1993b]). On the other hand, a real turbulent field is expected to exhibit multiaffine scaling at any time. Such a feature can be reproduced by using the hierarchical shell models discussed in section 3.2.3. As when constructing a field that exhibits multiaffine scaling after performing only a spatial average, one has to deal with a wavelet expansion and not simply with a generalization of the Mandelbrot–Weierstrass function [Benzi et al. 1993b].

3.5   A closure theory for the GOY model

At present, there is no first principle derivation of multifractality in three-dimensional turbulence. It is therefore important to understand whether a possible derivation could be found in the context of a shell model.

Benzi et al. [1993a] proposed a closure approach to determine the stationary probability $P[u]$, of the set $[u]$ of all the variables u_n of the GOY model, for large n and in the zero viscosity limit. In this region, one may suppose that the results are universal in the sense that they do not depend on the detailed form of the forcing. Fortunately the form of $P[u]$ is not arbitrary: it is strongly constrained by closure equations which can be written as $\langle dA[u]/dt \rangle = 0$, where A is a functional of u and we use the equation of motion (3.10) to compute the derivative of A. Different choices of the function A lead to different closure equations. For $P[u]$ one can, in the fully turbulent regime, make the following ansatz:

$$P[u] \sim \exp\left(-\sum_n H_n\right), \tag{3.34}$$

where H_n is given by $H(u_n, u_{n-1}, u_{n+1}, u_{n-2}, u_{n+2} \ldots)$. It is easy to verify that the above expression is compatible with the closure equations. In other words we assume that the Hamiltonian, $H = \sum_n H_n$, corresponding to the stationary distribution is invariant under scale transformations (i.e. translations with respect to n). It is natural to suppose, in agreement with what happens in normal phase transitions, that H is essentially short range, i.e. the dependence of H_n on u_{n+m}

can be neglected for m sufficiently large. The other crucial assumption is that H_n is a homogeneous function of degree zero: it depends only on the ratio between the u's and their angles. This assumption automatically leads to a scaling law for u. In order to understand the meaning of these hypotheses on $P[u]$, we recall a useful theorem which states that under some conditions the probability distribution (3.34) may be generated by a random (multiplicative) process. Let us consider the simple case where $H_n = H(u_n, u_{n+1})$. If the integral equation

$$\int dy\, e^{-H(x,y)} \psi(y) = \lambda \psi(x) \tag{3.35}$$

has a solution with positive λ, then the function

$$P(x, y) = \frac{e^{-H(x,y)} \psi(y)}{\lambda \psi(x)}$$

is normalized (i.e. $\int dy P(x, y) = 1$). Thus we can construct a Markov chain where the conditional probability of having u_{n+1}, for given u_n, is just given by $P(u_n|u_{n+1})$. It is a very simple computation to verify that the probability distribution of the u's generated by this process is given by (3.34).

The fact that the transition probability P is a function of degree zero in the u's tells us that this process is essentially a random multiplicative process. More generally we are supposing that there is a conditional probability $P(..., u_{n-2}, u_{n-1}, u_n|u_{n+1})$ which is a homogeneous function of zero degree in the u's, whose dependence on the far away u's may be neglected. Moreover, the process generated by it is assumed to approach an equilibrium distribution. If we suppose that $P(..., u_{n-2}, u_{n-1}, u_n|u_{n+1})$ depends only on m variables, we remain with a function of $m+1$ variables to be determined, by imposing the condition that the closure equations are satisfied as well as possible.

In order to implement this method, consider the variables u_n defined as $u_n = k_n^{-1/3}\phi_n$. The equations for ϕ_n are (in the absence of forcing):

$$\left(\frac{d}{dt} + \nu k_n^2\right) \phi_n = i k_n^{2/3} \left(\phi_{n+1}^* \phi_{n+2}^* - \frac{1}{2}\phi_{n-1}^* \phi_{n+1}^* - \frac{1}{2}\phi_{n-2}^* \phi_{n-1}^*\right). \tag{3.36}$$

Now, from the variables ϕ_n, we define $\phi_n = \rho_n \exp(i\theta_n)$, and look for the equations of moduli ρ_n:

$$\left(\frac{d}{dt} + \nu k_n^2\right) \rho_n = k_n^{2/3} \left[\rho_{n+1}\rho_{n+2} \sin(\theta_n + \theta_{n+1} + \theta_{n+2})\right.$$

$$-\frac{1}{2}\rho_{n-1}\rho_{n+1} \sin(\theta_n + \theta_{n+1} + \theta_{n-1})$$

$$\left. -\frac{1}{2}\rho_{n-2}\rho_{n-1} \sin(\theta_n + \theta_{n-1} + \theta_{n-2})\right]. \tag{3.37}$$

In the following the variables $\Delta_n = \theta_{n-2} + \theta_{n-1} + \theta_n$ are used in order to simplify the notation. With this choice, equation (3.36) becomes

$$\left(\frac{d}{dt} + vk_n^2\right)\rho_n = k_n^{2/3}\left[\rho_{n+1}\rho_{n+2}\sin(\Delta_{n+2})\right.$$
$$\left. -\frac{1}{2}\rho_{n-1}\rho_{n+1}\sin(\Delta_{n+1}) - \frac{1}{2}\rho_{n-2}\rho_{n-1}\sin(\Delta_n)\right]. \quad (3.38)$$

In the inertial range, we can neglect the effect of viscosity, so that the Kolmogorov solution corresponds to $\rho_n = $ const. The reader should notice that the first nonlinear term in (3.38) is the transfer of energy from small scales, the last term is the transfer of energy from large scales, while the second term could be a transfer of energy either from the large or from the small scales depending on the sign of $\sin(\Delta_{n+1})$. We argue that in order to have a cascade of energy from large to small scales, $\sin(\Delta_{n+1})$ should be negative, at least on the average.

It is interesting to note that one can prove that the probability distribution of θ_n is uniform in the interval $[0, 2\pi]$. Indeed, by simply taking

$$\theta_{3n} \rightarrow \theta_{3n} + 2\epsilon,$$
$$\theta_{3n-1} \rightarrow \theta_{3n-1} - \epsilon, \qquad\qquad (3.39)$$
$$\theta_{3n+1} \rightarrow \theta_{3n+1} - \epsilon,$$

for any n, we transform a solution of the equation of motion into another solution. Because of the existence of this $U(1)$ symmetry, ϵ can be chosen in a random uniform way, and therefore the variable θ_n must be uniformly distributed (another symmetry of the equation consists in changing the sign of the real part of the u's). The fact that the phases θ_n are uniformly distributed in $[0, 2\pi]$ does not imply that the Δ_n are uniformly distributed. Thus even in the Kolmogorov picture we should introduce some phase coherence in order to satisfy the requirement of an energy cascade. Next, consider the time average $\langle ... \rangle$ for the moment of order p of ρ_n. For n in the inertial range we obtain

$$0 = \langle \rho_n^p \frac{d\rho_n}{dt} \rangle = \langle \rho_n^p \rho_{n+1}\rho_{n+2}S_{n+2}\rangle$$
$$-\frac{1}{2}\langle \rho_n^p \rho_{n+1}\rho_{n-1}S_{n+1}\rangle - \frac{1}{2}\langle \rho_n^p \rho_{n-1}\rho_{n-2}S_n\rangle, \qquad (3.40)$$

where we have introduced the variables $S_n = \sin(\Delta_n)$. The aim is to solve equations (3.40) for all p by using the idea that a multiplicative process could represent a reasonable approximation of the equal time probability distribution of the real dynamical system.

The starting point is the hypothesis that

$$\rho_{n+1} = a_{n+1}\,\rho_n, \qquad\qquad (3.41)$$

where a_n is a random variable to be specified. By substituting equation (3.41) into (3.40) we obtain:

$$\langle a_{n+2}a_{n+1}^2 a_n^{p+2} a_{n-1}^{n+2} S_{n+2}\rangle - \frac{1}{2}\langle a_{n+1}a_n^{p+1} a_{n-1}^{p+2} S_{n+1}\rangle$$

$$- \frac{1}{2}\langle a_n^p a_{n-1}^{p+1} S_n\rangle = 0. \tag{3.42}$$

In order to solve these equations, one must specify the correlation among the a_j and the S_n. We first assume that a_n are uncorrelated variables (among themselves). This is quite a strong assumption which has been numerically tested. It turns out that the correlation $Q_n(m) = \langle a(n+m)a(n)\rangle - \langle a(n+m)\rangle\langle a(n)\rangle$ is nearly zero already at $m = 1$, for n in the inertial range.

Next, assume that

$$a_n = C(1 - \beta S_n). \tag{3.43}$$

As a consequence of these two assumptions the S_n are uncorrelated variables. Thus a_n could depend only on S_n or on S_{n+1}, in order to maintain statistical independence of a_n. The assumption (3.43) gives the Kolmogorov scaling law for $C = 1$ and $\beta = 0$. In order to find a different solution it is convenient to introduce the moment:

$$\Pi_p = \langle (1 - \beta S_n)^p\rangle, \tag{3.44}$$

where $\langle ...\rangle$ should be considered the average of the stochastic process βS_n. Using (3.43) in (3.42) one obtains

$$2C^6\Pi_{p+2}^2\Pi_2\langle(1-\beta S)S\rangle - C^3\Pi_{p+1}\Pi_{p+2}\langle(1-\beta S)S\rangle$$

$$- \Pi_{p+1}\langle(1-\beta S)^p S\rangle = 0. \tag{3.45}$$

Given the probability distribution of S we can regard (3.45) as a set of equations $F_p(\beta) = 0$. It is not clear at this stage whether this infinite set of equations can be simultaneously satisfied for the same values of β and C.

Let us now consider the equation for $p = 1$. We obtain

$$2C^6\Pi_3^2 - C^3\Pi_3 - 1 = 0. \tag{3.46}$$

The solutions are $C^3\Pi_3 = 1$ and $C^3\Pi_3 = -1/2$. Because the a_j are positive definite, the only physical solution is $C^3\Pi_3 = 1$, corresponding to $\zeta_3 = 1$. Equation (3.46) is a consequence of the assumption that we have made on the independence of the a_n among themselves. Furthermore, C is not an independent quantity but is fixed by $C^3 = 1/\Pi_3$. We can get a better insight into (3.45) by using the following trick. Assume that $\beta \neq 0$ and define $X = 1 - \beta S$, so that $S = (1 - X)/\beta$. By noticing that $\Pi_p = \langle X^p\rangle$, we obtain

$$2C^6\Pi_{p+2}^2\Pi_2(\Pi_1 - \Pi_2) - C^3\Pi_{p+1}\Pi_{p+2}(\Pi_1 - \Pi_2) - \Pi_{p+1}(\Pi_p - \Pi_{p+1}) = 0. \tag{3.47}$$

These equations show that the Kolmogorov solution $\Pi_p = 1 = C$. Also they can be used to compute Π_p as function of Π_1 and Π_2. Indeed from (3.47) for $p = 1$ we have an equation for C^3. Next from $p = 2$ we can compute Π_4 as a function of (Π_1, Π_2). From $p = 3$ we can compute Π_5 and so on. At this stage one can check whether or not the assumptions on the nature of the multiplicative process are at least consistent with the numerical results discussed in section 3.4. Notice that this check is independent of the probability distribution of S. In order to perform this check we compute the exponent ζ_p, linked to Π_p in the following way:

$$\zeta_p = p/3 - \log_2(C^p \Pi_p). \tag{3.48}$$

Figure 3.9 shows ζ_p computed from Π_1, Π_2 by (3.48) and ζ_p obtained numerically. Here Π_1 and Π_2 are chosen by imposing the condition that the values of ζ_1 and ζ_2 coincide with the numerical data of section 3.4. As can be seen, the numerical agreement is quite good. This indicates that the assumption on the multiplicative process can be considered to be a good first approximation.

3.6   Shell models for the advection of passive scalars

The equation for the time evolution of a scalar field Θ passively carried by the fluid is

$$\partial_t \Theta + (\mathbf{v} \cdot \nabla)\Theta = D\Delta\Theta + f_\Theta, \tag{3.49}$$

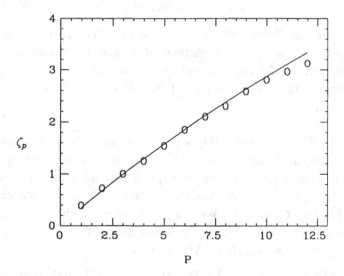

Figure 3.9 Circles are the values of ζ_p obtained from a numerical integration of (3.10). The solid line is computed from (3.48) by inserting in (3.47) the values of Π_1, Π_2 coincident with the corresponding numerical results [Benzi et al. 1993a].

where **v** is the velocity field, given by the Navier–Stokes equations with appropriate boundary conditions. Here f_Θ is an external forcing and D is the diffusion coefficient. A first characterization of the statistics of passive scalars is given by the correlation function

$$C(\mathbf{r}) = \langle \Theta(\mathbf{x})\,\Theta(\mathbf{x}+\mathbf{r}) \rangle,$$

where $\langle \ldots \rangle$ means spatial average. The theory of Obukhov [1949] and Corrsin [1951], which is the analogue of the K41 theory for the passive scalar, predicts that its power spectrum

$$\Gamma(k) = \int_{|\mathbf{k}|=k} \mathrm{d}\mathbf{k} \int C(\mathbf{r}) e^{-\mathrm{i}\,\mathbf{k}\cdot\mathbf{r}}\,\mathrm{d}\mathbf{r}$$

exhibits the power law decay $\Gamma(k) \sim k^{-5/3}$ in an inertial convective sub-range, where there is a nonlinear transfer of energy and molecular diffusion is negligible. For Prandtl number $Pr = \nu/D = O(1)$, this range coincides with the usual inertial range where inertia forces dominate and there exists an energy cascade towards small scales with a power law scaling for the energy spectrum $E(k)$. Hereafter we shall assume $Pr = 1$, for simplicity. The inertial range extends down to scales where the energy is dissipated by molecular viscosity.

The shell model for the passive scalar advection [Jensen et al. 1992] uses the velocity field of the GOY model.

The evolution equations are obtained according to the following criteria:

(a) The linear term for θ_n is given by $-Dk_n^2\theta_n$.

(b) The advection terms for θ_n are combinations of the form $k_n\theta_{n'}u_{n''}$.

(c) n' and n'' are nearest and next nearest neighbours of n.

(d) In the absence of forcing and damping one has conservation of volume in phase space (Liouville theorem) and the conservation of $\sum_n |\theta_n|^2$.

(e) The ('non-intermittent') scaling laws of Obukhov and Corrsin $\Gamma(k) \sim k^{-5/3}$, i.e. $\theta_n \sim k^{-1/3}$ is an unstable fixed point of the inviscid unforced evolution equations.

Properties (a), (b) and (d) are also valid for equations (3.49) in Fourier space, while (c) is an assumption on the locality of interactions among modes.

The equations of the cascade model with N shells can be written as

$$\left(\frac{\mathrm{d}}{\mathrm{d}t} + Dk_n^2\right)\theta_n = \mathrm{i}[e_n\,(u_{n-1}^*\theta_{n+1}^* - u_{n+1}^*\theta_{n-1}^*) + g_n\,(u_{n-2}^*\theta_{n-1}^* + u_{n-1}^*\theta_{n-2}^*)$$
$$+ h_n\,(u_{n+1}^*\theta_{n+2}^* + u_{n+2}^*\theta_{n+1}^*)] + \bar{f}\delta_{n,4}, \qquad (3.50)$$

where θ_n are complex variables. The u_n evolve according to the GOY model (3.10), D is the molecular diffusion and $\bar{f}$ is an external forcing (here on the fourth mode). The conservation of phase space (for $\nu = D = f = \bar{f} = 0$) is automatically satisfied by the absence of diagonal terms proportional to u_n, θ_n

on the right hand side of (3.50). The coefficients of the nonlinear terms follow from demanding the conservation of $\sum_n |\theta_n|^2$ in the absence of forcing and with $v = D = 0$. A possible choice is

$$e_n = \frac{k_n}{2}, \qquad g_n = -\frac{k_{n-1}}{2}, \qquad h_n = \frac{k_{n+1}}{2},$$

with

$$e_1 = e_N = g_1 = g_2 = h_{N-1} = h_N = 0.$$

The unstable fixed point of (3.10) and (3.50) when $v = D = f = \bar{f} = 0$ is given by the Obukhov–Corrsin scaling $u_n \sim \theta_n \sim k_n^{-1/3}$.

Let us briefly recall the predictions of the Obukhov–Corrsin theory, which makes use of arguments similar to those of the Kolmogorov K41 theory. By dimensional arguments, it is assumed that the time for the spectral transfer at $k \sim r^{-1}$ is $\tau(r) \sim r/\delta u(r)$, where $\delta u(r)$ is the velocity increment on a scale r. It follows that the transfer rate of Θ on a scale r is

$$N(r) = \frac{\delta \Theta^2(r)}{\tau(r)} \sim \frac{\delta u(r)\,\delta \Theta^2(r)}{r}; \tag{3.51a}$$

as in the K41 theory the energy transfer rate is

$$\epsilon(r) = \frac{\delta u^2(r)}{\tau(r)} \sim \frac{\delta u(r)^3}{r}. \tag{3.51b}$$

The classical theories ignore the intermittency in $N(r)$ and $\epsilon(r)$, which are assumed to be constant and equal to the corresponding dissipation (and injection) rates, N and ϵ, taken uniform in space and time. Under this hypothesis, the structure functions are easily derived to be

$$\langle \delta u^p \rangle \sim \epsilon^{p/3} r^{p/3} \sim r^{\zeta_p}, \qquad \zeta_p = p/3;$$

$$\langle \delta \Theta^p \rangle \sim N^{p/2} \epsilon^{-p/6} r^{p/3} \sim r^{H_p}, \qquad H_p = p/3.$$

However, there is clear evidence of strong fluctuations in the dissipation rates.

According the second revised similarity hypothesis of Kolmogorov [1962], the fluctuations of $\delta \Theta$ and δu are determined by the fluctuations of ϵ and N so that the scaling laws are given by

$$\langle \delta u^p \rangle \sim \langle \epsilon(r)^{p/3} \rangle r^{p/3}, \tag{3.52a}$$

$$\langle \delta \Theta^p \rangle \sim \langle N(r)^{p/2} \epsilon(r)^{-p/6} \rangle r^{p/3}. \tag{3.52b}$$

The problem of estimating the exponents H_p is quite complicated, even if one introduces the generalized dimensions d_q for the probability measure given by the passive scalar dissipation via the scaling

$$\langle N_x(r)^q \rangle \sim r^{(q-1)(d_q-3)}. \tag{3.53}$$

It is then simple to construct multifractal cascade models for N. However, knowledge of the two multifractal spectra is not sufficient to determine the scaling exponents H_p of the structure functions without taking into account the correlation between N and ϵ.

In order to continue our argument we should introduce the so-called joint multifractal distribution for N and ϵ. This goes beyond the scope of this presentation but we refer to the discussion in Jensen et al. [1992]. However, we shall give a simple expression that relates H_p to d_q and D_q, at least for not too large q's. Since $\epsilon(r)$ does not appear in combinations of the kind $\delta\Theta(r)^{2q}\delta u(r)^q$, we have

$$\langle \delta\Theta(r)^{2q}\delta u(r)^q \rangle \sim \langle N(r)^q \rangle r^q \sim r^{(q-1)(d_q-3)+q}. \tag{3.54}$$

Under the assumption that $\delta\Theta$ and δu are uncorrelated, i.e.

$$\langle \delta\Theta(r)^{2q}\delta u(r)^q \rangle \sim \langle \delta\Theta(r)^{2q} \rangle \langle \delta u(r)^q \rangle, \tag{3.55}$$

the structure function exponents become

$$H_{2q} = q - \zeta_q + (q-1)(d_q - 3). \tag{3.56}$$

Let us now discuss the results obtained with the shell model. The scaling $\langle |\theta_n|^p \rangle \sim k_n^{-H_p}$ does not follow the Obukhov–Corrsin prediction $H_p = p/3$, even for small p's, as shown in Fig. 3.10. In fact, H_p is a more strongly nonlinear

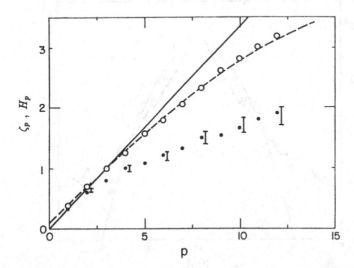

Figure 3.10 ζ_p (open circles) and H_p (dots) versus p; the data are obtained from a numerical integration of (3.10) and (3.50). The full line indicates the Obukhov–Corrsin prediction $\zeta_p = H_p = p/3$. The dashed line is the random β model prediction (2.62) with $x = 0.88$. The experimental data [Antonia and Van Atta 1978] are indicated by vertical bars, with a slight shift of p [Jensen et al. 1992].

function of p than ζ_p. Figure 3.10 also shows the experimental data of H_p. The agreement between the experimental values and the results obtained by the passive scalar model is good.

In particular, for $p = 2$ one finds $H_2 = 0.61 \pm 0.01$, and $\zeta_2 = 0.70 \pm 0.01$, so that the correction to the Obukhov–Corrsin exponent 5/3 for the power spectrum $\Gamma(k)$ has a sign opposite to that of the correction to the exponent for the energy spectrum $E(k)$, since $\Gamma(k) \sim k^{-(H_2+1)}$ and $E(k) \sim k^{-(\zeta_2+1)}$. This result has also been observed in real fluids [Antonia and Sreenevasan 1977]. The probability distribution function (PDF) of the passive scalar increments at different scales, i.e. the PDF of θ_n at varying n, is shown in Fig. 3.11.

For k_n at the beginning of the inertial range, that is, Θ differences on large scales, the PDF is quite close to a gaussian form. On the other hand, at the end of the inertial range, the PDF exhibits the typical exponential tails of velocity and Θ gradients in turbulence. Note that in model (3.50), for n at the end of the inertial range, k_n^{-1} is of the order of the Kolmogorov length and one can identify $k_n\theta_n$ with the gradient of Θ. The PDF of $\theta_{n=23}$ in Fig. 3.11 should thus be compared with the experimental PDF for the Θ gradients [Antonia et al. 1984, Castaing et al. 1990].

Moreover, the fractal dimensions d_q computed from the time signal of the Θ^2 dissipation are in good agreement with the experimental data of Antonia and

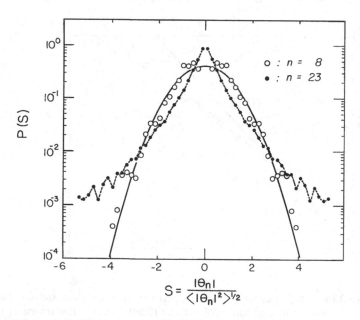

Figure 3.11 PDF of $s = \theta_n/\langle|\theta_n|^2\rangle^{1/2}$ for $n = 8$ (open circles) and $n = 23$ (dots). The full line is the standard gaussian. The PDF for negative s is drawn as an aid to the eye: it is taken equal to the PDF of the corresponding positive s-values [Jensen et al. 1992].

Sreenevasan [1977]. The numerical data indicate that the approximate relation (3.56) between H_q, d_q and ζ_q works rather well.

Let us conclude this section by mentioning that other shell models have also been introduced for many turbulent phenomena, such as turbulence in binary fluid mixtures [Ruiz and Nelson 1981], thermal convection [Brandenburg 1992, Suzuki and Toh 1995] and magnetohydrodynamics [Carbone 1994, 1995]. Their description is beyond the purpose of our book since it would require a detailed discussion of the phenomenology of those problems.

3.7   Shell models for two-dimensional turbulence

We have seen that for $\delta > 1$, in addition to energy conservation, the GOY model also conserves a generalized enstrophy $\sum k_n^\alpha |u_n|^2$, with $\alpha = 2$ for $\delta = 5/4$. If there is a forward transport of enstrophy, neglecting the intermittency fluctuation one has

$$\langle |W_n|^2 \rangle = \langle |k_n u_n|^2 \rangle \sim \epsilon_\omega^{\frac{2}{3}} k_n^0,$$

where ϵ_ω is the mean dissipation of enstrophy per unit time. Similarly in the inverse transport range one gets the estimate

$$\langle |W_n|^2 \rangle = \langle |k_n u_n|^2 \rangle \sim \epsilon^{\frac{2}{3}} k_n^{\frac{4}{3}}.$$

Let us introduce a new shell model motivated by the observation that the GOY model makes no difference between velocity and vorticity, which only differ by a scale factor. One could therefore hope that a shell model which preserves some trace of the vector structure of the velocity field will, in a qualitative sense, be closer to the Navier–Stokes equations.

In 2D one can write the vorticity equation in Fourier space as

$$\frac{d}{dt}\hat{\omega}(\mathbf{k}) = \sum_{\mathbf{k}+\mathbf{k}'+\mathbf{k}''=0} \frac{\mathbf{k}'\times\mathbf{k}''}{|\mathbf{k}'|^2}\,\hat{\omega}(\mathbf{k}')\,\hat{\omega}(\mathbf{k}'') - \nu|\mathbf{k}|^2\hat{\omega}(\mathbf{k}) + F(\mathbf{k}). \qquad (3.57)$$

One can model (3.57) by taking a few shell variables, $W_{n,j}$, $j = 0, 1, \ldots$, per shell, associating to each of them a wave vector $\mathbf{k}_{n,j}$ with length k_n and direction $\hat{e}_j$, e.g. in a hexagonal pattern, $\hat{e}_j = [\cos(2\pi j/3), \sin(2\pi j/3)]$ for $j = 0, 1, 2$. One possible set of interactions are then the same as in the 2D GOY model, but only between triples of shell variables having different directions:

$$\frac{d}{dt}W_{n,j} = (W^*_{n+1,j+1}W^*_{n+2,j+2} - 5W^*_{n+1,j+1}W^*_{n-1,j+2} + 4W^*_{n-2,j+1}W^*_{n-1,j+2})$$
$$+(W^*_{n+2,j+1}W^*_{n+1,j+2} - 5W^*_{n-1,j+1}W^*_{n+1,j+2} + 4W^*_{n-1,j+1}W^*_{n-2,j+2})$$
$$-\nu k_n^2 W_{n,j} - \nu' k_n^{-2\gamma} W_{n,j} + F_{n,j}. \qquad (3.58)$$

In (3.58) we have used the notation that $W_{n,j+3} = W_{n,j}$ and we shall call (3.58) the coupled GOY model. It has some moderate numerical advantages over the original GOY model. In the original GOY models with a steady force one observes that the shell energy spectrum in the inertial range exhibits oscillations superimposed on a mean power law [Pisarenko et al. 1993]. One has no trace of such oscillations in (3.58), so one concludes that three variables per shell are enough to sufficiently randomize the system and remove these undesired oscillations.

In the inviscid case the conservation laws and the Liouville theorem allow one to use an equilibrium statistical mechanics description, as discussed in section 2.1.1. Considering two cut-offs k_{min} and k_{max} in Fourier space, one has that the canonical ensemble distribution function is (see (2.14))

$$P(\hat{\omega}(\mathbf{k})) \sim \exp\left[-\left(\frac{\beta_1}{2}|\hat{\omega}(\mathbf{k})|^2 + \frac{\beta_2}{2}|\mathbf{k}|^{-2}|\hat{\omega}(\mathbf{k})|^2\right)\right], \qquad (3.59)$$

so that

$$\langle|\hat{\omega}(\mathbf{k})|^2\rangle = \frac{1}{\beta_1 + \beta_2|\mathbf{k}|^{-2}},$$

which, depending on $|\mathbf{k}|^2$ and the Lagrange multipliers β_1 and β_2, separates into two branches

$$\langle|\hat{\omega}(\mathbf{k})|^2\rangle \sim \begin{cases} |\mathbf{k}|^2/\beta_2, & |\mathbf{k}|^2 \ll \dfrac{\beta_2}{\beta_1}; \\[2mm] \beta_1^{-1}, & |\mathbf{k}|^2 \gg \dfrac{\beta_2}{\beta_1}. \end{cases} \qquad (3.60)$$

The corresponding result for the shell models previously considered is

$$\langle|W_n|^2\rangle \sim \begin{cases} k_n^2/\beta_2, & k_n^2 \ll \dfrac{\beta_2}{\beta_1}; \\[2mm] \beta_1^{-1}, & k_n^2 \gg \dfrac{\beta_2}{\beta_1}. \end{cases} \qquad (3.61)$$

The two branches in (3.61) simply correspond to shell energy equipartition and shell enstrophy equipartition. We thus have that in 2D shell models, but not in 2D hydrodynamics, a formal statistical equilibrium and a non-intermittent forward cascade of enstrophy give the same scaling law.

Let us briefly discuss the numerical results on the coupled GOY model (3.58); see Aurell et al. [1994b] for details. Figure 3.12 shows the second moment of shell vorticity vs. shell number.

The dominant overall feature is one branch at positive n's which is nearly flat, i.e. implies that $\langle|W_n|^2\rangle \sim k_n^0$, and one branch at negative n's which is closely fitted by $\langle|W_n|^2\rangle \sim k_n^2$. Both results are in agreement with the predictions of a statistical equilibrium.

Looking at Fig. 3.13, one recognizes that the spectrum in the range of forward transport is not quite flat, and that the 'k_n^0-range' seems to extend about five shells to the left of the forced shell $n = 0$. The second phenomenon is difficult to explain in the framework of the cascade picture. It is not a problem in the statistical equilibrium picture if we assume that the shell temperatures of enstrophy and energy (β_2^{-1}, β_1^{-1}) are such that the bend in the curve occurs somewhat to the left of the force.

The numerically observed distributions of the variables W_n are well fitted with a gaussian. Even the computation of the kurtosis for the shell enstrophy confirms the gaussian nature of the fluctuations of W_n. In addition one has that the mean values of the fluxes of both energy and enstrophy are always much smaller ($O(10^{-3}$–$10^{-4})$) than the standard deviations. The fluxes are thus always small corrections superimposed on a mean randomly fluctuating state. These results can be simply and coherently explained by a formal non-equilibrium statistical mechanics, close to local equilibrium [Aurell et al. 1994b].

In conclusion, the GOY shell model with $\delta = 5/4$ has a very poor connection with 2D turbulence. Therefore, a natural question is: why have shell models for 3D turbulence given reasonable results, while the 2D models have not? A qualitative answer goes as follows: the statistical equilibrium picture should be relevant if the time scales of relaxation to local equilibrium are faster than the time scales of transport of the conserved quantities to the viscous sinks.

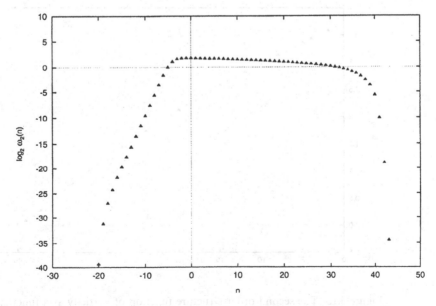

Figure 3.12 The second moment of shell vorticity as a function of shell number; logarithmic scale [Aurell et al. 1994b].

The time scale for relaxation to local equilibrium can be estimated as the local shell turnover times, which in 3D shell models decrease with shell number (as $t_n \sim k_n^{-\frac{2}{3}}$ at the Kolmogorov fixed point). The time scale for transport to the energy dissipation range from a shell in the forward transport range can then be estimated as the geometric sum of the local turn-over times up to the energy dissipation shell. Hence both time scales are of the same order, and it is unlikely that 3D shell models display statistical equilibrium. And indeed a cascade scenario is observed.

On the other hand, in 2D shell models the local turnover times in the forward range are constant, and the time to transport enstrophy to the dissipation range is proportional to how far away in shells that range is. Therefore, local statistical equilibrium has a chance to develop. It remains to explain why a similar argument does not work for the 2D Navier–Stokes equations, where, in the Batchelor–Kraichnan cascade picture, the time scales in the forward range are also constant. One possible answer is that one has to take into account the full non-linearities of the Navier–Stokes equations with the non-local transfer of energy and enstrophy.

On the other hand, our arguments can be pushed forward [Ditlevsen and Mogensen 1996] for the whole family of GOY models where the second invariant is the generalized enstrophy $\Omega_\alpha = \sum_n k_n^\alpha |u_n|^2$. Assuming a 'unifractal' scaling $|u_n| \sim k_n^{-h}$, the turn-over time on a scale k_n is $\tau_n \sim k_n^{h-1}$, by dimensional counting.

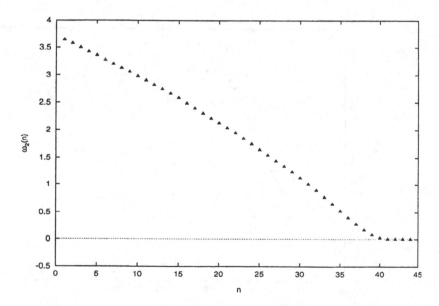

Figure 3.13 The second-order structure function of vorticity as a function of shell number in the range of forward transport of enstrophy, linear scale. The forcing is on the shell $n = 0$ [Aurell et al. 1994b].

In the framework of a generalized enstrophy cascade scenario, one should make a hypothesis of Kolmogorov type that the rate of generalized enstrophy transfer $k_n^\alpha |u_n|^2/\tau_n$ is constant, obtaining $h = (\alpha + 1)/3$, and therefore

$$\tau_n \sim k_n^{(\alpha-2)/3}. \tag{3.62}$$

Such a scaling is contradictory for $\alpha \geq 2$ (i.e. for non-singular velocity fields with $h \geq 1$), since it implies that the turn-over time τ_n of an eddy does not decrease on lowering its size. In this case, it is reasonable to expect the existence of a formal statistical equilibrium. In other words, the GOY shell models should exhibit a direct cascade of generalized enstrophy only for $\alpha < 2$, that is, in the range $5/4 < \delta < 2$. Neglecting the intermittency corrections, one obtains two different dimensional predictions on the exponents ζ_p controlling the structure functions of vorticity defined in the GOY model as

$$\langle |W_n|^p \rangle = \langle |k_n u_n|^p \rangle \sim k_n^{p+\zeta_p}.$$

In the presence of an enstrophy cascade with constant transfer rate, one has

$$\zeta_p = p \left(\frac{\alpha + 1}{3} \right) \qquad \text{for } \alpha < 2,$$

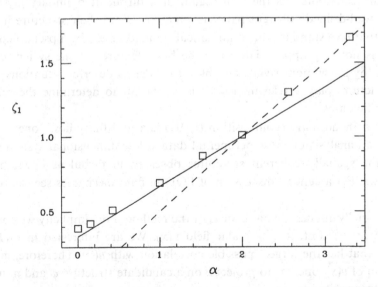

Figure 3.14 Value of the exponent ζ_1 versus $\alpha = -\log_2(\delta - 1)$ for GOY models where the second invariant is the generalized enstrophy $\sum_n k_n^\alpha |u_n|^2$. The squares are the numerical data of Ditlevsen and Mogensen [1996]. The full line is the result $\zeta_1 = (\alpha + 1)/3$ obtained by assuming a direct cascade of the generalized enstrophy; the dashed line is the statistical equilibrium prediction $\zeta_1 = \alpha/2$.

while for statistical equilibrium, repeating the same arguments that lead to (3.61), one has

$$\zeta_p = p \frac{\alpha}{2} \qquad \text{for } \alpha \geq 2.$$

The numerical simulations confirm these results, as shown in Fig. 3.14. They thus provide striking evidence that GOY models are described by a statistical equilibrium formalism for $1 < \delta \leq 5/4$.

3.8   Low-dimensional models for coherent structures

In section 2.3.4 we saw that the number of equations necessary to describe a turbulent flow increases as Re^b, where $b \simeq 9/4$, while the integration step decreases as Re^{-a}, where $a \simeq 1/2$. Present computers are able to perform a realistic simulation only up to $Re = O(10^3)$. On the other hand, dynamical systems with a few degrees of freedom, such as the GOY models, can reproduce the scaling of the structure functions and other statistical properties, e.g. the probability distribution of the velocity gradients, but they cannot describe any kind of coherent structures.

In many cases, such as the wall region of a turbulent boundary layer, one observes well-defined coherent patterns which maintain their structure in time. In this situation a standard direct numerical simulation, e.g. by a pseudo-spectral method, is not very appropriate since the basic features of the system can be described by a few, non-trivial, variables. In order to describe situations where large structures play a relevant role it is important to determine the relevant modes judiciously.

Proper orthonormal decomposition (POD) is a tool that allows one to build up, from an analysis of the experimental data, a low-dimensional system which models the spatially coherent structures observed in turbulent flows, mainly shear flows. For a general discussion of POD in fluid mechanics see Berkooz et al. [1993]

Let us briefly discuss the basic ideas of the method. For simplicity we consider the POD in the context of a scalar field $u(x)$. We are interested in finding a structure that has the largest possible correlation with $u(x)$. Therefore, given a realization of $u(x)$, one has to project u on a candidate structure ϕ and minimize the projection

$$u_\phi = \frac{(\phi, u)}{(\phi, \phi)^{1/2}} \tag{3.63}$$

in quadratic mean, where $(\,,\,)$ indicates the inner product in the Hilbert space

L_2. Denoting by $\langle \rangle$ the ensemble average, one thus has to maximize $\langle |u_\phi|^2 \rangle$. The calculus of variations reduces the problem to the following integral equation:

$$\int R(x, x')\phi(x')\mathrm{d}x' = \lambda\phi(x), \tag{3.64}$$

where

$$R(x, x') = \langle u(x)u^*(x') \rangle. \tag{3.65}$$

The theory of Hilbert–Schmidt operators guarantees that there exists a denumerable, complete, orthonormal set of eigenfunctions $\phi_k(x)$, called empirical eigenfunctions, such that

$$R(x, x') = \sum_k \lambda_k \phi_k(x)\phi_k^*(x'). \tag{3.66}$$

Moreover u can be reconstructed in the following way:

$$u(x) = \sum_k a_k \phi_k(x), \tag{3.67}$$

where

$$\langle a_k a_m^* \rangle = \lambda_k \delta_{km}. \tag{3.68}$$

The λ_k are ordered in such a way that the convergence of the representation is optimized. The mean square of a_1 is as large as possible, a_2 is the largest of the remainder of the series once the first term has been subtracted, and so on.

Using the POD and the Galerkin methods one obtains a system of ordinary differential equations which capture the maximum amount of kinetic energy, i.e. $\langle (u, u) \rangle$, among all the possible truncations of the same order. Taking into account N terms in the series (3.67) and inserting u in the partial differential equation which rules the evolution of u, e.g. the Navier–Stokes equation, one has a system of ordinary differential equations for the coefficients $a_1, a_2, \ldots$

$$\frac{\mathrm{d}\mathbf{a}}{\mathrm{d}t} = \mathbf{F}(\mathbf{a}).$$

The previous equation does not contain the interaction among a_k with $k \leq N$ and a_j with $j > N$, although these interactions can be included by some closure hypotheses. Therefore the term $\mathbf{F}(\mathbf{a})$ becomes $\mathbf{F}(\mathbf{a}) + \delta\mathbf{F}(\mathbf{a})$, where $\delta\mathbf{F}(\mathbf{a})$ mimics the effects of small scales.

This kind of approach has been used to model different phenomena, such as the jet–annular mixing layer [Zheng and Glauser 1990], 2D flows in complex geometries [Deane et al. 1991], and the Ginzburg–Landau equation [Sirovich 1990].

The best application of the POD is the one developed by Aubry et al. [1988] for the wall region in a turbulent boundary layer where organized structures, which manifest themselves as streamwise streaks in visualizations, are experimentally observed. The behaviour of these structures is intermittent in space and time. One has a succession of events in which their sudden breakup corresponds to large fluctuations in the turbulent energy production (bursting events).

Using the POD, the Galerkin method and a Heisenberg model to take into account the small-scale effects, one obtains a set of ordinary differential equations which is able to reproduce the bursting phenomenon. The model captures not only the static behaviour of the coherent structures, but also major features of their dynamics. It suggests that the bursts originate very close to the wall and shows how the pressure perturbation from the outer part of the boundary layer can trigger the occurrence of a burst, i.e. bursting frequency is determined by the pressure signal.

These results, which are in good agreement with experiments and direct simulations, show that a low-dimensional system, obtained in a judicious way, is able to describe complex situations in real flows.

Chapter 4

Turbulence and coupled map lattices

The precise modelling of turbulent phenomena generally requires the study of nonlinear systems of partial differential equations (PDE's). The numerical simulation of such PDE's is costly in computer time and does not always lead to much progress in terms of qualitative understanding of the phenomena involved. A class of much simpler models, with a correspondingly more distant connection to concrete physical phenomena, is provided by so-called coupled map lattices [Kaneko 1993], large systems of nonlinear maps connected by usually local interactions and situated on regular spatial lattices. These systems are thus discrete in both space and time. The maps that are coupled together are already, or can at least be, chaotic on their own, and thus these coupled map lattices will not provide any understanding of the origin of chaos in simple models. We shall see, however, that many features of turbulent systems, especially those involving coherence versus disorder, the spreading of turbulent regions and their characterization in terms of Lyapunov exponents and fractal dimensions, can be addressed very simply in terms of such coupled chaotic maps, which form a very direct transition from 'small' to 'large' systems, as discussed in Chapter 1. For these coupled map lattices, there is no simple continuum limit, since the chaotic nature of the maps depends crucially on the discreteness of the dynamics. On the other hand, we shall mainly discuss aspects that are related to large scales and thus they should not depend strongly on the details of the map lattices. We shall therefore not enter into a discussion of the many ordered patterns with length scales down to the lattice spacing, which have been studied by several authors. For this, we can refer the interested reader to the reviews by Crutchfield and Kaneko [1987] and Kaneko [1989b, 1993].

4.1   Introduction to coupled chaotic maps

Let us consider a d-dimensional hypercubic lattice of volume L^d where the sites are labelled by an index $\mathbf{i} = (i_1, i_2, \ldots, i_d)$, with $i_k = 1, 2, \ldots, L$. On each site one has a field $u_{\mathbf{i}}(n)$, where n indicates the discrete time variable. The dynamics of the field is given by [Kaneko 1984, 1993]

$$u_{\mathbf{i}}(n+1) = (1 - \epsilon)f(u_{\mathbf{i}}(n)) + \frac{\epsilon}{2d} \sum_{\mathbf{i}'} f(u_{\mathbf{i}'}(n)), \tag{4.1}$$

where $\mathbf{i}'$ are the 2D nearest neighbours of $\mathbf{i}$. Here and in the following, periodic boundary conditions are assumed if not otherwise stated. The function $f(x)$ is a nonlinear map that can sustain chaotic motion. The logistic map $f(x) = R x (1 - x)$ is the most popular choice. As is well known, this mapping of the unit interval into itself has a period-doubling cascade, i.e. a stable periodic orbit with period 2^n for $R_n < R < R_{n+1}$ with accumulation point $R_\infty = 3.569\,9456\ldots$ For $R_\infty < R < 4$ the attractor is either chaotic or periodic.

The map (4.1) has a homogeneous solution $u_{\mathbf{i}}(n) = M(n)$. This *mean field* solution is obtained by iterating the map $f : M(n+1) = f(M(n))$. As we shall see later, the fact that the function f appears also in the coupling term makes the linear stability analysis easy and ensures that a sensible 'continuum limit' exists above the onset of chaos, which occurs at the same R-value as for the single map f.

Another possible field dynamics is given by the linearly coupled lattice

$$u_{\mathbf{i}}(n+1) = (1 - \epsilon) f(u_{\mathbf{i}}(n)) + \frac{\epsilon}{2d} \sum_{\mathbf{i}'} u_{\mathbf{i}'}(n) \tag{4.2}$$

and here the onset of turbulence can occur below the onset of chaos for the single map f (see, for example, Kook et al. [1991]). As we shall see in section 4.2.4 the critical properties of the linearly coupled maps are different, but it is also not clear whether a sensible 'continuum limit' exists in this case [Kuznetsov 1993].

At first we shall restrict our attention to cases such as (4.1) and (4.2) which are symmetric in the spatial couplings. In section 4.5, we shall consider asymmetrical couplings in order to describe open flows, boundary layers and convection phenomena.

Let us recall that the transition to chaos for low-dimensional maps via period doubling is well understood by renormalization group techniques [Feigenbaum 1978, 1979]. We briefly give the main results of the theory that shall be used through this chapter without entering into the details. A fairly complete exposition in the context of one-dimensional maps can be found in the monograph by Collet and Eckmann [1980], and in Cvitanović [1989], a collection of original papers with a short but useful introduction. Feigenbaum discovered, at first numerically, an amazing universality in nonlinear maps of the interval which undergo period-doubling bifurcations, looking at the bifurcation points R_n, where the period 2^{n-1} cycle becomes unstable and a new stable period 2^n cycle appears. Feigenbaum realized that the ratio

$$\delta_n = \frac{R_{n-1} - R_n}{R_n - R_{n+1}} \qquad (4.3)$$

rapidly converges to the value $\delta = 4.669\,2016\ldots$ irrespective of the map f chosen, if f has a parabolic maximum.

In order to understand this result, one looks at the scaling properties after many iterations of f, by introducing a series of limiting functions

$$g_i(x) = \lim_{n \to \infty} \alpha^n f_{R_{n+i}}^{(2^n)}(x/\alpha^n), \qquad (4.4)$$

where $f_R^{(k)}$ denotes the kth iterate of f with control parameter equal to R, and a period k cycle is a trajectory $(x, f(x), \ldots, f^{(k)}(x))$ that satisfies $f^{(k)}(x) = x$. The doubling operator $\mathbf{T}$ is defined as

$$\mathbf{T}g_i(x) \equiv \alpha g_i[g_i(x/\alpha)] = g_{i-1}(x). \qquad (4.5)$$

It is then possible to show that there exists a universal scaling function

$$g(x) = \lim_{i \to \infty} g_i(x) \qquad (4.6)$$

which is a fixed point of the doubling operator. Thus it satisfies the so-called Feigenbaum–Cvitanović equation

$$g(x) = \mathbf{T}g(x) \equiv \alpha g[g(x/\alpha)], \qquad (4.7)$$

and this equation uniquely determines the universal value α. Note that the precise details of the starting map f are lost under the renormalization group transformation and one ends up with the same results within each universality class, determined only by the form of the maximum of f. Here and in the following, we consider the class of maps with a quadratic maximum, which is the generic case, and for these maps, $\alpha = -2.502\,907\ldots$ The constant δ appears as the relevant, unstable eigenvalue controlling small perturbations around the fixed point g of the doubling operator.

4.1.1 Linear stability of the coherent state

It is easy to investigate the linear stability of the flat chaotic state $u_i(n) = M(n)$, where $M(n+1) = f(M(n))$ is the mean field solution obtained by neglecting the spatial couplings. In the chaotic state long waves are unstable. A state which is almost uniform, i.e. of the form $u_i(n) = M(n) + z_i(n)$, will, to linear order in z, change to $M(n+1) + z_i(n+1)$, where

$$z_i(n+1) = f'(M(n)) \left((1 - \epsilon)z_i(n) + \frac{\epsilon}{2d} \sum_{i'} z_i(n) \right). \qquad (4.8)$$

By Fourier transformation, i.e. expanding z as

$$z_i(n) = \sum_q \hat{z}_q(n) e^{i q \cdot i}, \tag{4.9}$$

where $\mathbf{q} = (q_1, q_2, ..., q_d)$, we get

$$\hat{z}_q(n+1) = f'(M(n)) \left(1 - \epsilon + \frac{\epsilon}{d} \sum_{i=1}^d \cos q_i \right) \hat{z}_q(n). \tag{4.10}$$

A perturbation $z(0) e^{i q \cdot i}$ of the flat state therefore grows as $z(0) e^{[\lambda_0 + \delta \lambda_\epsilon(\mathbf{q})] n}$, where λ_0 is the Lyapunov exponent of the single map f, i.e.

$$\lambda_0 = \lim_{N \to \infty} \frac{1}{N} \sum_{n=1}^N \log |f'(M(n))| \tag{4.11}$$

and

$$\delta \lambda_\epsilon(\mathbf{q}) = \ln \left| 1 - \epsilon + \frac{\epsilon}{d} \left(\sum_{i=1}^d \cos q_i \right) \right|, \tag{4.12}$$

which, for small q, becomes

$$\delta \lambda_\epsilon(\mathbf{q}) \simeq -\frac{\epsilon}{2d} q^2. \tag{4.13}$$

The allowed wave vectors are of the form $\mathbf{q} = \frac{2\pi}{L}(n_1, n_2, ..., n_d)$, where n_k are integers. Thus, the first wave vector $\mathbf{q} = \frac{2\pi}{L}(1, 0, ..., 0)$ (and, because of the symmetry, similarly for any permutation of the coordinates) becomes unstable at system size L_1 determined by $\lambda_\epsilon(\mathbf{q} = (2\pi/L_1, 0, ..., 0)) = 0$, which for large L_1 means that $L_1 \simeq 2\pi \sqrt{\epsilon/(2d\lambda_0)}$.

4.1.2 Spreading of perturbations

Let us look at the spreading of a local perturbation in the coupled map lattice (4.1) by going back to (4.8). As a crude approximation, assume that $|f'|$ in (4.8) is constant and equal to or larger than unity. Let $\lambda = \ln |f'|$ and assume that the disturbance at $n = 0$ is completely localized, i.e. $z_i(0) = z(0) \delta_{i, i_0}$. Then one has

$$z_i(n) = z(0) e^{n\lambda} \sum_q e^{i q \cdot i + n \delta \lambda_\epsilon(\mathbf{q})}, \tag{4.14}$$

where $\delta \lambda_\epsilon(\mathbf{q})$ is given by (4.13). To study the speed of propagation of the disturbance we let the point $\mathbf{i}$ move through the lattice with velocity $\mathbf{v}$. Inserting $\mathbf{i} = \mathbf{i}_0 + \mathbf{v} n$ and looking at a very large system, where the spacing between q-vectors is very small, we get

$$z_i = z_{i_0 + v n} = z(0) e^{i q \cdot i_0} e^{n\lambda} \int e^{(i q \cdot v + \delta \lambda_\epsilon(\mathbf{q})) n} d\mathbf{q} \tag{4.15}$$

which, for large n, is dominated by the saddle point of the exponent, determined by the equation

$$i\mathbf{v} = -\nabla_\mathbf{q} \delta\lambda_\epsilon(\mathbf{q}).$$ (4.16)

The saddle point $\widetilde{\mathbf{q}}$ is in general complex and the exponential growth rate of (4.15) is then the real part of

$$\lambda(\mathbf{v}) \equiv \lambda + i\widetilde{\mathbf{q}} \cdot \mathbf{v} + \delta\lambda_\epsilon(\widetilde{\mathbf{q}}).$$ (4.17)

Small velocities $\mathbf{v}$ correspond to small $\widetilde{\mathbf{q}}$ since $i\mathbf{v} = \frac{\epsilon}{2}\widetilde{\mathbf{q}}$ in this limit. As a consequence, for small velocities one has

$$\lambda(\mathbf{v}) = \lambda - \frac{v^2}{\epsilon},$$ (4.18)

showing that there is exponential growth in any frame travelling with speed less than

$$v_s = \sqrt{\epsilon\lambda},$$ (4.19)

whereas the disturbance decays exponentially in frames moving more rapidly. Arguments of this type recur several times in this book, so we have included a more detailed treatment in appendix D.

Returning to the argument of section 1.5.4 that a generic large chaotic system has finite coherence length ξ, we expect the velocity c appearing in (1.16) to be related to v_s and thus to depend on the maximum Lyapunov exponent λ as (4.19). Therefore, in the limit of small λ, one has

$$\xi \leq \text{const } \sqrt{\frac{\epsilon}{\lambda}}.$$ (4.20)

Moreover, going back to (4.15) and using the approximation for $\delta\lambda_\epsilon(\mathbf{q}) \simeq -\frac{\epsilon}{2d}q^2$, valid for small $\mathbf{q}$, we can carry out the integral over $\mathbf{q}$ and find for a disturbance centred at $\mathbf{i_0}$:

$$z_\mathbf{i}(n) = z(0)\,e^{\lambda n - \frac{|\mathbf{i}-\mathbf{i_0}|^2}{\epsilon n}},$$ (4.21)

from which one can again read off (4.19) and the fact that the disturbance spreads by exponential growth of a gaussian envelope. To obtain these results we made the very rough approximation of neglecting spatial variations in f'. Temporally varying, spatially uniform choices of f' can easily be included. We simply have to replace λ in (4.21) by the (mean field) Lyapunov exponent.

Approaching the critical point $R = R_\infty$, we expect the coupled map lattice to become flatter and thus (4.21) to be better. We shall check this conjecture in the next section and see that, although the true Lyapunov exponent scales like the mean field one (the value obtained by iterating the single map f), their absolute values can be quite different owing to the spatial inhomogeneity of f'.

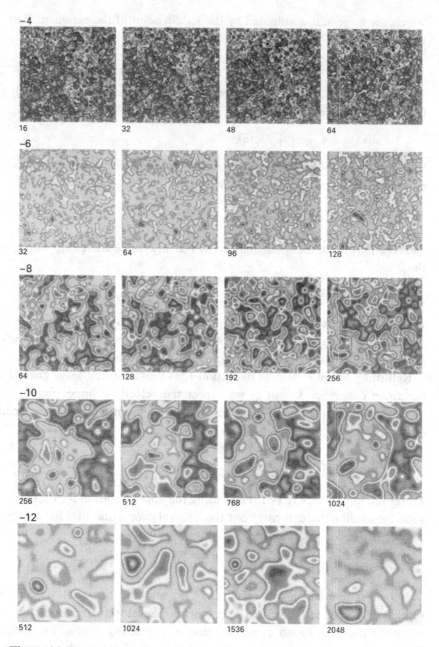

Figure 4.1 Evolution of the 2D CML (4.1) of logistic maps with $\epsilon = 0.4$ at different values of the non-linearity parameter $\rho(= -4, -6, -8, -10, -12)$ from the top to the bottom row. Each row shows four snapshots of the lattice, taken at multiples of the basic periodicity due to the chaotic bands. In each row the lattice is allowed to equilibrate for around 30 000 iterates [van de Water and Bohr 1993].

4.2 Scaling at the critical point

Many of the conjectures made above have been confirmed in 2D coupled logistic map lattices. Thus the scaling properties of Lyapunov exponents, length scales and spreading velocities have been investigated in the critical regime where each map is slightly above the onset of chaos ($R = R_\infty$) in van de Water and Bohr [1993], which we shall follow below. The reason for choosing 2D instead of 1D lattices is to avoid 'getting stuck' in special states. Coupled maps have a huge phase space and thus, in principle, a huge number of basins of attraction. In 2D, however the chances of getting stuck in a non-turbulent state is much smaller since the curvature of the borders of 'islands' (corresponding, for example, to different elements of a periodic cycle) causes them to shrink. As the critical point R_∞ is approached from above, the iterates of a single map move between chaotic bands. In terms of the rescaled non-linearity parameter $\rho = \log[(R - R_\infty)/R_\infty]$, the chaotic bands accumulate at $\rho = -\infty$ with the number of bands p scaling as $\ln_2 p = \rho/\ln \delta$, where δ is the Feigenbaum constant. This property is carried over to the coupled map lattice, although the bifurcations are displaced in ρ. As the number of bands diverges, their width vanishes. In the simulations by van de Water and Bohr [1993], the coupled map lattices were always started with random initial conditions in a single chaotic band, $u_i(0) = 0.3 + 10^{-6}\sigma_i$, where σ_i are pseudo-random numbers uniformly distributed in [0,1]. For most of the computations the lattice size was $L = 128$ and the boundary conditions were always periodic.

Figure 4.1 shows the state of the lattice at various values of ρ, many iterates after it has been started. Clearly the relevant length scales are different and the coherence length seems to increase with $|\rho|$. In the following we shall quantify this dependence. The frames at a particular value of ρ in Fig. 4.1 are taken at times n that are multiples of the basic period of the system coming from the chaotic bands of the map f. Therefore, they show the temporal evolution of the lattice within a particular chaotic band. The near periodicity of the system causes problems for the precise computation of a coherence length scale of the coupled map lattice. The problem is that the width of the coupled map lattice is set by the chaotic bands of the single maps f, and thus the scale of the field variable u varies with time.

4.2.1 Scaling of the Lyapunov exponents

In the 2D coupled logistic maps, there is clear numerical evidence that the maximum Lyapunov exponent scales as

$$\lambda \simeq A (R - R_\infty)^\beta, \tag{4.22}$$

where $\beta \approx 0.448$ (see Fig. 4.2). Note that the Lyapunov exponent of a single logistic map scales with $\beta = \ln 2 / \ln \delta = 0.4498\ldots$ [Huberman and Rudnick 1980]. The reason is that the scaling by the Feigenbaum δ, as discussed in section 4.1, is achieved by the doubling operator **T**, which combines two maps into one and thus reduces the number of iterations by a factor of two. Thus the maximum Lyapunov exponent of the coupled map lattice seems to scale with the same exponent β as found for a single map, although the non-universal amplitudes are different ($A \approx 0.725$ for the coupled maps and $A \approx 1.371$ for a single logistic map).

For a single map, β is defined by the envelope of an extremely spiky curve $\lambda(R)$. In each periodic window (forming a dense subset of the interval $[R_\infty, 4]$) there is an R such that $\lambda \to -\infty$. However, in the coupled map case, for sufficiently large L, these coherent periodic windows are never seen. Presumably many of the chaotic states observed in a system like (4.1) are transient in a technical sense (i.e. they are so-called chaotic repellers), but the transient times grow at least exponentially in L.

In this CML system, one can also test the conjecture, discussed in chapter 1, that the Lyapunov density exists in the thermodynamic limit. Figure 4.3 shows the number N_+ of positive Lyapunov exponents as a function of the linear system size L. The plot shows that indeed, for large L, the number of active modes is an extensive quantity, i.e.

$$N_+ \sim L^2. \tag{4.23}$$

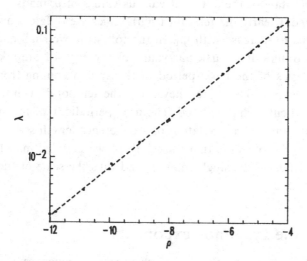

Figure 4.2 The maximum Lyapunov exponent λ of a 128×128 CML (4.1) with $\epsilon = 0.4$ as a function of the non-linearity parameter ρ. The broken line shows the fit $\ln \lambda = a + b\rho$ with $a = 0.725$ and $b = 0.448$ [van de Water and Bohr 1993].

4.2.2   Scaling of the correlation length

The standard way of defining a coherence length is via the correlation function

$$G(\mathbf{r}) = \langle u_i u_{i+r} \rangle - \langle u \rangle^2. \tag{4.24}$$

where $\langle \ldots \rangle$ denotes a spatial and temporal average. The hope is that there is an exponential decrease of correlations, i.e. $G(r) \sim \exp(-r/\xi)$, so that the decay rate defines a correlation length ξ. The trouble with the time average is that the scale of u at different times is very different. A time series of u registered at a point (or series of points) on the lattice is p-periodic with small chaotic modulations. The size of these modulations at different times modulo p is very different. Therefore, a straightforward time average in (4.24) does not work. To overcome this difficulty one can use the so-called *mutual information function* $I(\mathbf{i}_1, \mathbf{i}_2)$, which is directly expressed in terms of the joint probability $P(u_1, u_2)$ that two points, at position $\mathbf{i}_1$ and $\mathbf{i}_2$, have their field values equal to u_1 and u_2 respectively [Fraser and Swinney 1986]:

$$I(\mathbf{i}_1, \mathbf{i}_2) = \int du_1 du_2 P(u_1, u_2) \log_2 \left(\frac{P(u_1, u_2)}{P(u_1)P(u_2)} \right). \tag{4.25}$$

In contrast to the correlation function the information function is invariant under a (spatially uniform) transformation of u.

The details of the algorithm to compute the information function can be found in van de Water and Bohr [1993]. As shown in Fig. 4.4, $I(r)$ behaves as

$$I(r) = I_0 e^{-r/\xi} + I_\infty, \tag{4.26}$$

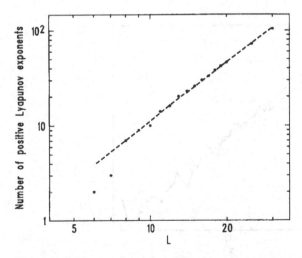

Figure 4.3 The number of positive Lyapunov exponents N_+ as a function of the lattice size L at $p = -4$, for the 2D CML (4.1). The broken line shows the fit $N_+ \simeq aL^b$ with $a = 0.097$ and $b = 2.06$ (consistent with the value $b = 2$) [van de Water and Bohr 1993].

where $r = |\mathbf{i}_1 - \mathbf{i}_2|$. The non-zero value I_∞ is due to the basic periodicity of the coupled map system. Even points very far away on the lattice remain within the same band, and the bands become more numerous (and thinner) as ρ decreases.

We can now use (4.26) to extract a correlation length ξ. In Fig. 4.5 we show the dependence of ξ on the non-linearity parameter ρ. The measured curve exhibits large fluctuations, parts of which are systematic, i.e. reproducible with

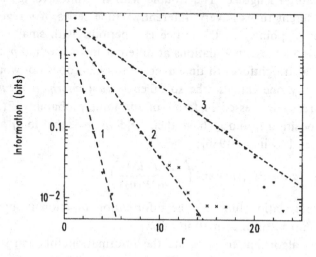

Figure 4.4 The mutual information function $I(r) - I_\infty$ for the 2D CML (4.1), with $\epsilon = 0.4$, as a function of r for $\rho = -4.3$, -8.4 and -12 in curves 1–3 respectively. The broken lines are fits $I(r) - I_\infty = I_0\, e^{-r/\xi}$ [van de Water and Bohr 1993].

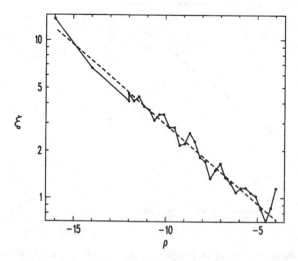

Figure 4.5 Correlation length ξ, computed from mutual information, for the 2D CML (4.1), with $\epsilon = 0.4$. The broken line shows the fit $\ln \xi = a - b\rho$ with $a = -1.25$ and $b = -0.23$ [van de Water and Bohr 1993].

different random initial conditions of **u**. However, the average value, as indicated by the dashed line, follows the power law

$$\xi \sim (R - R_\infty)^{-b}, \qquad (4.27)$$

with $b \approx 0.23$. Combining (4.27) with the scaling of the Lyapunov exponent (4.22) we obtain

$$\xi \sim \lambda^{-\nu}, \qquad (4.28)$$

with $\nu \approx 0.51$, which is very close to the value $\nu = \frac{1}{2}$ predicted earlier in (4.20) (see also Kaspar and Schuster [1986], Bohr and Christensen [1989]).

4.2.3   Spreading of localized perturbations

In section 4.1.2 we argued that relation (4.20) is due to the gaussian distribution for the spreading of localized perturbations combined with the exponential increase in time. Let us assume a gaussian spreading, i.e. that an initial localized disturbance of size Δ_0, at time t, behaves as

$$\Delta(r,t) = \Delta_0 \exp\left(\lambda t - \frac{r^2}{4Dt} \right), \qquad (4.29)$$

where D is the effective diffusion coefficient. Naively, taking the continuum limit of (4.1) one has $D = \epsilon/2d$. In a system of coordinates moving along a ray with speed v, i.e. where $r = vt$, the size of the perturbation would grow as

$$\Delta(r = vt, t) = \Delta_0 \, e^{(\lambda - \frac{v^2}{4D})t}, \qquad (4.30)$$

and thus the velocity v_s of the front separating the exponentially increasing region from the exponentially decaying region would be

$$v_s = \sqrt{4D\lambda}, \qquad (4.31)$$

analogously to (4.19).

To check the assumption of gaussian diffusive spreading, one may iterate the coupled map lattice with two different initial states. One of them is a well-equilibrated turbulent state $u_i(0)$ and the other, $u'_i(0)$, differs from it only at one site, which is slightly perturbed. Now both the systems are iterated forward and we keep track of the difference

$$\Delta u_i(n) = |u_i(n) - u'_i(n)|. \qquad (4.32)$$

Figure 4.6 shows the evolution of Δu for different values of ρ. The frames of the figure are snapshots taken at times n that are a multiple of the basic period due to the chaotic bands. A point perturbation is seen to spread to an approximately circular blob, but the deterministic dynamics induces sharply delineated coherent structures.

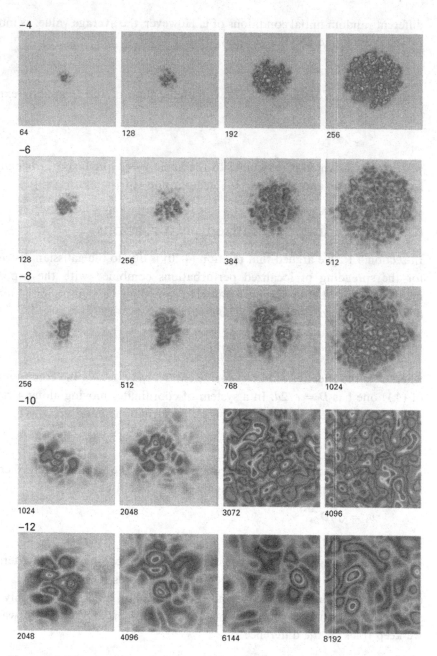

Figure 4.6 The spreading of a point perturbation as a function of time for various values of ρ, for the 2D CML (4.1), with $\epsilon = 0.4$. Time runs horizontally and increases proportionally to the basic period due to the chaotic bands. ρ decreases downward from -4 (top row) to -12 (bottom row) [van de Water and Bohr 1993].

The most direct way of quantifying the distribution of disturbances is by computing the radial distribution function $P(r, t)$. First the difference field values $\Delta u_i(t)$ are averaged over (discrete) circles to get $\Delta u_r(t) = \frac{1}{n_r} \sum_i \Delta u_i(t)$, where $i_1^2 + i_2^2 = r^2$ and n_r is the number of pairs in the summation. Thus

$$P(r, t) = \left\langle\!\!\left\langle \frac{\Delta u_r(t)}{\sum_r r \Delta u_r(t)} \right\rangle\!\!\right\rangle, \tag{4.33}$$

where the double brackets denote the average over many samples. Figure 4.7 shows the radial distribution function $P(r, t)$ at different t and $\rho = -4$. The distribution function is very nearly gaussian, although, at large times, its central part flattens considerably, which is to be expected since the disturbance in the central parts has to saturate. The width of the gaussian grows as $\sqrt{4Dt}$ with $D = 0.11$, which differs only slightly from $\epsilon/4 = 0.10$.

Finally we show results for the spreading velocity, which, in the gaussian approximation should be given by (4.31). By computing the expansion rate $\lambda(v)$ on circles that move out radially from the initial perturbation with velocity v, the spreading velocity v_s can be found from $\lambda(v_s) = 0$. Figure 4.8 shows the dependence of v_s on ρ. In terms of the Lyapunov exponent this leads to

$$v_s \sim \lambda^\alpha, \tag{4.34}$$

where $\alpha \approx 0.54$, very close to the expected value $1/2$.

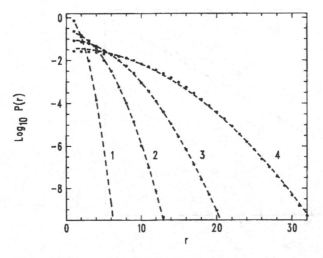

Figure 4.7 Average radial distribution function $P(r, t)$ as a function of r for different t and $\rho = -4$. Curves 1–4 show times $4, 18, 48$ and 128 respectively. The dashed lines show gaussian fits [van de Water and Bohr 1993].

4.2.4   Renormalization group results

Many of the results for the critical regime can be obtained very naturally from a renormalization group approach, which gives very detailed information about the scaling of the bifurcations and the structure of the periodic states for parameter values R slightly below R_∞. We shall not go into detail about these procedures here, but only give a brief review and refer the reader to the literature, especially the recent review by Kuznetsov [1993]. For earlier work see Kuznetsov and Pikovsky [1986] and Kook et al. [1991] for the case of coupled logistic maps and Stassinopoulos and Alstrøm [1992] for coupled circle maps.

The starting point is the Feigenbaum–Cvitanović equation,

$$g(u) = \alpha g[g(u/\alpha)], \qquad\qquad (4.35)$$

already introduced in section 4.1. In a coupled map lattice there is an obvious fixed point of the renormalization group, for which each map has the same value and the coupling never appears. If the actual state is very flat, we can assume that we are close to this fixed point and determine how perturbations around it will behave. Thus we assume that the map lattice close to the critical point has the form (which corresponds to nearest neighbour interactions)

$$u_i' = \phi_i(u_i, u_{i+1}, u_{i-1}) = g(u_i) + \epsilon\psi(u_{i-1}, u_i, u_{i+1}), \qquad (4.36)$$

where $u = u(n)$ and $u' = u(n+1)$ and where $u = 0$ has been chosen as the maximum for the maps. Applying the doubling operator **T**, at the lowest order

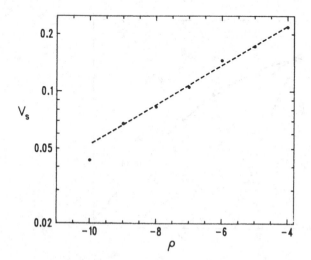

Figure 4.8 Spreading velocity v_s as function of ρ found from the exponents $\lambda(v)$. The dashed line is a fit with $\ln v = a\rho - b$, with $a = 0.24$ and $b = 0.56$ [van de Water and Bohr 1993].

in ϵ, one finds that

$$\mathbf{T}\phi_i = \alpha g[g(u_i/\alpha)] + \epsilon\alpha g'(u_i/\alpha)\psi(u_{i-1}/\alpha, u_i/\alpha, u_{i+1}/\alpha)$$
$$+ \alpha\epsilon\psi(g(u_{i-1})/\alpha, g(u_i)/\alpha, g(u_{i+1})/\alpha), \tag{4.37}$$

which leads to the eigenvalue equation

$$\epsilon\alpha g'(u_i/\alpha)\psi(u_{i-1}/\alpha, u_i/\alpha, u_{i+1}/\alpha) + \alpha\epsilon\psi(g(u_{i-1})/\alpha, g(u_i)/\alpha, g(u_{i+1})/\alpha)$$
$$= \lambda\psi(u_{i-1}, u_i, u_{i+1}) \tag{4.38}$$

for the deviations around the critical point. Further, differentiating (4.38), using (4.35) and introducing the function [Kook et al. 1991]

$$F(u) = \frac{\partial\psi(u_{i-1}, u_i, u_{i+1})}{\partial u_{i+1}} \tag{4.39}$$

evaluated on the flat solution $u_j = u$ for all j, we find that

$$g'[g(u/\alpha)]F(u/\alpha) + F[g(u/\alpha)]g'(u/\alpha) = \lambda F(u). \tag{4.40}$$

From (4.35), one sees that one solution of (4.40) is given by

$$F(u) = g'(u), \tag{4.41}$$

with eigenvalue $\lambda = 2$. There is, however, another solution with eigenvalue $\lambda = \alpha$, as can be seen by evaluating (2.40) at $u = 0$, using $g'(0) = 0$ and $g'(1) = 1$ and assuming that $F(0) \neq 0$. Since $|\alpha| > 2$ one would expect that the latter always dominates asymptotically, but if the coupling is of the form chosen in (4.1), often called forward coupling, it can be shown [Kuznetsov 1993] that the perturbation has (and acquires) no component in the direction of the eigenvector corresponding to $\lambda = \alpha$ (because of the quadratic nature of the maximum of the universal function g), and that the relevant eigenvalue therefore is $\lambda = 2$. For linear coupling $\lambda = \alpha$ dominates, as expected.

Thus, for the purely diffusive case (4.1) close to the critical point, one has $\epsilon \to \epsilon 2^n$ after n applications of the doubling operator (4.37). If we start very close to the critical point, say at $R - R_\infty = A\delta^{-n}$, n doubling transformations will take us out of the critical regime and generate an effective diffusive coupling $\epsilon \sim 2^n \sim (R - R_\infty)^{\ln 2/\ln \delta}$. For a diffusive system with length scales much larger than the lattice spacing, we assume that the relevant length scale is the diffusion length of the order of $\sqrt{\epsilon}$. The scaling of the coherence length is then

$$\xi \sim \sqrt{\epsilon} \sim (R - R_\infty)^{-b}, \tag{4.42}$$

where the exponent has the value $b = \ln 2/2\ln \delta = 0.2249...$ in good agreement with (4.27). For linear coupling $\lambda = \alpha$ dominates, as expected, and using the same arguments would lead to a violation of (2.42). But as mentioned earlier, this coupling also introduces non-trivial spatial structure and the introduction of the diffusion length would not be justified.

There exist very few exact results on turbulent coupled map lattices. In an interesting dynamical multicomponent model where the local couplings are random (in the spirit of spin-glasses), Hansel and Sompolinsky [1993] were able to write down a self-consistent nonlinear partial differential equation for the correlations in space and time in the limit of infinitely many components. Near the onset of turbulence, for low spatial dimensions ($d < 3$), they found exponentially decaying spatial correlations and again the relation (4.28) with $v = 1/2$. For $d > 3$ the correlations decay only algebraically, even after the onset of turbulence. In Bhagavatula et al. [1992] examples are also given of coupled map lattices with algebraically decaying correlations, for which the exponents can be calculated from noisy, non-chaotic models. Superficially these results would seem to violate inequality (1.16) (since $\xi = \infty$), but one has to keep in mind that the correlation length of relevance for the argument leading to (1.16) is the correlation length for the chaotic fluctuations, which one cannot in general hope to capture through the correlation function of the field. In the logistic coupled map case such correlation functions in fact do not decay to zero at all because of the ordering in bands, which coexists with the turbulent dynamics. Similar problems are discussed again in section 5.11.1.

4.3 Lyapunov spectra

The Lyapunov spectrum of an extended dynamical system is the complete set λ_i of Lyapunov exponents. As discussed in section 1.5.2, in systems with many degrees of freedom, a rather natural question is whether the thermodynamic limit $L \to \infty$ for the Lyapunov spectrum , i.e. a limiting function $\Lambda(x)$, with $x = i/L^d \in [0, 1]$, exists such that $\lambda_i = \Lambda(x)$ as $L \to \infty$. The existence of such a limit implies that the Kolmogorov–Sinai entropy H is proportional to the size of the system. Indeed, from the Pesin formula (1.10) one has

$$H \simeq hL^d, \tag{4.43}$$

where h is the entropy per degree of freedom:

$$h = \int_0^1 \Lambda(x)\theta(\Lambda(x))\,dx. \tag{4.44}$$

In addition, for large L, the number of positive Lyapunov exponents N_+ is proportional to the number of degrees of freedom:

$$N_+ \sim L^d. \tag{4.45}$$

From the Kaplan–Yorke conjecture (1.13) one then infers that the dimension of
the attractor D_λ in the phase space of an extended system is also proportional
to L^d, and thus there exists the dimension density

$$\delta_\lambda = D_\lambda / L^d. \tag{4.46}$$

In some coupled map lattices [Bunimovich and Sinai 1988, Gundlach and
Rand 1993], the existence of the thermodynamic limit can be proved. Let us
remark that a Lyapunov density has also been found numerically in some
dissipative systems [Pomeau et al. 1984], symplectic systems [Livi et al. 1986,
1987, Eckmann and Wayne 1988, 1989] and in products of random matrices
[Paladin and Vulpiani 1986].

Moreover it has been conjectured by Ruelle [1982] that the Lyapunov spectrum for a strongly turbulent fluid should be singular near $\lambda = 0$ where the
density $n(\lambda)$ of Lyapunov exponents would be infinite. In section 3.3.3 we saw
that, for the GOY shell model, there exists an asymptotic limit density $\Lambda(x)$ with
a finite fraction of almost vanishing Lyapunov exponents in agreement with the
Ruelle conjecture.

In this section we shall see that Lyapunov spectra in one-dimensional dynamical systems should generically show singularities if the dynamics has a local
conservation law [Bohr et al. 1995]. A well-known case [Manneville 1988] is the
1D Kuramoto–Sivashinsky equation, described in chapters 5 and 7. Simulations
of Rayleigh–Bénard convection in 3D shows clearly the dependence on dimension: the Lyapunov spectra [Deane and Sirovich 1991a,b] exhibit a singularity
only at small Rayleigh numbers (Fig. 4.9a), where the rolls are aligned along
one axis for long periods of time and the system is thus almost one dimensional.
In Fig. 4.9b the one-dimensional character of the system has disappeared and
the singularity is absent.

4.3.1 Coupled map lattices with conservation laws

It is easy to modify model (4.1) in order to include local conservation laws
[Grinstein et al. 1991, Bourzutschky and Cross 1992, Bhagavatula et al. 1992].
Any model where $u_i(n+1) - u_i(n)$ can be written as a discrete gradient conserves
the quantity $Q = \sum_i u_i(n)$. We shall again assume symmetry between i and $-i$,
and thus allow only even spatial derivatives. The map

$$u_i(n+1) = u_i(n) + \epsilon_1 \Delta_i f(u(n)) + \epsilon_2 \Delta_i u(n) , \tag{4.47}$$

where

$$\Delta_i f(u) \equiv \frac{1}{2d} \sum_{i'} f(u_{i'}) - f(u_i) \tag{4.48}$$

and $\mathbf{i}'$ are the nearest neighbours of the site $\mathbf{i}$, for positive constants ϵ_1 and ϵ_2, is a simple example in which Q is conserved. Another possibility is to include additional fields, e.g.

$$u_i(n+1) = u_i(n) + f(u_i(n)) + \epsilon_1 \Delta_i u(n) + b v_i(n),$$
$$v_i(n+1) = v_i(n) + \epsilon_2 \Delta_i v(n) + \epsilon_3 \Delta_i u(n). \tag{4.49}$$

Here $Q = \sum_i u_i(n)$ is not conserved, but $P = \sum_i v_i(n)$ clearly is; b and the ϵ_i (> 0) are constants.

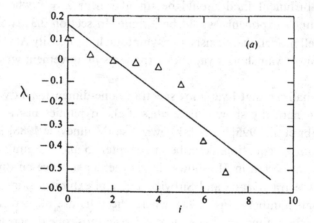

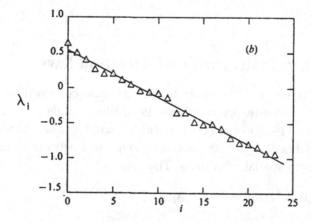

Figure 4.9 Spectrum of Lyapunov exponents from a numerical simulation of Rayleigh–Bénard convection at small Rayleigh numbers: (a) $Ra = 5$, where the system is quasi-one-dimensional and there is a singularity at $\lambda = 0$; (b) $Ra = 15$, where no singularity is visible [Deane and Sirovich 1991a].

Bohr et al. [1995] studied a modified version of (4.47), which was used as a model of a roughening interface (like the models discussed in section 7.4.3). The modified equation reads

$$u_i(n+1) = u_i(n) + \epsilon_1 \Delta f(u_i(n)) + \epsilon_2 \Delta^3 u_i(n) \,. \tag{4.50}$$

In Fig. 4.10 we show the Lyapunov spectra for the two models (4.49) and (4.50) in 1D. Here the function f was chosen as $f(u) = Rx(1-x)$, where x is the fractional part of u, i.e. the left-over when the integer part is subtracted. This special choice of f was made in order to let u be an unbounded variable

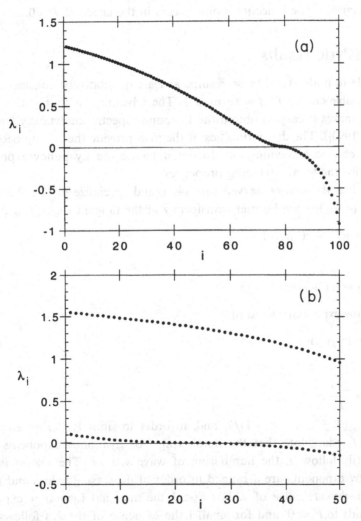

Figure 4.10 Lyapunov spectrum for the 1D conserving coupled map lattices: (a) equation (4.50), with $R = 5$, $\epsilon_1 = 1$, $\epsilon_2 = 0.000\,625$ and $L = 100$: (b) equation (4.49), with $R = 10$, $\epsilon_1 = 0.2$, $\epsilon_2 = 0$, $\epsilon_3 = 0.2$, $b = 1$ and $L = 50$. Note the two distinct 'bands', the lower containing $\lambda_{51} - \lambda_{100}$ [Bohr et al. 1995].

and thus to give the models an approximate 'interface symmetry', as discussed in chapter 7, but has no bearing on the singularities of the Lyapunov spectra.

For the non-conserved case (4.1) we know from Fig. 1.9, that the spectrum passes through zero smoothly, with non-zero slope. For the conserving, single-field case (4.50), there is a strong indication in Fig. 4.10a that the curve has vanishing slope at $\lambda = 0$, i.e. that the density $n(\lambda)$ diverges at $\lambda = 0$. In the conserving case with two fields (4.49) there is a marked discontinuity in the Lyapunov spectrum (Fig. 4.10b); one can therefore think of the spectrum as separated into two distinct bands, one without and one with an inflection point. These bands correspond roughly to the non-conserved u-field and the conserved v-field respectively. The inflection point occurs in the latter, at $\lambda = 0$.

4.3.2   Analytic results

It is possible to understand these results, at least qualitatively, considering the simplest possible choice $f(u) = ru$ mod 1. The advantage is that $f'(u) = r$ for all u, which makes it easy to obtain the Lyapunov spectra analytically (see also [Isola et al. 1990]). The discontinuities of the map provide the mixing necessary to generate chaotic behaviour but do not influence the Lyapunov exponents, which describe the local stretching properties.

Consider first the non-conserved case (4.1), and specialize to 1D. We return to equation (4.10) for the Fourier transform $\hat{z}$ of the tangent vector (for $d = 1$):

$$\hat{z}_k(n+1) = \chi(k)\hat{z}_k(n), \tag{4.51}$$

where

$$\chi(k) = r(1 + \epsilon(\cos k - 1)). \tag{4.52}$$

The Lyapunov exponents are simply

$$\lambda_i = \ln |\chi(k_j)| . \tag{4.53}$$

Here

$$k_j = \frac{2\pi j}{L}, \tag{4.54}$$

where $j = 0, \pm 1, \pm 2, ..., \pm(L-1)/2$, and, in order to simplify the notation, we have taken L odd. Note that the numbering i of Lyapunov exponents need not necessarily follow j, the numbering of wave vectors. The reason is that the Lyapunov exponents are arranged in order of decreasing values and (4.53) contains the *absolute value* of χ. For $\epsilon < 1$ the maximal Lyapunov exponent λ_0 corresponds to $k = 0$, and for small i the sequence of the λ_i's follows that of the k_i's, because the lowest values of χ/r are around $1 - 2\epsilon > -1$. If $\chi(k)$ becomes negative (i.e. if $\epsilon > 1/2$), the ordering in k starts to differ from the ordering in size since both a positive and a negative χ leads to the same λ.

This happens below the value $\lambda_c \equiv \lambda(k_c) = \lambda(\pi)$ corresponding to the minimal χ, where $\cos k_c = (3\epsilon - 2)/\epsilon$. The behaviour around $k = k_c$ (or $i \sim i_c = k_c L/\pi$) is different when approaching $i = i_c$ from above and from below. From above, the curve is non-singular, passing linearly through λ_c; the branch corresponding to negative χ does not contribute. Above i_c, however, taking into account the quadratic minimum in $\chi(k)$ at $k = \pi$, we find that $\lambda(i) \approx \lambda_c - a(i - i_c)^2$, where a is a constant. Thus, the curve approaches λ_c with zero slope from below. The corresponding density of Lyapunov exponents $n(\lambda)$ diverges as $(\lambda_c - \lambda)^{-1/2}$ from below.

For $\epsilon < 1/2$ this singularity is still present; now, however, it appears at the bottom of the band, $i = (L - 1)/2$, i.e. at $\lambda = \lambda_b \equiv \ln r(1 - 2\epsilon)$. Note too that, owing to the quadratic maximum in λ at $k = 0$, $n(\lambda)$ has an additional singularity at the top of the band, i.e. at $\lambda = \lambda_t \equiv \ln r$; $n(\lambda)$ of course vanishes for $\lambda > \lambda_t$, but diverges as $(\lambda_t - \lambda)^{-1/2}$ as $\lambda \to \lambda_t$ from below.

We conclude that the positions of the singularities in the spectra for the non-conserving case move with the parameters. Thus, for example, the value λ_c depends on ϵ. In contrast, for the conserving case (4.47) one has

$$\chi(k) = 1 + 2(\epsilon_1 r + \epsilon_2)(\cos k - 1) . \tag{4.55}$$

Analysis similar to the previous one shows that the singularity must always be at $\lambda = 0$. Here the singular contribution to $\lambda = 0$ comes from $k \approx 0$. Again the singularity in the density of Lyapunov exponents is $n(\lambda) \approx (-\lambda)^{-1/2}$, and occurs only as $\lambda = 0$ is approached from below; the density approaches a finite constant as $\lambda \to 0^+$.

For the model (4.49), again with $f = rx \bmod 1$, a new feature appears. Now, as in Fig. 4.10b, we get two distinct bands – one with a singularity at $\lambda = 0$ and one without. They originate from the two solutions for $\chi(k)$:

$$\chi_\pm(k) = \frac{1}{2}(T \pm \sqrt{\Delta}), \tag{4.56}$$

where

$$T(k) = r + 2 + 2(\epsilon_1 + \epsilon_2)z, \tag{4.57}$$

with $z = \cos k - 1$, and

$$\Delta(k) = r^2 + [4r(\epsilon_1 - \epsilon_2) + 8b\epsilon_3]z + [2(\epsilon_1 - \epsilon_2)z]^2 . \tag{4.58}$$

Around $k = 0$ the two solutions behave as

$$\chi_+(k) \approx 1 + r - \left(\epsilon_1 + \frac{b\epsilon_3}{r}\right)k^2 \tag{4.59}$$

and

$$\chi_-(k) \approx 1 + \left(\frac{b\epsilon_3}{r} - \epsilon_2\right)k^2 , \tag{4.60}$$

respectively. Thus, depending on the sign of $b\epsilon_3/r - \epsilon_2$, the singularity occurs as $\lambda = 0$ is approached from above or below.

The analogous expressions for χ in d space dimensions are found by letting $\cos k - 1 \rightarrow \sum_{i=1}^{d}(\cos k_i - 1)$, which, for small k, becomes $-\frac{1}{2}|\mathbf{k}|^2$. As a consequence the number of Lyapunov exponents in an interval $[0, \lambda]$ is

$$N(\lambda) \approx A\lambda^{d/2} + B\lambda . \tag{4.61}$$

The first term comes from the branch with $k \approx 0$, and the last term is the non-singular piece coming from finite k. For $d \geq 2$ the last term actually dominates and the singularity is basically invisible, though for $d = 3$ it should in principle show up in the derivative $n'(\lambda)$.

4.4   Coupled maps with laminar states

4.4.1   Spatio-temporal intermittency

The coupled map lattices considered up to now employ maps that are always chaotic in their temporal development. In turbulent phenomena, it is often observed that there is an alternation between turbulent and laminar states usually called *intermittency*. Intermittent phenomena have already been discussed in detail in chapter 3, where, e.g. the time series of the energy dissipation of the GOY shell model exhibits alternations between violent energy bursts and laminar quiescent periods. Just as important is the alternation between turbulent and laminar regions in space, which is usually called *spatio-temporal intermittency*. In this section, we shall focus on a CML which captures these intermittent behaviours in a simple way. These CML's use maps which exhibit both chaotic and laminar states and were introduced by Chaté and Manneville [1988, 1989]:

$$f(x) = \begin{cases} rx, & \text{if } x \in [0, 1/2]; \\ r(1-x), & \text{if } x \in [1/2, 1]; \\ x, & \text{if } x \in \,]1, r/2]. \end{cases} \tag{4.62}$$

The chaotic motion of f for $x \leq 1$ is governed by a standard tent map of slope r. However, when r exceeds the value 2, the trajectory may escape to a 'laminar' state with $x > 1$, and this state is marginally stable, because the slope in the line of fixed points is one. The laminar state is absorbing, i.e. the trajectory cannot be pulled back into the chaotic state. This is no longer the case when the maps are coupled, since the interactions with its neighbours may pull a laminar site back into chaotic motion, thus causing an interesting interplay between laminar and turbulent regions, as shown in Fig. 4.11. Using the f given by (4.62), the 1D Chaté and Manneville CML has the form (4.1)

$$u_i(n+1) = f(u_i(n)) + \frac{\epsilon}{2}[f(u_{i-1}(n)) - 2f(u_i(n)) + f(u_{i+1}(n))]. \tag{4.63}$$

In the following we use the more compact form

$$\mathbf{u}(n+1) = \mathbf{F}(\mathbf{u}(n)) \ , \tag{4.64}$$

where $\mathbf{u} = (u_1, \ldots, u_L)$. For coupling strength ϵ exceeding a critical value ϵ_c, a turbulent site percolates through the system in a way that closely resembles the statistical mechanics of *directed percolation*, a simple probabilistic process which has been intensively studied and seems to underlie a large class of (directed) non-equilibrium statistical processes. Directed percolation will play an important role later (in chapter 7), and in appendix H we introduce it, explain the critical properties and guide the reader to some important references. It describes for instance the wetting of a porous medium by water slowly percolating through it in a gravitational field. At a given time, slices through the medium look like snapshots of the 2D CML, so that increasing time in the CML corresponds to increasing depth of the medium. Directed percolation is characterized by an 'absorbing' (empty) state: when no water exists in a given layer of the porous medium, it will not appear in layers below. Likewise, in the CML, a state in which all sites are laminar is absorbing. When ϵ is below ϵ_c any initial state eventually reaches such a laminar state, where it freezes.

Similar types of spatio-temporal intermittency patterns are observed experimentally for quasi-one-dimensional Rayleigh–Bénard systems with a large aspect ratio. Ciliberto and Bigazzi [1988] and Daviaud et al. [1989] performed experiments in a circular cell and a straight cell, respectively. Their control parameter was the temperature difference between top and bottom and they measured the

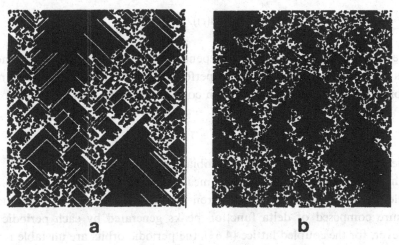

a **b**

Figure 4.11 Spatio-temporal intermittency patterns generated by the CML (4.63). Time is running upwards during 250 iterations after long transients. Laminar sites ($u_i > 1$) are in black, turbulent sites in white. (a) $r = 3$, $\epsilon = \epsilon_c = 0.360$. (b) $r = 2.1, \epsilon = 0.360$ (here $\epsilon_c = 0.0047$) [Chaté and Manneville 1988].

turbulent 'spots' appearing in the cells. These correspond to the white regimes of the model shown in Fig. 4.11. Ciliberto and Bigazzi [1989] found that the distribution of turbulent spots follows a power law close to the transition to spatio-temporal intermittency and changes to an exponential distribution as they move above the transition. A similar pattern is found for the CML model (4.63). Of course, one cannot expect that a CML captures the precise quantitative description of the hydrodynamic Rayleigh–Bénard convection, but the critical properties appear to be very similar.

4.4.2   Invariant measures and Perron–Frobenius equation of CML

The invariant measure produced by iterations of low-dimensional maps is described by the Perron–Frobenius equation (see for instance Ott [1993]) and can be straightforwardly extended to CML. The probability density $\rho_{\mathbf{u}}(t)$ of $\mathbf{u}$ at time t can be obtained from the $\rho_{\mathbf{u}}(0)$ at the initial time using the following formulation of the Perron–Frobenius equation:

$$\rho_{\mathbf{u}}(t+1) = \int_0^{r/2} \rho_{\mathbf{v}}(t)\delta(\mathbf{u} - \mathbf{F}(\mathbf{v})) \prod_{i=1}^{L} dv_i \,. \tag{4.65}$$

Of particular interest is the invariant measure $\rho^{(I)}(\mathbf{u})$, which is the stationary solution of (4.65), i.e. it is the eigenfunction of the Perron–Frobenius operator corresponding to the eigenvalue one. The invariant measure can also be defined from the time average over a single trajectory

$$\rho^{(I)}(\mathbf{u}) = \lim_{M \to \infty} \frac{1}{M} \sum_{t=0}^{M} \delta(\mathbf{u} - \mathbf{u}(t)). \tag{4.66}$$

If the CML is ergodic, $\rho^{(I)}(\mathbf{u})$ is independent of the initial conditions, and time averages are equivalent to averages performed over the invariant measure. The Perron–Frobenius equation (4.65) can conveniently be rewritten in the form

$$\rho_{\mathbf{u}}(t+1) = \sum_{\mathbf{v}=\mathbf{F}^{-1}(\mathbf{u})} \frac{1}{J(\mathbf{v})} \rho_{\mathbf{v}}(t) \,, \tag{4.67}$$

where $J(\mathbf{u}(t)) = |\det\mathbf{J}(\mathbf{u}(t))|$ is the Jacobian of the CML mapping.

Whether or not a unique invariant measure exists is not known a priori. There are clearly many solutions of the Perron–Frobenius equation, e.g. the stationary measure composed of delta function peaks generated by each periodic orbit. However, for the coupled lattice (4.63), the periodic orbits are unstable repellers and the probability of observing the corresponding singular measures starting from random initial conditions is zero.

For the CML (4.63) it is clear that once a trajectory reaches the absorbing part of phase space given by $1 < u_i \le r/2$, $i = 1,\ldots,L$, it has to remain there. The escape rate from the complementary phase space volume can be defined as the asymptotic rate by which probability density is transferred to the absorbing region. The average lifetime of a chaotic orbit is then the inverse of the escape rate. Another consequence of an absorbing region is that there is no contribution to the invariant measure outside this region. Nevertheless, the lifetimes of orbits can become very large as the system size is increased. During this 'intermediate' time, an interesting steady-state measure can develop before the trivial invariant measure is eventually reached.

4.4.3 Mean field approximation and phase diagram

A mean field Perron–Frobenius equation [Kaneko 1989a, Houlrik et al. 1990] can be formulated assuming that the probability measure $\rho_{\mathbf{u}}(t)$ is a product of single-site probabilities

$$\rho_{\mathbf{u}}(t) = \prod_i \tilde{p}_{u_i}(t) . \tag{4.68}$$

Because of the translational invariance of the lattice, the measures $\tilde{p}_{u_i}(t)$ are identical. As in statistical mechanics, this approximation corresponds to the neglect of spatial correlations. The introduction of a single-site distribution is motivated by direct lattice calculations, where it has been observed [Chaté and Manneville 1988] that the iterates of a single map in a large lattice can be described by a quasi-invariant measure. The single-site measure $\tilde{p}$ can be determined self-consistently from the recursion relation

$$\tilde{p}_u(t+1) = \int_0^{r/2} \delta \left(u - (1-\epsilon)f(u_2) - \frac{\epsilon}{2}[f(u_1) + f(u_3)] \right) \prod_{i=1}^{3} \tilde{p}_{u_i}(t) du_i .$$
$$\tag{4.69}$$

When $\epsilon = 0$, the single-map Perron–Frobenius equation is recovered.

In general, (4.69) must be solved numerically using a finite resolution, e.g. with the following procedure. The integration over phase space is replaced by a triple summation over N^3 grid points, where N is the number of points in the interval $[0, r/2]$. One defines $N_1 = [N/(r/2)]$, where the square brackets stand for the integer part, and for each of the variables u_1, u_2, u_3, a grid is defined by

$$u_k^{(j)} = \frac{j + 0.5}{N_1}, \quad j = 0, 1, ..., N-1, \quad k = 1, 2, 3, \tag{4.70}$$

where $u_k^{(j)} < 1$ (i.e. a turbulent site) if $0 \le j_i \le N_1 - 1$ and $u_k^{(j)} > 1$, i.e. a laminar site, if $N_1 \le j_i \le N - 1$. On each of these grid points, the argument of the delta function in (4.69) is evaluated, and the measures are defined on N grid points in the interval $[0, r/2]$, i.e.

$$u = \frac{[uN_1] + 0.5}{N_1} . \tag{4.71}$$

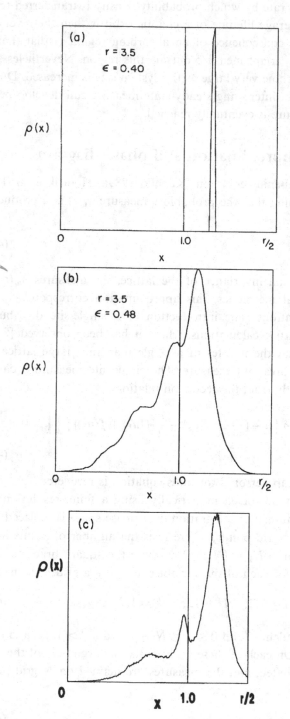

Figure 4.12 The self-consistent measure calculated from equation (4.69). (a) When the parameters are chosen in the non-chaotic region of the phase diagram ($r = 3.5$ and $\epsilon = 0.40$), all of the measure is sharply peaked in the laminar regime. (b) In the turbulent part of the phase diagram ($r = 3.5$ and $\epsilon = 0.48$), the measure is distributed over the whole interval. (c) Steady-state probability distribution obtained from direct iteration with the same parameters as in (b) [Houlrik et al. 1990].

For each of the N^3 grid points, the following operation is performed:

$$\tilde{\rho}_u = \alpha \tilde{\rho}_{u_1} \tilde{\rho}_{u_2} \tilde{\rho}_{u_3} , \qquad\qquad (4.72)$$

where α is introduced in order to have a normalized $\tilde{\rho}_u$. Then the function $\tilde{\rho}_u$ is put back into (4.69) and the procedure is continued until the measure has converged with a certain accuracy. The asymptotic solution of (4.69) is not unique but depends on the initial distribution. The calculations can be initiated by a uniform distribution, but when the initial distribution is concentrated in the laminar regime, it stays there even if $\epsilon > \epsilon_c$. This corresponds to the fact that the laminar phase is absorbing. The initial distribution should therefore always have some contribution in the u-interval $[0, 1]$. A somewhat non-uniform initial distribution does not change the final result, i.e. the same $\tilde{\rho}_u$ appears.

Figure 4.12 shows the result of this procedure after 80 steps for $r = 3.5$ and $\epsilon = 0.40 < \epsilon_c = 0.42$ with $N = 100$ grid points. The entire measure has settled on a delta peak in the laminar regime. This is not surprising, since the parameter value is below the transition point and any state will turn into a laminar phase. Figure 4.12b shows the result for the same value of r and N with $\epsilon = 0.48$ and 400 iterations. Clearly, there is a large part of the distribution below $u = 1.0$. Since the values $u < 1$ correspond to the turbulent sites, this signals that the system is now above the critical transition line for the onset of spatio-temporal chaos, i.e. the value of ϵ is above the critical value ϵ_c (within the given approximation). For comparison, Fig. 4.12c shows a distribution obtained by direct iteration with the same parameters as in Fig. 4.12b. The transition

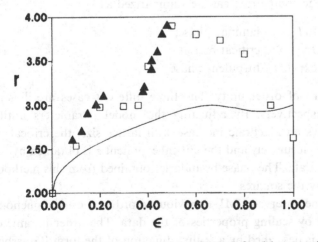

Figure 4.13 Phase diagram of the CML defined by equation (4.63). The solid triangles are obtained using the mean field Perron–Frobenius approximation, the solid line is a cellular automata approximation, while the squares are the critical points obtained from direct iterations [Houlrik et al. 1990].

points resulting from this calculation at selected parameter values are plotted in Fig. 4.13. For comparison, the results of a cellular automata approximation and the transition points obtained by direct iterations are also shown. The mean field criterion based on a self-consistent measure described in this section is in good agreement (within a few per cent) with the finite-lattice iterations. Despite its simplicity, this method therefore seems to capture the essential features of the CML quite well.

4.4.4   Direct iterates and finite size scaling

In this section, finite size scaling approaches are applied to a systematic study of the dependence of the properties of the CML on the lattice size L. The main quantities to be calculated are the escape time τ and the order parameter m. Both quantities are averaged over an ensemble of N_s random initial conditions. The escape time is the average number of elapsed time steps before the absorbing state, where all sites are laminar, is reached. The order parameter is defined as the fraction of non-laminar sites, i.e. sites with $u_i < 1$, and is a decreasing function of time.

The critical point is located by calculating the escape time as a function of the parameters r and ϵ and the system size. An example is given in Fig. 4.14a, where τ is shown as a function of ϵ for $r = 3$. The escape time increases rapidly as ϵ approaches 0.36 and 0.91. This behaviour is consistent with the mean field phase diagram calculated earlier. When the behaviour around these two critical points is analysed in detail, one finds that finite-size scaling relations of the same form as used for directed percolation describe the data rather accurately. The size dependence of the escape rate can be summarized as

$$\tau(r,\epsilon) \sim \begin{cases} \ln L & \text{laminar phase,} \\ L^z & \text{critical region,} \\ \exp(L^c) & \text{turbulent phase,} \end{cases} \tag{4.73}$$

where c is a constant of order unity. The three different cases are illustrated in Figs. 4.14a–c, respectively. By adjusting the model parameters until the absorbing time shows an algebraic increase with lattice size, the critical point can be located rather accurately and the critical exponent z is obtained by fitting a straight line to the data. The phase boundaries obtained from this method are shown in Fig. 4.13 by the squares.

The similarity of the observed CML behaviour to ordinary critical phenomena is further supported by scaling properties of the data. The order parameter in directed percolation is described by a scaling function of the form [Grassberger and de la Torre 1979]

$$m(L,t,\delta) \sim \xi^{-\beta/\nu} f_m(L/\xi, t/\xi^z) , \tag{4.74}$$

where δ is the deviation from the critical point. The assumption is that this scaling function also describes the behaviour of the CML. The spatial correlation length ξ diverges as $\delta^{-\nu}$ when $\delta \to 0$ at criticality. At the same time, the temporal correlation length diverges as ξ^z, where z is the dynamic critical exponent. If the problem is viewed as a $(1+1)$-dimensional anisotropic static phenomena, which is discussed in chapter 7, then two separate exponents $\nu_\perp$ and $\nu_\parallel$ are defined for the spatial and temporal directions, respectively, and the dynamic exponent is written $z = \nu_\parallel / \nu_\perp$. In this case, the absorbing time measures the correlation length in the time direction and we have $\tau \sim \xi_\parallel$. An estimate of the exponent

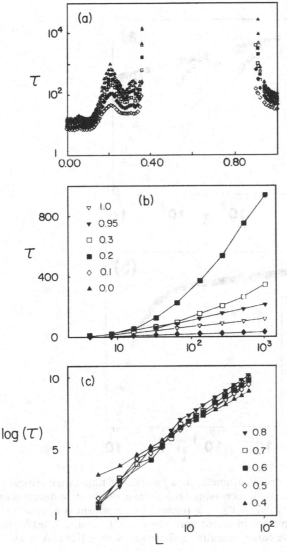

Figure 4.14 (a) Escape time τ as a function of ϵ for $r = 3$ and lattice sizes $L = 16$ (diamonds), $L = 32$ (solid diamonds), $L = 64$ (squares), $L = 128$ (solid squares), $L = 256$ (triangles), $L = 512$ (solid triangles), $L = 1024$ (circles). (b) Semi-logarithmic plot of τ versus L in the laminar phase. (c) Logarithmic plot of τ versus L in the turbulent regime [Houlrik et al. 1990].

$v_\parallel$ can thus be obtained directly from a plot of τ versus δ if the system size is sufficiently large, such that finite-size effects can be neglected.

In order to test the scaling assumption (4.74), the order parameter m can be plotted as a function of L and t at the critical point. If the first argument of the scaling function f_m is kept constant, at $\delta = 0$ then

$$m(L,t) \sim L^{-\beta/\nu} g_1(t/L^z) , \tag{4.75}$$

where g_1 is a new scaling function. If the second argument is constant, then instead

$$m(L,t) \sim t^{-\beta/\nu z} g_2(L/t^{1/z}) . \tag{4.76}$$

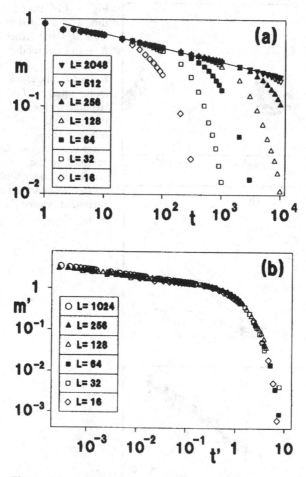

Figure 4.15 (a) The order parameter as a function of time at the critical point $\epsilon = 1$ and $r_c = 2.628$. With increasing lattice size a steady state distribution with power law decay $m(t) \sim t^{-\beta/\nu z}$ is formed. The exponent is $\beta/\nu z = 0.16$. (b) When the data are plotted in scaled variables $t' = t/L^z$ and $m' = mL^{\beta/\nu}$ the points fall on a single curve indicating a finite-size scaling [Houlrik et al. 1990].

At early times $t \ll L^z$ the system does not feel the constraint of a finite L and the above scaling functions approach the limits $g_1(x) \to x^{\beta/\nu z}$ and $g_2(x^{-1/z}) \to 1$. This scaling behaviour is verified in Fig. 4.15. From a log–log plot of m versus time (Fig. 4.15a), one directly obtains the exponent describing the decay $m \sim t^{-\beta/\nu z}$. Inserting the values for z and $\beta/\nu z$, the order parameter can be replotted, Fig. 4.15b, using the scaled coordinates $t' = t/L^z$ and $m' = L^{\beta/\nu} m$. The fact that the data collapse onto a single curve is a strong indication of dynamical scaling.

4.4.5 Spatial correlations and hyperscaling

Spatial correlations can be obtained directly by calculating the pair correlation function

$$C_j(t) = \frac{1}{L} \sum_{i=1}^{L} \langle u_i(t) u_{i+j}(t) \rangle - \langle u(t) \rangle^2. \tag{4.77}$$

The brackets now stand for an ensemble average over different initial conditions

$$\langle u(t) \rangle = \frac{1}{N_s} \sum_{s=1}^{N_s} u^{(s)}(t), \tag{4.78}$$

where $u(t)$ denotes any quantity defined on the lattice at time step t and N_s is the number of initial conditions. If there is only weak coupling between sites, i.e. small ϵ, one might expect the spatial correlations to fall off rapidly with distance. If this is the case, (4.77) can be written as $C_j(t) \sim \exp(-j/\xi(t))$ for large j. The function $\xi(t)$ can be interpreted as a time-dependent non-equilibrium correlation length.

If the model, on the other hand, is at criticality $\delta = 0$ one may expect a scaling form similar to the behaviour observed in Hamiltonian lattice models approaching thermodynamic equilibrium [Gunton et al. 1983]

$$C_j(t) = j^{1-\eta} \, \psi(j/\xi(t)), \tag{4.79}$$

where η is the exponent describing the algebraic decay of correlations in the static case [Houlrik and Jensen 1992]. As the lattice relaxes towards a steady state, correlations are induced over a length scale $\xi(t) \sim t^{1/z}$. An example of the time-dependent correlations calculated at the critical points (as defined above) $\epsilon = 0.7$ and $r = 3.735$ is shown in Fig. 4.16a. Using the value of z in Table 4.1, it is possible to determine the value of η such that the data approximately collapse onto a single curve. The pair correlation function thus provides additional evidence of critical correlations in both space and time.

It is interesting to compare the value of η to the other critical indices. In ordinary critical phenomena, the exponents are expected to satisfy the (hyper) scaling relation

$$2\beta/v = d - 2 + \eta , \qquad (4.80)$$

where $d = 1$ is the spatial dimension. When η is calculated from this relation, values in the range $\eta = 1.51$ to $\eta = 1.58$ are obtained, with no apparent variation as a function of the model parameters r and ϵ. Comparing to Table 4.1, the scaling relation (4.80) holds within the uncertainty for all entries except one ($r = 3$ and $\epsilon = 0.3598$).

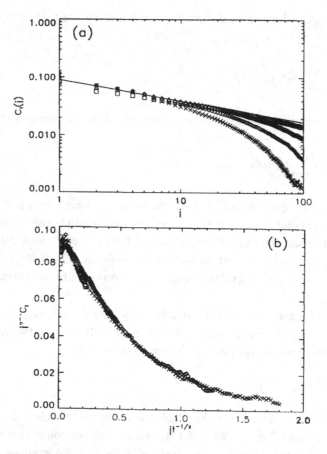

Figure 4.16 (a) The spatial pair correlation function defined in equation (4.77) as a function of distance j at different times t. As time proceeds, the pair correlation function develops a power law decay of the form $C(j) \sim j^{1-\eta}$. The parameters are $\epsilon = 0.7$ and $r = 3.735$ and the exponent is $\eta \simeq 1.52$. (b) Pair correlations in terms of the scaling variables $j/t^{1/z}$ and $j^{\eta-1}C_j(t)$. When $z = 1.70$ and $\eta \simeq 1.52$, the data points fall approximately on a single curve, indicating dynamical scaling of the form (4.79) [Houlrik and Jensen 1992].

All the critical exponents calculated for the CML system (4.63) are listed in Table 4.1 for various values of the model parameters ϵ and r. There is a clear dependence on the parameters, which shows that the exponents are not universal in the sense of ordinary critical phenomena. On the other hand, this variation is weak for small ϵ and for ϵ close to one. It was conjectured by Pomeau, see Chaté and Manneville [1988], that the system (4.63) belongs to the directed percolation universality class. The exponents for that class are shown in the bottom line of Table 4.1. The results indicate that the system is not in that class even though, for some parameter values, the exponents are quite close. It is indeed surprising that the CML does not belong to the directed percolation class, because the propagation of the turbulent sites follows a directed branching process (see Fig. 4.11) with an absorbing state, namely the globally laminar phase.

The origin of this discrepancy should probably be sought in the complicated spatial correlations, which are built up due to the completely deterministic dynamics represented by the synchronous update [Marcq et al. 1996]. It has thus recently been shown numerically [Rolf et al. 1998] that if the model (4.63) is updated in a random, asynchronous fashion, the system indeed falls in the class of directed percolation. In this approach a random site on the lattice is chosen at each time step and iterated forward according to (4.63). The randomness which is introduced in the process of selecting which site to update next, seems to be sufficient to make the phase transition universal and bring it into the universality class of directed percolation.

4.5 Coupled map lattices with anisotropic couplings

4.5.1 Convective instabilities and turbulent spots

The coupled map lattice (4.1) describes a closed system with no preferred direction. Many hydrodynamical experiments are, however, performed under

Table 4.1 *Critical exponents as a function of selected model parameters r and ϵ. The last entry contains the exponents for directed percolation.*

ϵ	r	z	β	ν	η
0.1	2.539(1)	1.44(2)	0.30(2)	1.04(2)	1.63(5)
0.3598(5)	3	1.42(2)	0.24(2)	0.92(2)	1.36(5)
0.7	3.735(5)	1.70(5)	0.24(2)	0.84(3)	1.52(5)
0.9083(5)	3	1.65(2)	0.22(2)	0.85(2)	1.50(5)
1	2.628(5)	1.60(2)	0.22(2)	0.86(2)	1.50(5)
DP		1.57	0.28	1.10	

open flow conditions, where the fluid comes in from one side and leaves through the other, as in a pipe flow. Let us, for simplicity, consider one spatial dimension. One can break the symmetry between j and $-j$ by admitting gradient terms into (4.1). Neglecting the Laplacian term the simplest 1D model is

$$u_j(n+1) = f(u_j(n)) - c(f(u_j(n)) - f(u_{j-1}(n))) \tag{4.81}$$

which includes coupling between sites j and $j-1$ but not between sites j and $j+1$. The coefficient c is analogous to the mean flow velocity. Models like this were studied in Crutchfield and Kaneko [1987], Deissler and Kaneko [1987], Bohr and Rand [1991]. Assume that the map f has an unstable fixed point M with $|f'(M)| = a > 1$ and consider the boundary condition $u_0(n) = M$ for all n. One should think of M as the laminar state and thus the incoming fluid is laminar. Letting $u_j(n) = M + z_j(n)$ and linearizing in the deviation z from the mean field solution we find

$$z_j(n+1) = (1-c)a\, z_j(n) + ca\, z_{j-1}(n) \tag{4.82}$$

for $j > 1$, and

$$z_1(n+1) = (1-c)\, a\, z_1(n) \tag{4.83}$$

for $j = 1$. These equations can be written in matrix form as

$$
\begin{pmatrix} z_1(n+1) \\ \cdot \\ \cdot \\ \cdot \\ z_L(n+1) \end{pmatrix}
=
\begin{pmatrix}
(1-c)a & 0 & 0 & \cdot & \cdot \\
ca & (1-c)a & 0 & 0 & \cdot \\
\cdot & \cdot & \cdot & \cdot & \cdot \\
0 & ca & (1-c)a & 0 & \\
\cdot & \cdot & \cdot & 0 & (1-c)a
\end{pmatrix}
\begin{pmatrix} z_1(n) \\ \cdot \\ \cdot \\ \cdot \\ z_L(n) \end{pmatrix}.
\tag{4.84}
$$

Since this matrix is triangular, its eigenvalues are the diagonal elements. All of them have the value $\lambda = (1-c)\, a$. So, even if $a > 1$, λ can be smaller than unity, making the laminar state linearly stable.

The system is, however, convectively unstable. An easy way to study the linear system (4.84) is to introduce a generating function $g(\xi, n)$ well known from the theory of random walks [Feller 1971]. Thus, we let

$$g(\xi, n) = \sum_{j=1}^{L} z_j \xi^{j-1}, \tag{4.85}$$

which, using (4.82) and (4.83), must satisfy the recursion relation

$$g(\xi, n+1) = ((1-c)a + ac\xi)\, g(\xi, n). \tag{4.86}$$

If, at time $n = 0$, only the left edge u_1 is perturbed, we can take as initial condition the vector $z_j(0) = \delta_{j,1}$ and find

$$g(\xi, n) = ((1-c)a + ac\xi)^n. \tag{4.87}$$

Thus

$$z_j(n) = \binom{n}{n-j} [(1-c)a]^{n-j} [c\,a]^j. \tag{4.88}$$

Looking at $z_j(n)$ in a frame of reference moving down the chain with speed v

$$z_{vn}(n) \sim e^{n\lambda(v)} \tag{4.89}$$

for $n \gg 1$ where Stirling's formula is applicable, we get [Crutchfield and Kaneko 1987]:

$$\lambda(v) = \ln a - (1-v)\ln\frac{(1-v)}{(1-c)} - v\ln\frac{v}{c}. \tag{4.90}$$

The maximal value of $\lambda(v)$ is the positive number $\ln a > 0$ and this occurs at $v = c$. There exists an interval $[v_{\min}, v_{\max}]$ around $v = c$ in which $\lambda(v)$ is positive; outside of it $\lambda(v)$ is negative.

The initial disturbance becomes a wave-packet with exponentially growing amplitude, whose front propagates downstream with leading velocity $v_{\max}$ and whose rear has a trailing velocity $v_{\min}$. Consequently, the spatial extent of the packet grows linearly in time.

If we want to consider more complicated problems, e.g. adding higher order spatial derivatives to (4.81) the method presented above becomes very cumbersome. We shall therefore give a more direct method based on the dispersion relation for the fields [Bohr and Rand 1991]. In fact the exponent $\lambda(v)$ turns out to be the Legendre transform of the analytical continuation of the dispersion relation, i.e. the spectrum for linear perturbations. This is discussed in greater detail in appendix D for linear perturbations of partial differential equations.

Let us, for simplicity, return to (4.82). In terms of the spatial Fourier transform of z, replacing ξ by e^{-ik} in (4.86), we find

$$\widehat{z}_k(n+1) = L(k)\widehat{z}_k(n) \tag{4.91}$$

where

$$L(k) = \left((1-c)a + ace^{-ik}\right). \tag{4.92}$$

Thus we can, at least formally, solve (4.82) by

$$z_j(n) \sim \int dk\, e^{n\ln L(k) + ijk} \widehat{z}_k(0). \tag{4.93}$$

Since we are looking for spatially growing perturbations we let $\alpha = ik$ and

$$S(\alpha) \equiv \ln L(k) = \ln[a(1-c) + ace^{-\alpha}]. \tag{4.94}$$

The solution, in a reference frame moving with velocity v, is then

$$z_{vn}(n) \sim \int d\alpha\, e^{n(S(\alpha) + \alpha v)} \widehat{z}_{-i\alpha}(0). \tag{4.95}$$

By a standard saddle point argument, as used earlier in (4.15) and (4.18), one finds that the dominant contribution comes from the saddle point of $S(\alpha) + \alpha v$ given by $\alpha = \alpha(v)$ where

$$S'(\alpha) = -v, \tag{4.96}$$

which gives us, for the exponent in (4.89),

$$\lambda(v) = S(\alpha(v)) + \alpha(v)v. \tag{4.97}$$

Further

$$\lambda'(v) = \alpha \tag{4.98}$$

so $S(\alpha)$ and $\lambda(v)$ are Legendre transform pairs and the maximal exponent is found, when v satisfies $\alpha(v) = 0$. In this example a simple computation gives

$$v = -S'(\alpha) = \frac{ace^{-\alpha}}{a(1-c) + ace^{-\alpha}} \tag{4.99}$$

therefore

$$S(\alpha(v)) = \ln\left(\frac{a(1-c)}{1-v}\right) \tag{4.100}$$

corresponding to (4.90) for $\lambda(v)$. This procedure can easily be used for more complicated coupled map systems or partial differential equations, although, in general, the saddle point will not correspond to purely imaginary k (real α).

We now discuss a simple example of a coupled map lattice generating localized turbulent spots which move downstream [Bohr and Rand 1991]. Consider the two-dimensional lattice maps

$$u_{i,j}(n+1) = f(u_{i,j}(n)) - c(f(u_{i,j}(n)) - f(u_{i-1,j}(n))) + \epsilon\Delta f(u_{i,j}(n)) \quad (4.101)$$

where

$$\Delta f(u_{i,j}(n)) = \frac{1}{4}(f(u_{i+1,j}(n)) + f(u_{i-1,j}(n)) + f(u_{i,j+1}(n)) + f(u_{i,j-1}(n)) - 4f(u_{i,j}(n))) \tag{4.102}$$

(as defined by (4.48)) and f is the piecewise linear map

$$f(x) = \begin{cases} 3x + 2, & \text{if } x \in [-1, -1/3] \\ ax, & \text{if } x \in [-1/3, 1/3] \\ 3x - 2, & \text{if } x \in]1/3, 1] \end{cases} \tag{4.103}$$

shown in Fig. 4.17. The constants a and c are adjusted such that $a > 1$, and $(1-c)a < 1$. Thus the system is absolutely stable, but convectively unstable, when the left edge is fixed at the unstable fixpoint (i.e. $u_{0,j}(n) = 0$ for all j and n).

Figure 4.18 shows the evolution of a localized perturbation around the unstable fixed point $u_* = 0$ of the map (4.102), when the system is perturbed slightly at the initial time at the centre of the left-hand edge. An initial small disturbance grows downstream and finally the field enters the region outside of the interval $[-1/3, 1/3]$, where it can undergo chaotic motion until it gets reinjected into the laminar region $[-1/3, 1/3]$. The velocities v_{min} and v_{max} fit nicely with the predictions from linear stability theory sketched above.

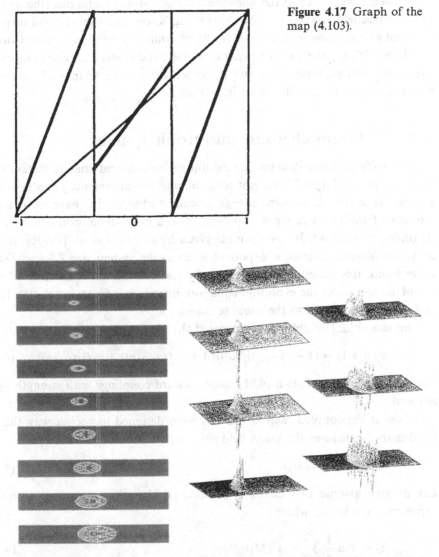

Figure 4.17 Graph of the map (4.103).

Figure 4.18 Evolution of a turbulent spot in the model (4.102–103). An initial disturbance of amplitude 10^{-6} was applied to the centre of the left-hand boundary. The plots on the left were extracted at 'times' $n = 191, 193, 197, 201, 205, 219, 229, 248, 260$. The plots on the right, at times $n = 192, 193, 195, 200, 220, 240, 260$. The parameters are $a = 1.5$, $c = 0.5$ and $\epsilon = 0.4$ [Bohr and Rand 1991].

In two-dimensional systems without spatial isotropy such turbulent spots can take on interesting forms. It is thus well known that the onset of turbulence in a boundary layer formed on a flat plate can happen via the formation of turbulent spots, the so-called Emmons spots shown in Fig. 1.2, of a characteristic horseshoe or boomerang-like form, which is maintained in time, while the spot becomes larger. This was explained by Gaster [1962], who studied the form of a localized disturbance in the linearized Navier–Stokes equations (the so-called Orr–Sommerfeld equations). The same thing is seen in coupled maps or partial differential equations with various forms of symmetry breaking terms. Conrado and Bohr [1995] gave a rather general analysis and showed how to compute the shape of spots perturbatively and discussed the generic shapes of such spots. We shall briefly discuss this issue in section 5.10.3.

4.5.2   Coherent chaos in anisotropic systems

Convectively unstable systems can exhibit stable coherent chaotic states even if their size is very large. This seems to go against the arguments given in chapter 1, that large chaotic systems are generically turbulent, i.e. have an extensive number of positive Lyapunov exponents and a finite dimension density in the thermodynamic limit. In the example given by Aranson et al. [1992b], there is only one positive Lyapunov exponent, even as the system size $L \to \infty$. On the other hand, this system is not structurally stable, i.e. a small amount of noise, or small changes in the evolution equation, introduce a finite correlation length and thus take us back to the usual scenario.

Aranson et al. [1992b] have considered the 1D coupled map lattice

$$u_j(n+1) = (1 - \gamma_1 - \gamma_2)f(u_j(n)) + \gamma_1 f(u_{j-1}(n)) + \gamma_2 f(u_{j+1}(n)), \quad (4.104)$$

which is more general than (4.81) since forward couplings with strength γ_2 are allowed.

As usual the coupled map lattice has been designed in such a way that the local map f generates the mean field solution

$$M(n+1) = f(M(n)). \tag{4.105}$$

Let us also assume that chaos is present, i.e. that the mean field Lyapunov exponent $\lambda_0 = \ln|\chi_0|$, where

$$\lambda_0 = \lim_{n \to \infty} \frac{1}{n} \sum_{i=1}^{n} \ln |f'(M(i))| \tag{4.106}$$

is positive. Let us now study the stability of this uniform solution (4.105). If $u_j(n) = M(n) + z_j(n)$, the linearized equation for z is

$$z_j(n+1) = f'(M(n)) \left[(1 - \gamma_1 - \gamma_2)z_j(n) + \gamma_1 z_{j-1}(n) + \gamma_2 z_{j+1}(n) \right]. \tag{4.107}$$

Surprisingly at first sight, the result depends strongly on the type of boundary condition chosen. For periodic boundary conditions, utilizing

$$z_j(n) = \sum_k e^{ikj} z_k(n) \tag{4.108}$$

with $k = 2\pi m/L$ and $m = 0, \pm 1, \pm 2, \ldots, \pm L/2$, each Fourier component $z_k(n)$ obviously grows/decays as $(\chi^{(p)}(k))^n$, where

$$\chi^{(p)}(k) = \chi_0[(1 - \gamma_1(1 - e^{-ik}) - \gamma_2(1 - e^{ik})]. \tag{4.109}$$

For small values of k one has

$$\ln(\chi^{(p)}(k_j)) \simeq \lambda_0 + i(\gamma_2 - \gamma_1)k_j - \frac{1}{2}(\gamma_2 + \gamma_1)k^2, \tag{4.110}$$

and thus the real part has a quadratic maximum at $k = 0$, where the value is λ_0.

If we choose free boundary conditions, the result is rather different. A way to implement free boundary conditions numerically is to introduce additional sites 0 and $L + 1$, such that $u(0) = u(1)$ and $u(L + 1) = u(L)$. As shown in [Aranson et al. 1992b], this changes the spectrum to

$$\chi^{(f)}(k_m) = \chi_0(1 - \gamma_1 - \gamma_2 + 2\sqrt{\gamma_1\gamma_2}\cos k_m), \tag{4.111}$$

where now $k = \pi m/L$ and $m = 0, 1, 2, \ldots, L - 1$.

This result can be found in a straightforward way using the form

$$z_j(n) = \sum_m (\chi^{(f)}(k_m))^n e^{\delta j} \cos(k_m j + \phi_m) \tag{4.112}$$

where

$$e^{2\delta} = \frac{\gamma_1}{\gamma_2}. \tag{4.113}$$

The phases ϕ_m are fixed by the boundary conditions, giving:

$$\tan \phi_m = \frac{\cos k_m - e^{-\delta}}{\sin k_m}. \tag{4.114}$$

The spectrum has a gap in the sense that, for $k \to 0$, $\chi^{(f)}(k) \to \chi_0[1 - (\gamma_1^{1/2} - \gamma_2^{1/2})^2]$, which is different from the result χ_0 at $k = 0$.

In fact (4.111) can be obtained directly by analytical continuation of the spectrum for the periodic case. First note, in analogy with the last section, that there exist a saddle point k_* of $\chi^{(p)}$ satisfying

$$e^{2ik_*} = \frac{\gamma_1}{\gamma_2}, \quad \text{and} \quad \chi^{(p)}(k_*) = \chi_0[1 - (\gamma_1^{1/2} - \gamma_2^{1/2})^2]. \tag{4.115}$$

As seen in the last section the saddle point value of $\chi^{(p)}(k_*)$ controls the absolute stability of the state and therefore exponentially growing states like (4.112) must have $\delta = ik_*$, in agreement with (4.113) and (4.115). If we shift the values of k in the complex plane as $k = k_* + q$ we find

$$\chi^{(p)}(k) = \chi^{(f)}(q), \qquad\qquad (4.116)$$

whereby one can see that the form of the spectrum (4.111) does not only apply to free boundary conditions, but to any boundary conditions allowing exponentially growing solutions. Of course, these boundary conditions should be consistent with the mean field solution, which in the case of a flat state singles out the free boundary conditions.

If γ_1 and γ_2 are sufficiently different then $|\chi^{(p)}(k_*)| < 1$ and the state is absolutely stable. Under such conditions one does indeed find that the flat, chaotic state is stable and can be generated even from an initially disordered state. The ordered state develops by the motion of a synchronizing front through the lattice, whose velocity can be calculated by the methods discussed in the previous section. Aranson et al. [1992b] noted that the synchronization wave does not propagate through the whole lattice if L is sufficiently large. They showed that this is due to lack of numerical accuracy and that the length l_c of the flat region scales as

$$l_c \sim -\ln(\sigma) \qquad\qquad (4.117)$$

where σ is the strength of an imposed noise or the numerical inaccuracy.

To understand this, let us consider a particular form of noise consisting of a defect at the origin. That is, we slightly change the boundary condition at the origin, to say

$$u_0(n) = u_1(n) + \sigma. \qquad\qquad (4.118)$$

This creates an exponentially growing disturbance

$$\delta u_j(n) \sim \sigma e^{\delta j}, \qquad\qquad (4.119)$$

and this is $0(1)$ when

$$j = l_c \sim -\frac{1}{\delta} \log \sigma. \qquad\qquad (4.120)$$

We conclude this section by stressing the fact that even a small amount of noise has a strong influence on the system (4.107), and takes us back to the generic situation in which the number of positive Lyapunov exponents grows with system size. For example the number of positive Lyapunov exponents depends on the noise level, or the number of digits, in the numerical computation [Biferale et al. 1993]. This is a clear evidence that this system is non generic. Moreover, the stability eigenvector corresponding to the first Lyapunov exponent has support only inside the flat state [Biferale et al. 1993], except for bursts created on the interface between the flat and disordered part, that are swept downstream while being exponentially damped. The stability eigenvectors for the higher Lyapunov exponents have support in the disordered part towards the right end of the system.

4.5.3   A boundary layer instability in an anisotropic system

In this section we discuss a simple coupled map lattice model of a convective system with a boundary layer. The model is constructed to mimic a fluid in a container that is heated from below. As discussed in section 1.2.1, a convective state commences when the temperature gradient exceeds a critical value. If the convective effects exceed the diffusive effects the boundary layer becomes unstable and emits 'plumes' from the boundary layer, as shown in Fig. 1.3a.

In the convective state, the hot fluid moves upwards and the cold fluid downwards by means of convective rolls. The number of convective rolls is determined by the aspect ratio of the container. In the experiments developed by Libchaber and co-workers [Heslot et al. 1987, Castaing et al. 1989], the aspect ratio is close to 1 and only few convective rolls are present. Boundary layers play important roles for Rayleigh–Bénard convection: close to the bottom and top plates there is a small region, a thermal boundary layer, where heat is not transported by convection but only by diffusion, e.g. see Landau and Lifshitz [1987]. As the temperature gradient is increased further to high values of the Rayleigh number ($\sim 10^8$), these boundary layers become unstable, at first against small amplitude travelling waves on the top of the layers. At an even higher value of the Rayleigh number, the amplitude of these waves has grown so large that they begin to emit patches ('hot plumes') from the boundary layer into the centre of the cell [Castaing et al. 1989]. The motion of these plumes causes large temperature fluctuations, which gives rise to the exponential distributions of the temperature fluctuations, characteristic of the 'hard' turbulent state.

The essential feature of the CML model, which appears to capture the same phenomenology, is the presence of a competition between diffusion and convection leading to an unstable diffusive boundary layer, emitting plumes into the laminar regime [Jensen 1989a,b]. It is important to note that by using CML one loses some essential features of the turbulence as, for instance, conservation laws. Therefore one can only compare with experiments on a qualitative level. In Rayleigh–Bénard convection, one always has a mixture of regular laminar motion given by the large-scale circulation, with strongly turbulent motion around the boundary layers. The maps should therefore exhibit both turbulent and laminar behaviour, and such maps were introduced in section 4.3.1, (4.62). The motion is chaotic when $u \leq 1$, which plays the role of a turbulent (or 'hot') state, and the motion is laminar for $u > 1$. When coupled diffusively in 1D, (4.63), Chaté and Manneville [1988, 1989] found that as the coupling strength ϵ exceeds a critical value ϵ_c, a turbulent site percolates through the system. When ϵ is below ϵ_c any initial state turns into a laminar state.

To include an anisotropic coupling, one can introduce a convective term of strength v that displaces 'hot' fluid (i.e. sites with $u \leq 1$) upwards, which leads to a model, in 2D, of the form (4.101) (except that the convection occurs in the vertical direction):

$$u_{i,j}(n+1) = f(u_{i,j}(n)) + \frac{\epsilon}{4}(f(u_{i-1,j}(n)) + f(u_{i+1,j}(n)) + f(u_{i,j-1}(n))$$
$$+ f(u_{i,j+1}(n)) - 4f(u_{i,j}(n))) + v(f(u_{i,j-1}(n)) - f(u_{i,j}(n))). \quad (4.121)$$

This ensures that the model employs a competition between a diffusive term of strength ϵ and a convective term of strength v.

Next, a boundary condition at the bottom is introduced which plays the role of a constant hot temperature at the bottom plate. Since the sites with $u \leq 1$ are the hot sites, we enforce this condition by the constraint

$$u_{i,1} = u_B < 1. \quad (4.122)$$

This means the u-value in the first row is kept fixed at u_B (in the following, we set $u_B = 0$). Simulations of the model are initiated in a laminar state $u_{i,j}(0) = 1.1 + \eta_{i,j}$, where $\eta_{i,j}$ are random terms. The time evolution is visualized by marking the hot sites, i.e. sites where $u_{i,j} < 1$. In this way, it is easy to observe the hot plumes that travel through the laminar regime. For small values of the convection term, i.e. v less than a critical value v_c, the hot boundary condition at the bottom introduces a hot boundary layer of a depth of a few lattice lengths. Above the boundary layer, the system is in its laminar state. This thermal boundary layer is identified by calculating the average number of hot sites in each layer above the bottom. This number shows a sharp gradient over the width of the boundary layer and then goes to zero in the laminar state, just like the sharp temperature gradient in experimentally observed thermal boundary layers.

When the strength of the gradient term exceeds the critical value v_c, the boundary layer becomes unstable and starts to emit patches (hot plumes) into the laminar regime. The shapes and sizes of the plumes can vary a lot. Fig. 4.19 shows a snapshot of the simulation. For $\epsilon = 0.12$ used in Fig. 4.19, the critical value is found to be $v_c \simeq 0.018$ by monitoring the value of v for which the boundary layer begins to emit plumes. These observations are qualitatively similar to the experimental findings. In the visualization of the experiment, one observes that the boundary layer becomes unstable against small amplitude traveling waves and as the Rayleigh number is increased, the waves may detach as convective plumes [Castaing et al. 1989], see Fig. 1.3a.

The boundary layer instability can be understood in more detail, using a simple mean field approximation. A mean field value m_j of $u_{i,j}(n)$ is defined at each layer j in the lattice and is calculated in the following way (for specific values of r, ϵ and v):

$$m_j = \frac{1}{TN} \sum_{i=1}^{N} \sum_{n=1}^{T} u_{i,j}(n). \tag{4.123}$$

By definition, $0 \le m_j \le \frac{r}{2}$, and of course $m_1 = u_B$. Within this mean field, one can compare the diffusive effects against the convective effects. At a specific layer j in the lattice, j chosen in the top of the boundary layer, the following two quantities are calculated

$$D_d = \frac{\epsilon}{4}(m_{j-1} - 2m_j + m_{j+1}), \tag{4.124a}$$

$$D_c = v(m_{j-1} - m_j). \tag{4.124b}$$

These two quantities are the diffusive and convective terms in the local mean field. For $\epsilon = 0.12$, the top of the boundary layer is around $j = 3$ which is used in (4.124). Figure 4.20 shows D_d and D_c plotted versus v. The two curves cross at $v \simeq 0.018$. This is the point where, for an increasing value of v, the convective term becomes larger than the diffusive term. Above this point, the boundary layer will not only be a diffusive layer but will also be unstable against convection such that patches from the top of the boundary layer will be released

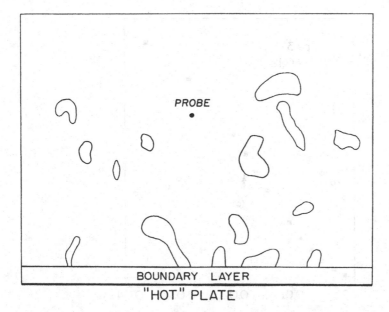

Figure 4.19 A snapshot of a simulation of model (4.121) with the parameters $\epsilon = 0.12$, $v = 0.04 > v_c \simeq 0.018$, $r = 3.0$, and $u_B = 0$. The patches are the hot plumes for which $u_{(i,j)} < 1$. The plumes are released from the boundary layer and drift upwards by convection [Jensen 1989a].

into the laminar regime. Therefore, this crossing point is the critical point for the boundary layer instability, v_c.

As in the experiments, for example, see Castaing et al. [1989], the turbulent state is quantitatively characterized by the temperature fluctuations in the laminar regime. These are measured by placing a probe at a specific point in the centre of the cell, see Fig. 4.19b. The number of time steps, t_p, for each plume to pass the probe is then measured. This time plays the role of a temperature fluctuation, i.e. a long time (large plume) will likely give rise to a large fluctuation in the temperature. As the system evolves, many plumes sweep intermittently across the probe. The corresponding probability distribution, $D(t_p)$, is plotted in Fig. 4.21 and clearly shows an exponential shape: $D(t_p) \sim \exp(-at_p)$ which appears to be robust to changes in the parameter values. Similar behaviour has been observed experimentally in the 'hard' turbulent regime [Castaing et al. 1989].

4.5.4   A coupled map lattice for a convective system

Yanagita and Kaneko [1993] extended the model discussed in the previous section in order to have a more detailed description of a convective system. Three numbers are important: the Rayleigh number proportional to δT, the

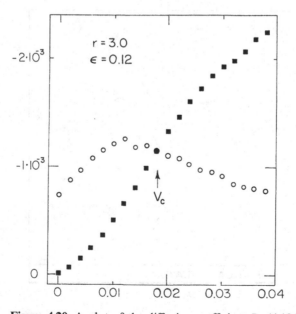

Figure 4.20 A plot of the diffusive coefficient D_d (4.124a), shown by circles, and the convective coefficient D_c (4.124b), shown by squares, versus the gradient strength v. Both are computed at site $j = 3$ (near the top of the boundary layer) and parameters $r = 3$ and $\epsilon = 0.12$ [Jensen 1989b].

Prandtl number v/κ and the aspect ratio L_x/L_y, where L_x and L_y are the horizontal and vertical size of the system respectively.

Two fields are introduced, the velocity field $\mathbf{u}_{i,j}(n) = (v_{i,j}, w_{i,j})$ and the internal energy $E_{i,j}(n)$. The corresponding evolution law is decomposed into Eulerian and Lagrangian parts to include advective motion. The Eulerian part has the following prescriptions: (I) a site with higher temperature receives a force in the upwards direction; (II) heat diffusion leads to diffusion for $E_{i,j}(n)$; (III) the velocity field $\mathbf{u}_{i,j}(n)$ diffuses due to the viscosity. The incompressibility condition is not strictly fulfilled but a term is added to avoid $\nabla \cdot \mathbf{u}$ being too large. Yanagita and Kaneko [1993] proposed the following model:

Buoyancy:

$$w_{i,j}(n)^* = w_{i,j}(n) + \frac{1}{2}c[2E_{i,j}(n) - E_{i+1,j}(n) - E_{i-1,j}(n)]$$

$$v_{i,j}(n)^* = v_{i,j}(n). \tag{4.125a}$$

Heat diffusion:

$$E_{i,j}(n)' = E_{i,j}(n) + \kappa\Delta E_{i,j}(n). \tag{4.125b}$$

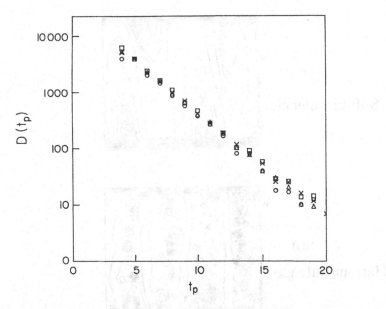

Figure 4.21 The distribution of passage times, $D(t_p)$, versus t_p on a logarithmic scale. Circles: $\epsilon = 0.12$, $v = 0.035$; triangles: $\epsilon = 0.12$, $v = 0.05$; squares: $\epsilon = 0.12$, $v = 0.08$; crosses: $\epsilon = 0.14$, $v = 0.05$. The different curves are normalized to the same value of $D(t_p)$ at $t_p = 5$. Measurements for $t_p \leq 3$ are disregarded. Each calculation is performed over $\sim 10^6$ time steps [Jensen 1989b].

Viscosity and pressure effects:

$$v_{i,j}(n)' = v_{i,j}(n)^* + v\Delta v_{i,j}(n)^* + \eta\left(\frac{1}{2}[v_{i+1,j}(n)^* + v_{i-1,j}(n)^*] - v_{i,j}(n)^*\right.$$

$$\left. +\frac{1}{4}[w_{i+1,j+1}(n)^* + w_{i-1,j-1}(n)^* - w_{i-1,j+1}(n)^* - w_{i+1,j-1}(n)^*]\right). \qquad (4.125c)$$

with a similar equation where $v \leftrightarrow w$. These three steps completes the Eulerian scheme. In the above equations c and η are constants, Δ indicates the discrete Laplacian in 2D, v is the kinematic viscosity and the fields with asterisks and primes are intermediate quantities generated by the Eularian dynamics.

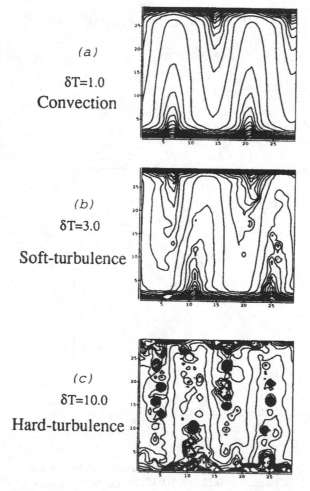

(a)

$\delta T = 1.0$

Convection

(b)

$\delta T = 3.0$

Soft-turbulence

(c)

$\delta T = 10.0$

Hard-turbulence

Figure 4.22 Snapshot of equi-energy contours obtained by a numerical simulations of model (4.125) for the parameter values $L_x = L_y = 20$, $v = \kappa = 0.2$, and three temperature differences: $\delta T = 1, 3, 10$ [Yanagita and Kaneko 1993].

The Lagrangian scheme completes the time step by advecting the velocity and the temperature. One places a test particle on the site (i, j), which then moves to $(i + v, j + w)$. All the variables, i.e. the velocity and the internal energy, are carried by this particle. In general there is no site in the position $(i+v, j+w)$, so one allocates the field variables on its four nearest neighbour sites, see Yanagita and Kaneko [1993] for details.

The full dynamics of the model can therefore be described by successive applications of the following procedure: $\{u_{i,j}(n), E_{i,j}(n)\} \rightarrow \{u_{i,j}(n)^*, E_{i,j}(n)\} \rightarrow \{u_{i,j}(n)', E_{i,j}(n)'\} - (\text{Lagrangian}) \rightarrow \{u_{i,j}(n+1), E_{i,j}(n+1)\}$. The temperature difference over the cell is set by: $E_{i,0} = \delta T = -E_{i,L_y}$, where L_y is the vertical size of the system.

For small values of the Rayleigh number various patterns of convective rolls appear. Usual oscillatory instabilities are observed in the roll pattern as the Rayleigh number is increased. As in the model described in the previous section, the boundary layers become unstable and emit plumes into the laminar regime. Figure 4.22 shows equi-energy (equi-temperature) lines for three different values of the Rayleigh number, corresponding to the convective, soft-turbulence and hard-turbulence regimes, respectively (cf. Castaing et al. [1989]). The corresponding distribution functions for the energy fluctuations in the centre of the system are shown in Fig. 4.23. Qualitatively, this is similar to what is observed in experimental Rayleigh–Bénard convection. Many interesting spatio-temporal patterns can be found in the simulations of this model which is easily generalized to 3D [Yanagita and Kaneko 1993].

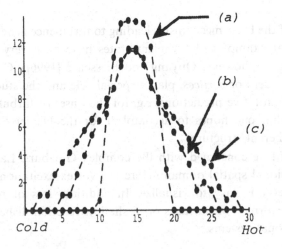

Figure 4.23 The distribution functions of $E_{i,j}$ for $j = L_y/2$ sampled over time and space for the same parameter values as in Fig. 4.22 [Yanagita and Kaneko 1993].

Chapter 5

Turbulence in the complex Ginzburg–Landau equation

In chapter 1 we have anticipated that it is possible to find turbulent states in nonlinear field theories, whose structure is much simpler than the Navier–Stokes equations. In the following, we shall discuss an important case, the so-called complex Ginzburg–Landau equation, which has been intensively studied in recent years, both theoretically and experimentally. Experimental works include oscillatory chemical reactions [Zaikin and Zhabotinsky 1970, Winfree 1972, 1987, Kuramoto 1984, Ouyang and Flesselles 1996], surface catalysis [Jakubith et al. 1990], multimode lasers [Arecchi et al. 1990, 1991], intracellular waves [Lechleiter 1991], colonies of social amoebae [Gerisch and Hess 1974], and cardiac arrhythmia [Winfree 1987, 1989, Davidenko et al. 1991].

For this system many of the basic mechanisms leading to turbulence have been understood, although clear examples of turbulent states have been very hard to observe experimentally (see however Ouyang and Flesselles [1996]). Certain excitations, namely *spiral waves* or *vortices*, play a special role and the study of their properties allows quantitative predictions, e.g. for the onset of turbulence. This is very similar to what one hopes to accomplish for the Navier–Stokes equations by studying 'coherent structures'.

Most of the chapter will be concerned with the complex Ginzburg–Landau equation in a two-dimensional spatial domain. Here the vortex excitations are point-like and the geometry is easy to visualize. In addition, lots of recent experiments have been performed on surfaces or in shallow reaction dishes, i.e. in essentially two dimensional systems.

5.1 The complex Ginzburg–Landau equation

The complex Ginzburg–Landau equation is an example of an *amplitude equation*. Close to a bifurcation point, where one or more modes become unstable, one assumes that the new structure selected by the dynamics can be described in terms of these modes. In large systems, slow spatial variations of such modes have to be taken into account and the result is an equation for the amplitude(s) of the mode(s) in question, valid close to the bifurcation. This concept goes back to Landau's work on equilibrium systems, in particular to the generic behaviour of a thermodynamic system close to continuous phase transitions [Landau 1937] and the structure of inhomogenous superconductors [Ginzburg and Landau 1950]. For non-equilibrium systems, pioneering work was done by Newell and Whitehead [1969], Segel [1969] and Stewartson and Stuart [1971]. A recent review of the application of amplitude equations to pattern formation can be found in Cross and Hohenberg [1993]. For the case of the complex Ginzburg–Landau equation a lot of valuable material including a thorough derivation can be found in the book of Kuramoto [1984].

There exist several systems, the most famous is probably the Belousov–Zhabotinsky chemical reaction, in which a global Hopf bifurcation occurs, and a homogenous periodic attractor appears out of a spatially homogenous steady state (see appendix A). If the system is kept homogenous, e.g. by stirring, many other bifurcations can be observed as well as beautiful examples of strange attractors [Roux et al. 1983]. In the following we shall however consider unstirred systems where inhomogeneities can form, but where we assume that the necessary chemicals can be provided in a homogeneous way.

Close to a Hopf bifurcation there is a characteristic frequency ω_0: in the bifurcated state, slightly above the bifurcation, it is the frequency of the periodic attractor. Slightly below, it characterizes the transients to the steady state. Let us now describe the state variables of the system, e.g. the densities of chemicals for the Belousov–Zhabotinsky reaction, by assuming that the only rapid time variation comes from the Fourier mode pertaining to the characteristic frequency ω_0. All other time scales are assumed to be slow compared to it, and we thus treat the amplitude of this mode, the *order parameter*, as a slowly varying field. This field, $A(\mathbf{x}, t)$ is a complex variable, since one must consider the Fourier coefficients for both ω_0 and $-\omega_0$. Assuming moreover that the modulus $|A|$ is small, which is true close to the bifurcation point as long as we restrict our attention to supercritical (i.e. continuous) bifurcations, we get the complex Ginzburg–Landau equation

$$\partial_t A = \mu A - (1 + i\alpha)|A|^2 A + (1 + i\beta)\nabla^2 A, \tag{5.1}$$

where we have rescaled A and $\mathbf{x}$ in order to make two of the coefficients equal to unity. The meaning of the coefficient α can be understood by neglecting spatial variations. For a homogenous state $A_0(t)$ we get the solution

$$A_0(t) = \sqrt{\mu}e^{-i\alpha\mu t}. \tag{5.2}$$

It is easy to verify that this solution is stable towards spatially homogenous perturbations when $\mu > 0$, whereas the solution $A = 0$ is unstable. For $\mu < 0$ the situation is opposite. The solution $A_0(t)$ has the frequency $\omega = \alpha\mu$, showing that α controls the variation of the characteristic frequency with the order parameter, i.e. that the local Hopf frequency varies like

$$\Omega = \omega_0 + \omega = \omega_0 + \alpha\rho^2, \tag{5.3}$$

where ρ is the modulus of the field $\rho = \sqrt{\mu}$. Thus, for small positive μ, one has that ρ and $\delta\omega$ are small, ensuring the validity of the assumptions leading to (5.1). We can now rescale μ to 1 by letting $x \to x/\sqrt{\mu}$, $A \to A\sqrt{\mu}$ and $t \to t/\mu$ obtaining

$$\partial_t A = A - (1 + i\alpha)|A|^2 A + (1 + i\beta)\nabla^2 A \tag{5.4}$$

which is the form of the complex Ginzburg–Landau equation used in the following.

The parameter β is related to differences in the diffusivity of the species involved in the chemical reaction. Indeed, assume that the reaction-diffusion equation describing the reaction is

$$\partial_t \mathbf{c} = \mathbf{f}(\mathbf{c}; \mu) + \mathbf{D}_c \nabla^2 \mathbf{c}, \tag{5.5}$$

where $\mathbf{c}$ is the concentration vector and $\mathbf{D}_c$ is the diffusion matrix. Assume further that the spatially uniform solution of (5.5) satisfying

$$\frac{d\mathbf{c}}{dt} = \mathbf{f}(\mathbf{c}; \mu) \tag{5.6}$$

undergoes a Hopf bifurcation at $\mu = \mu_0$. Thus, two of the eigenvalues of

$$(\mathbf{J}_0)_{nm} = \frac{\partial f_n}{\partial c_m} \tag{5.7}$$

are purely imaginary $= \pm i\omega_0$, while the rest are assumed to have negative real parts. These neutral eigenvectors of $\mathbf{J}_0$ provide geometrical information on how the emerging limit cycle is situated in concentration space. The Hopf plane is shown schematically in Fig. 5.1. One should notice that $\mathbf{J}_0$ is not a Hermitian matrix, and thus the eigenvalues need not be real. Thus left and right eigenvectors are in general different and will be indicated as $\mathbf{u}^{\pm}$ and $\mathbf{u}_{\pm}$, satisfying

$$\mathbf{J}_0\mathbf{u}_+ = i\omega_0\mathbf{u}_+ , \qquad \mathbf{u}^+\mathbf{J}_0 = i\omega_0\mathbf{u}^+$$

and

$$\mathbf{J}_0\mathbf{u}_- = -i\omega_0\mathbf{u}_- , \qquad \mathbf{u}^- \mathbf{J}_0 = -i\omega_0\mathbf{u}^-.$$

By proper normalization

$$\mathbf{u}^+ \cdot \mathbf{u}_+ = \mathbf{u}^- \cdot \mathbf{u}_- = 1,$$

whereas

$$\mathbf{u}^+ \cdot \mathbf{u}_- = \mathbf{u}^- \cdot \mathbf{u}_+ = 0.$$

Let us remark that in the Dirac notation $\mathbf{u}^\pm$ is the bra $\langle\mathbf{u}^\pm|$ and $\mathbf{u}_\pm$ is the ket $|\mathbf{u}_\pm\rangle$. Moreover, since it is assumed that $\mathbf{J}_0$ is real,

$$\mathbf{u}_- = (\mathbf{u}_+)^* \qquad \text{and} \qquad \mathbf{u}^- = (\mathbf{u}^+)^*.$$

The above assumptions leading to the complex Ginzburg–Landau equation are equivalent to the approximation

$$\mathbf{c}(\mathbf{x}, t) \approx A(\mathbf{x}, t)\, e^{i\omega_0 t}\mathbf{u}_+ + A^*(\mathbf{x}, t)\, e^{-i\omega_0 t}\mathbf{u}_-, \tag{5.8}$$

which, by expanding the reaction diffusion equation (5.5) to linear order in A, leads to

$$e^{i\omega_0 t}\mathbf{u}_+\partial_t A + e^{-i\omega_0 t}\mathbf{u}_-\partial_t A^* = e^{i\omega_0 t}\mathbf{D}_c\mathbf{u}_+\nabla^2 A + e^{-i\omega_0 t}\mathbf{D}_c\mathbf{u}_-\nabla^2 A^*. \tag{5.9}$$

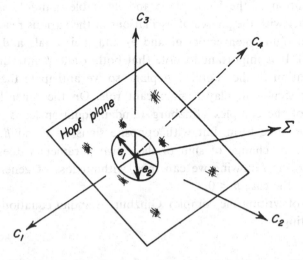

Figure 5.1 The Hopf plane in concentration space with origin in the stationary state, which loses stability at the Hopf bifurcation. The vectors $\mathbf{e}_1$ and $\mathbf{e}_2$ in the Hopf plane are shown, as well as the stable manifold Σ of the stationary state. The dual vector $\mathbf{e}^1$ is orthogonal to $\mathbf{e}_2$ and Σ and $\mathbf{e}^2$ is orthogonal to $\mathbf{e}_1$ and Σ, but $\mathbf{e}_1$, $\mathbf{e}_2$ and Σ are not in general orthogonal.

Multiplying on the left by $\mathbf{u}^+$ and using the orthonormality conditions, one gets

$$\partial_t A = \mathbf{u}^+ \mathbf{D}_c \mathbf{u}_+ \nabla^2 A + e^{-2i\omega_0 t} \mathbf{u}^+ \mathbf{D}_c \mathbf{u}_- \nabla^2 A^*. \tag{5.10}$$

Now, if A only varies on time scales much larger than $2\pi/\omega_0$ the last term will average out and we are left with a diffusion equation

$$\partial_t A = D\nabla^2 A, \tag{5.11}$$

where the diffusion coefficient is the matrix element

$$D = \mathbf{u}^+ \mathbf{D}_c \mathbf{u}_+ \tag{5.12}$$

and in general is complex. We can rewrite (5.12) in terms of real eigenvectors, such that $\mathbf{u}_+ = \mathbf{e}_1 + i\mathbf{e}_2$ and $\mathbf{u}^+ = \mathbf{e}^1 - i\mathbf{e}^2$, where, due to the orthonormality conditions for $\mathbf{u}^\pm$, we must have $\mathbf{e}^1 \cdot \mathbf{e}_2 = \mathbf{e}^2 \cdot \mathbf{e}_1 = 0$. Consequently we have

$$D = \mathbf{e}^1\mathbf{D}_c\mathbf{e}_1 + \mathbf{e}^2\mathbf{D}_c\mathbf{e}_2 + i(\mathbf{e}^1 \mathbf{D}_c \mathbf{e}_2 - \mathbf{e}^2 \mathbf{D}_c \mathbf{e}_1). \tag{5.13}$$

If, in particular, $\mathbf{D}_c$ is proportional to the unit matrix we see that D is purely real. On the other hand, even a diagonal $\mathbf{D}_c$, containing different diffusivities for different species, is enough to create a non-zero imaginary part.

It has recently been shown how the parameters α and β can be computed directly from an experiment [Hynne and Graae Sørensen 1993] using a so-called quenching technique [Hynne and Graae Sørensen 1990]. Near the Hopf bifurcation the system is systematically perturbed, thereby determining the change of concentrations necessary to put the system (temporarily) at the unstable fixed point, i.e. to hit a point on the stable manifold. In this way one can determine the location of the Hopf plane and its stable manifold. From this information together with the phases of oscillations of the various reactants one can determine the right eigenvectors $\mathbf{e}_1$ and $\mathbf{e}_2$ and their duals and thus the coefficients of (5.1). It is important to note that both α and β introduce a preferred sense of rotation in the complex A-plane so we anticipate that the relation between their signs will play a significant role. On the other hand, complex conjugation of the complex Ginzburg–Landau equation leads to an equation for A^* of the same form, but with opposite signs of α and β, and therefore the simultaneous change of sign of these two parameters does not affect the dynamics. In the following we can thus, without loss of generality, restrict our attention to the case $\beta \leq 0$.

Another useful way of writing the complex Ginzburg–Landau equation is in polar coordinates, putting

$$A = Re^{i\Phi}. \tag{5.14}$$

When the real and imaginary terms are separately equated in (5.4) one has

$$R\,\partial_t\Phi = -\alpha R^3 + R\nabla^2\Phi + 2\nabla R \cdot \nabla\Phi - \beta R(\nabla\Phi)^2 + \beta\nabla^2 R,$$
$$\partial_t R = R - R^3 - \beta R\nabla^2\Phi - 2\beta\nabla R \cdot \nabla\Phi - R(\nabla\Phi)^2 + \nabla^2 R. \tag{5.15}$$

It is important to stress that Φ itself never occurs in these equations, only its derivatives. Indeed, the complex Ginzburg–Landau equation is invariant under an overall shift of the phase $\Phi \to \Phi + C$ corresponding to time translational invariance of the reaction diffusion system.

5.2 The stability of the homogeneous periodic state and the phase equation

The homogeneous periodic state $A_0(t)$ is not necessarily stable towards spatially varying perturbations. If we look at weak perturbations of $A_0(t)$ in the form

$$A = (1 + \rho(\mathbf{x}, t))e^{i(-\alpha t + \phi(\mathbf{x}, t))}, \tag{5.16}$$

we find that ρ and ϕ will decay exponentially as long as the so-called Newell criterion

$$1 + \alpha\beta > 0 \tag{5.17}$$

is fulfilled, whereas long wavelength modes (with $|\mathbf{k}| < k_c \approx |1 + \alpha\beta|^{1/2}$) will be exponentially enhanced if $1 + \alpha\beta < 0$. In particular, if α and β have the same sign, $A_0(t)$ is always linearly stable.

Let us briefly sketch a derivation. If we insert (5.16) into (5.15) and retain only linear terms in ρ and ϕ we get

$$\partial_t \rho \approx -2\rho + \nabla^2 \rho - \beta\nabla^2 \phi, \tag{5.18}$$

$$\partial_t \phi \approx \nabla^2 \phi - 2\alpha\rho + \beta\nabla^2 \rho. \tag{5.19}$$

As these equations are linear, we can treat each Fourier mode independently. Thus assuming $\phi(\mathbf{x}, t) = e^{i(\omega t - \mathbf{k} \cdot \mathbf{x})}\phi_0$ and $\rho(\mathbf{x}, t) = e^{i(\omega t - \mathbf{k} \cdot \mathbf{x})}\rho_0$

$$(i\omega + 2 + k^2)\rho_0 = \beta k^2 \phi_0, \tag{5.20}$$

$$(i\omega + k^2)\phi_0 = -(2\alpha + \beta k^2)\rho_0, \tag{5.21}$$

which implies that

$$(i\omega + 2 + k^2)(i\omega + k^2) = -\beta k^2 (2\alpha + \beta k^2). \tag{5.22}$$

The two solutions are

$$i\omega = -(1 + k^2) \pm \sqrt{1 - 2\alpha\beta k^2 - \beta^2 k^4}. \tag{5.23}$$

Choosing the solution with the positive sign, we get, for small k,

$$i\omega = -(1 + \alpha\beta)k^2 - \frac{1}{2}\beta^2(1 + \alpha^2)k^4 + o(k^6). \tag{5.24}$$

When $1 + \alpha\beta < 0$, one branch of ω has a negative imaginary part for

$$k < k_c \approx \sqrt{\frac{2|1 + \alpha\beta|}{\beta^2(1 + \alpha^2)}},$$

implying exponential divergence for this band of modes. The branch corresponding to choosing the minus sign in (5.23) will always have a negative imaginary part, even for $k \to 0$, and thus always be stable.

The instability occurring in this system is mathematically analogous to the instability of long ocean waves found by Benjamin and Feir [1967] and is called the Benjamin–Feir instability in many papers. We shall refer to it as the Benjamin–Feir–Newell (BFN) instability criterion since the condition (5.17) was derived by Newell [1974].

We can then solve (5.20) and (5.21) for the ratios of the Fourier coefficients:

$$\left(\frac{\rho_0}{\phi_0}\right)^2 = -\frac{\beta k^2(i\omega + k^2)}{(i\omega + 2 + k^2)(2\alpha + \beta k^2)}. \tag{5.25}$$

For the 'soft branch' with $\omega(k) \to 0$ for $k \to 0$, we get

$$\frac{\rho_0}{\phi_0} \approx \frac{|\beta|}{2} k^2 \tag{5.26}$$

which shows that $\rho_0 \to 0$ as $k \to 0$ in the large wavelength limit. Thus the disturbances leading to instability are phase-like and affect only weakly the modulus of the field.

For the soft branch the dispersion relation is

$$i\omega \approx -(1 + \alpha\beta)k^2, \tag{5.27}$$

for small k, which means that long wavelength fluctuations decay very slowly. This is a consequence of the invariance of the complex Ginzburg–Landau equation under global shifts in the phase implying that the equation for phase disturbances has the form of a diffusion equation

$$\partial_t \phi = \nu \nabla^2 \phi, \tag{5.28}$$

with $\nu = 1 + \alpha\beta$.

This linear equation describes the relaxation of the phase when the initial state is inhomogeneous. It misses, however, some very important phenomena which determine, for example, the interaction between phase singularities, to be discussed later. It is thus important to take into account nonlinear terms in the phase equation [Kuramoto 1984]. First, let us rewrite (5.15) in terms of the deviations: $\rho = 1 - R$ and $\phi = \Phi + \alpha t$.

$$\partial_t \rho = -2\rho - 3\rho^2 - \rho^3 + \nabla^2\rho - \beta(1 + \rho)\nabla^2\phi - 2\beta\nabla\rho \cdot \nabla\phi - (1 + \rho)(\nabla\phi)^2$$

$$(1+\rho)\partial_t\phi = -2\alpha\rho - 3\alpha\rho^2 - \alpha\rho^3 + \beta\nabla^2\rho + (1+\rho)\nabla^2\phi$$
$$+2\nabla\rho\cdot\nabla\phi - \beta(1+\rho)(\nabla\phi)^2. \tag{5.29}$$

Now, since we are assuming slow spatial and temporal variations, we shall scale space and time according to $\mathbf{X} = \epsilon\mathbf{x}$ and $T = \epsilon^2 t$, where ϵ is a small parameter. The difference in scaling for $\mathbf{x}$ and t comes from the diffusive dispersion relation (5.27). Now, using (5.26), we see that it is reasonable to scale $\rho \to \epsilon^2\rho$ and retain ϕ unscaled. Inserting in (5.29) and equating all terms of order ϵ^2, we get

$$0 = -2\rho - \beta\nabla^2\phi - (\nabla\phi)^2$$
$$\partial_t\phi = -2\alpha\rho + \nabla^2\phi - \beta(\nabla\phi)^2 \tag{5.30}$$

or

$$\partial_t\phi = \nu\nabla^2\phi + \lambda(\nabla\phi)^2, \tag{5.31}$$

where $\lambda = \alpha - \beta$.

Crossing the BFN instability, the diffusion coefficient ν becomes negative. This will lead to a diverging phase equation even including the nonlinear term. We therefore have to include the stabilizing fourth-order term in (5.24) and find that

$$\partial_t\phi = -|\nu|\nabla^2\phi - \mu\nabla^4\phi + \lambda(\nabla\phi)^2, \tag{5.32}$$

where $\mu = \frac{1}{2}\beta^2(1+\alpha^2)$. Without the nonlinear term this equation would still be exponentially unstable. But the fourth-order term restricts the instability to the interval $[-\sqrt{|\nu|/\mu}, \sqrt{|\nu|/\mu}]$ in k (except $k = 0$, which is neutral). And, miraculously, the nonlinear term is precisely sufficient to stop ϕ from diverging [Nepomnyashchy 1974, Kuramoto 1984, Sivashinsky 1977]. For large enough system size and typical initial conditions it never settles down, but becomes turbulent [Kuramoto 1984, Sivashinsky 1977], as we shall discuss later. Note that for fourth-order terms to be comparable in size with the second-order ones, the diffusion coefficient $|\nu|$ must be very small, such that $|\nu|k^2 \sim k^4$ or $|\nu| \sim k^2$ consistent with having the unstable wave vectors well inside the allowed regime.

The Kuramoto–Sivashinsky equation is perhaps the simplest turbulent field theory and has thus been intensely studied in recent years. We shall return to it later in this chapter in the discussion of phase turbulence, in chapter 7 in the context of interface dynamics and finally in chapter 9 on turbulent diffusion. It should be noted that very little is known about the addition of other nonlinear terms into the Kuramoto–Sivashinsky equation. Such terms will appear if the expansion leading to the phase equation is continued. This is an important issue, which needs to be clarified in the future. Some work has been done by Sakagutchi [1990a,b] in which it is shown that certain terms can lead to finite time singularities.

5.3   Plane waves and their stability

Aside from the homogenous solution $A_0(t)$ (5.2) the complex Ginzburg–Landau equation has a whole family of harmonic travelling wave solutions. It is easily seen that the ansatz

$$A(\mathbf{x}, t) = A_{\mathbf{k}}(\mathbf{x}, t) = R_k e^{i(\mathbf{k}\cdot\mathbf{x} - \omega t)} \tag{5.33}$$

is a solution when $R_k^2 = 1 - k^2$ and $\omega = \alpha R_k^2 + \beta k^2$. Thus

$$\omega(k) = \alpha + (\beta - \alpha)k^2 \tag{5.34}$$

(where $k = |\mathbf{k}|$). Note that this agrees with the phase equation (5.31) when we take $\phi = (\alpha - \omega)t + f(\mathbf{x})$. Thus, for any $k < 1$ there is a plane wave solution and these waves are clearly nonlinear, since the dispersion relation involves the amplitude R_k, which is fixed for a given harmonic wave. The group velocity $\mathbf{v}_g = \nabla_{\mathbf{k}}\omega = 2(\beta - \alpha)\mathbf{k}$ is not a priori well defined, since superpositions of waves with different wave vectors do not, in general, solve the complex Ginzburg–Landau equation. However, we shall see below that the group velocity naturally occurs when we consider long wavelength perturbations on plane wave states.

5.3.1   The stability of plane waves

Let us briefly discuss the stability of the travelling wave solutions. It is convenient to write the perturbed field as $A(\mathbf{x}, t) = (R_k + u)e^{i(\mathbf{k}\cdot\mathbf{x} - \omega t)}$ where $u(\mathbf{x}, t)$ is complex. Inserting this ansatz into (5.4), we obtain, to linear order in u,

$$\partial_t u = -R^2(1 + i\alpha)(u + u^*) + 2i(1 + i\beta)\mathbf{k} \cdot \nabla u + (1 + i\beta)\nabla^2 u \tag{5.35}$$

together with the complex conjugate equation for u^*. In terms of the Fourier components $u_{\mathbf{q}}(t) = \int d\mathbf{x}\, e^{-i\mathbf{q}\cdot\mathbf{x}} u(\mathbf{x}, t)$, we obtain two equations which can be written in matrix form

$$\frac{\partial \mathbf{U}_{\mathbf{q}}}{\partial t} = \mathbf{L}\mathbf{U}_{\mathbf{q}},$$

where $\mathbf{U}_{\mathbf{q}}(t) = (u_{\mathbf{q}}(t), u_{\mathbf{q}}^*(t))$ and $\mathbf{L}$ is the following 2×2 matrix

$$\begin{pmatrix} -R_k^2(1 + i\alpha) - (2\mathbf{k}\cdot\mathbf{q} + q^2)(1 + i\beta) & -R_k^2(1 + i\alpha) \\ -R_k^2(1 - i\alpha) & -R_k^2(1 - i\alpha) - (-2\mathbf{k}\cdot\mathbf{q} + q^2)(1 - i\beta) \end{pmatrix}. \tag{5.36}$$

The problem is then reduced to the diagonalization of a 2×2 matrix with characteristic equation

$$v^2 + (a_1 + ia_2)v + b_1 + ib_2 = 0, \tag{5.37}$$

where

$$a_1 = 2(R_k^2 + q^2),$$
$$a_2 = 4\beta \mathbf{k} \cdot \mathbf{q},$$
$$b_1 = 2(1 + \alpha\beta)R_k^2 q^2 + (1 + \beta^2)(q^4 - 4(\mathbf{k} \cdot \mathbf{q})^2),$$
$$b_2 = 4(\beta - \alpha)R_k^2 \mathbf{k} \cdot \mathbf{q}, \tag{5.38}$$

and we obtain the two eigenvalues

$$\nu_\pm = -R_k^2 - q^2 - \mathrm{i}2\beta\mathbf{k} \cdot \mathbf{q} \pm R_k^2 \sqrt{1 + x + \mathrm{i}y}, \tag{5.39}$$

where x and y are the polynomials

$$x = -\frac{2\alpha\beta}{R_k^2}q^2 - \frac{\beta^2}{R_k^4}q^4 + \frac{4}{R_k^4}(\mathbf{k} \cdot \mathbf{q})^2,$$
$$y = \frac{4\alpha}{R_k^2}(\mathbf{k} \cdot \mathbf{q}) + \frac{4\beta}{R_k^4}(\mathbf{k} \cdot \mathbf{q})q^2. \tag{5.40}$$

For $k \to 0$, one has $R_k \to 1$ and we recover (5.23) (with $\nu = \mathrm{i}\omega$). For $\mathbf{k} \perp \mathbf{q}$ the results of the previous section are still valid, provided q is renormalized to q/R_k. However, in the general case, long wavelength disturbances can be unstable, even for $1 + \alpha\beta > 0$.

To see this, expand ν_+ for small q:

$$\nu_+(q) \approx -\mathrm{i}\nu_g q - Dq^2 + \mathrm{i}Cq^3 - Aq^4 + O(q^5), \tag{5.41}$$

where

$$\nu_g = 2(\beta - \alpha)k_q,$$
$$D = (1 + \alpha\beta) - 2(1 + \alpha^2)\frac{k_q^2}{R_k^2},$$
$$C = \frac{1}{R_k^4}(2\beta k_q R_k^2 (1 - 8\alpha^2) - 4\alpha^3 k_q^3),$$
$$A = \frac{1}{2R_k^6}(\beta^2(1 + \alpha^2)R_k^4 + 12\alpha\beta(1 + \alpha^2)k_q^2 R_k^2 - 4k_q^4(1 + 5\alpha^4)), \tag{5.42}$$

and $k_q = \mathbf{k} \cdot \mathbf{q}/q$. The expression for D shows that the least stable perturbations at a given $\mathbf{k}$ are those parallel to $\mathbf{k}$ and, in the long wavelength limit, they will be unstable if

$$k > k_c \approx \sqrt{\frac{R_k^2(1 + \alpha\beta)}{2(1 + \alpha^2)}}.$$

Finally, at the BFN transition all plane wave states become unstable.

In addition, the imaginary part of (5.41) shows that there is a drift velocity $\nu = 2(\beta - \alpha)\mathbf{k}$, which is precisely the group velocity $\mathbf{v} = \nabla_\mathbf{k}\omega$. Thus long wavelength disturbances are advected with the group velocity.

5.3.2 Convective versus absolute stability

As noted in chapter 4, systems with asymmetric couplings can have instabilities which are only convective. This means that the perturbation is convected so rapidly downstream that the instability is not felt at the originally perturbed site. The existence of the term $2i(1 + i\beta)\mathbf{k} \cdot \nabla u$ in (5.34) breaks the radial symmetry and selects a direction specified by the wave vector $\mathbf{k}$. In analogy with chapter 4 we must consequently determine the velocity-dependent stability exponent $\lambda(\mathbf{v})$. The method for doing this is reviewed in appendix D, but in the following we shall briefly discuss the concepts involved.

We assume that the system is unstable in the sense that the real part of v_+ is positive for some range of $\mathbf{q}$. A small wave-packet $u_\mathbf{q}(t = 0)$ with wave vectors centred around $\mathbf{q}$ will at time t be amplified to

$$u(\mathbf{x}, t) = \int d\mathbf{q} e^{(i\mathbf{q}\cdot\mathbf{x} + v_+ t)} u_\mathbf{q}(t = 0), \tag{5.43}$$

where the limits of integration are $-\infty$ to $+\infty$ for each component of $\mathbf{q}$. If we want to asses the stability in a moving frame, we must replace $\mathbf{x}$ by $\mathbf{x_0} + \mathbf{v}t$ and, in the limit of large t we evaluate the integral (5.43) in the saddle point approximation. We then have to look for the saddle points of $f(\mathbf{q}) = v_+(\mathbf{q}) + i\mathbf{q} \cdot \mathbf{v}$ and choose only the contributions from those for which the contour can be deformed back to the real axis without encountering singularities. If the saddle point (or points) occur for $\mathbf{q} = \mathbf{q}^\star(\mathbf{v})$, the velocity-dependent stability exponent $\lambda(\mathbf{v})$ is equal to $f(\mathbf{q}^\star)$, where the saddle point $\mathbf{q}^\star$ giving the largest real part of $f(\mathbf{q}^\star)$ is chosen. If $\lambda(\mathbf{v} = \mathbf{0}) > 0$ the system becomes *absolutely* unstable: perturbations will grow even in a non-moving frame.

As we have seen earlier, the most unstable wave vectors are those parallel to $\mathbf{k}$. Thus we can simplify the calculation by considering the one-dimensional subspace, where $\mathbf{k}$, $\mathbf{q}$ and $\mathbf{v}$ are all parallel and we only need the complex function $\lambda(v)$, where v is a real scalar. To repeat, the zeros v_- and v_+ of $\lambda(v)$ give the velocities of the two edges of a localized perturbation. If the interval $[v_-, v_+]$ contains the point $v = 0$, the state is absolutely unstable. The maximal growth rate is for v somewhere between v_- and v_+. For $C = 0$ in (5.41), it will be at the group velocity, although in general it depends on C.

5.4 Large–scale simulations and the coupled map approximation

In order to be able to study very large systems, it is convenient to use a 'coupled map' approach to the complex Ginzburg–Landau equation [Bohr et al. 1989]. It can be regarded as a rough approximation of (5.1) (which can be made exact by taking certain limits of the parameters) or as an interesting dynamical system in its own right. It has been used intensively in recent years, and the results seem to correspond well to those obtained by more conventional approximations, notably spectral methods [Coullet et al. 1989, Bodenschatz et al. 1989, Chaté and Manneville 1995].

We split the coupled map lattice into two parts: a local map $A' = F(A)$ representing the two first terms of (5.1) and a non-local part representing the complex heat (or diffusion) equation which results from omitting the local terms. The heat equation has the solution

$$A(t + \tau_0) = e^{\tau_0(1+i\beta)\nabla^2} A(t). \tag{5.44}$$

On the lattice we approximate the Laplacian by an average ΔA over neighbours. Thus on a two-dimensional square lattice one has

$$\Delta A_{i,j} = \frac{1}{6} \sum_{\sigma_1 = \pm 1, \sigma_2 = \pm 1} (2A_{i+\sigma_1,j} + 2A_{i,j+\sigma_2} + A_{i+\sigma_1,j+\sigma_2} - 5A_{i,j}). \tag{5.45}$$

The non-local map is then taken as

$$A^{\mathrm{nl}} = (1 + \frac{\tau_0}{M}(1 + i\beta)\Delta)^M A. \tag{5.46}$$

Here M is an integer that determines the range of the effective interaction. The limit $M \to \infty$ reproduces the exponential above (except, of course, that Δ and ∇^2 are not exactly the same). M is typically around 5, large enough to ensure that short wavelength instabilities do not occur.

The properties of the local map F are very simple. If we look at the complex Ginzburg–Landau equation without the last (Laplace) term it can be written as

$$\frac{dr}{dt} = \mu r - r^3,$$
$$\frac{d\phi}{dt} = -\alpha r^2, \tag{5.47}$$

which can also be seen by going back to (5.15) and reinserting μ, since the lattice spacing introduces a new length scale. The qualitative features can be easily reproduced by maps of the form

$$r(n+1) = f(r(n))$$
$$\phi(n+1) = \phi(n) - g(r(n)) \tag{5.48}$$

where the map f has an unstable fixed point in 0 and a stable one in $r = \sqrt{\mu}$. Specifically, one can integrate (5.47) obtaining

$$r(t + \tau) = \frac{\sqrt{\mu} r(t)}{\sqrt{\lambda\mu + (1 - \lambda)r(t)^2}} \tag{5.49}$$

and

$$\phi(t + \tau) = \phi(t) - \frac{\alpha}{2} \left(2\mu\tau + \log \left((1 - \frac{r^2(t)}{\mu})\lambda + \frac{r^2(t)}{\mu} \right) \right), \tag{5.50}$$

where $\lambda = e^{-2\mu\tau}$. From (5.49) and (5.50), the functions f and g are fixed. In most simulations, one uses the approximation

$$\phi(n+1) = \phi(n) - \tau\alpha\, r(n)^2, \tag{5.51}$$

valid for small τ.

The full map lattice can now be written in the form

$$A_{i,j}(n+1) = F(A_{i,j}^{\text{nl}}(n)). \tag{5.52}$$

The travelling waves and linear stability properties for this discretized version of the complex Ginzburg–Landau equation have been worked out in [Bohr et al. 1990b]. They closely resemble those of the continuous system. For the stability of the homogeneously rotating state $A(n) = \sqrt{\mu}e^{-i\alpha\mu\tau n}$ the criterion replacing $1 + \alpha\beta > 0$ is

$$1 + \frac{2\mu}{1-s}\tau\alpha\beta > 0, \tag{5.53}$$

where s is the stability parameter for the stable fixed point of f, i.e. $s = f'(\sqrt{\mu})$. For the map (5.49) $s = e^{-2\mu\tau}$, and it is seen that the new stability criterion approaches the old one as $\tau \to 0$.

5.5 Spirals and wave number selection

The instabilities studied above, occurring in the simple states of the complex Ginzburg–Landau equation, often lead to a new kind of excitation containing phase singularities. Spiral waves or vortices typically appear due to instability of wave fronts or due to noisy initial conditions. Vortices are defined as singularities of the phase field. The total variation of the phase over a closed loop, i.e. $\Delta\phi = \oint d\phi$, does not necessarily vanish if the loop encloses vortex centres. Instead, $\Delta\phi = 2\pi n$, where the integer n is the total 'vorticity' of the region enclosed, i.e. the difference between the numbers of positive and negative spirals. In the vortex centre the angle is not defined so for the order parameter itself $(A = Re^{i\phi})$ to remain well defined R must vanish. Typical vortex $(\Delta\phi = 2\pi)$ and antivortex $(\Delta\phi = -2\pi)$ configurations are shown in Fig. 5.2. The spiral field has the form

$$A_s(r,\theta) = R_s(r)e^{i(\sigma\theta + \psi(r))}e^{-i\Omega t}, \tag{5.54}$$

where σ is the 'topological charge' $= \pm 1$. For large distances $\psi(r) \approx qr + \text{const}$, where q is the pitch, $R_s \to R_q = \sqrt{1 - q^2}$ and Ω is consequently the frequency of the asymptotic plane wave $\Omega = \alpha + (\beta - \alpha)q^2$. Asymptotically the spirals are thus Archimedian. For small distances $R_s \to r$ and $\psi(r) \to \frac{1}{2}r^2 + \text{const}$, which describes the spiral core.

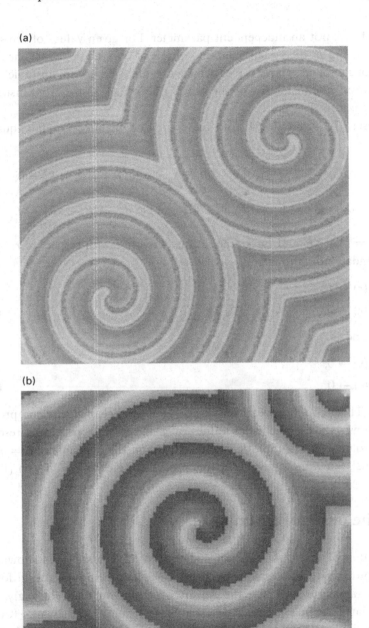

Figure 5.2 Spirals. (a) Spiral waves in the Belousov–Zhabotinsky reaction [Müller et al. 1987] (b) Simulation of complex Ginzburg–Landau equation with $\alpha = 0.4$ and $\beta = -1.25$. Shading corresponds to the phase of the complex field [Huber 1993].

The pitch q is not an independent parameter. For given values of the system parameters α and β, the system contains spirals of only one pitch q. The spirals can be thought of as plane waves which are matched to an unstable centre emitting waves. The instability of the centre traces back to the Hopf instability, since at the centre $A \to 0$.

If we insert the ansatz (5.54) into the complex Ginzburg–Landau equations we get

$$\Delta_r R_s - \frac{1}{r^2} R_s - (\psi')^2 R_s - \beta \left(R_s \Delta_r \psi + 2\psi' R_s' \right) + R_s - R_s^3 = 0,$$

$$\beta \left(\Delta_r R_s - \frac{1}{r^2} R_s - (\psi')^2 R_s \right) + R_s \Delta_r \psi + 2\psi' R_s' - \Omega R_s - \alpha R_s^3 = 0, \quad (5.55)$$

where $\Delta_r f = f'' + \frac{1}{r} f'$ and the prime means derivative with respect to r.

The boundary conditions are

$$R_s(r) \to R_q,$$
$$\psi'(r) \to q, \qquad\qquad\qquad (5.56)$$

for $r \to \infty$ and

$$R_s(r) \to 0,$$
$$\psi'(r) \to 0, \qquad\qquad\qquad (5.57)$$

for $r \to 0$. These equations can be viewed as a nonlinear eigenvalue problem with solutions only for particular values of q. In general one has to resort to numerical solution or asymptotic expansion to find the dependence $q = q(\alpha, \beta)$ [Hagan 1982]. For a more geometrical approach see Keener and Tyson [1986].

5.6   The onset of turbulence

The formation of spiral waves can lead to a turbulent state, as shown first by Kuramoto [1984]. In this state small spiral pairs are spontaneously formed and others annihilate such that the density of spirals fluctuates strongly. Such 'defect-mediated' turbulence has been studied in several recent works [Coullet et al. 1989, Bodenschatz et al. 1989, Bohr et al. 1990a, Huber et al. 1992]. Figure 5.3 shows the real part of the complex field A in a turbulent state. How do we know whether a state is turbulent? According to our definition in chapter 1 we must show that the system has positive Lyapunov exponents. The calculation of Lyapunov exponents is described in chapter 1 and in more detail in appendix C. In a chaotic system, the length of the tangent vector grows exponentially and the slope of the curve in Fig. 5.4a gives the value of the maximum Lyapunov exponent. One can easily extend the calculation to find several Lyapunov exponents and verify that, for a large system, the number of positive ones will grow proportionally to the available volume. However, the fact that the chaotic attractor has a high dimensionality is almost obvious from the lack of coherence in the system.

The inset in Fig. 5.4a shows another case, where turbulence dies out after a time T. After this time, the running maximum 'Lyapunov exponent' is zero and the state becomes periodic. If we, for example, fix the parameter β and vary α we can study the onset of turbulence. As such, this might not be a very well-defined concept, because the system might have many coexisting attractors. In the following we shall assume that the initial conditions are chosen at random, i.e. that the real and imaginary parts of A are randomly distributed in some interval around zero. Initially we thus have a large number of vortices: places where both the real and the imaginary parts vanish. For such initial conditions it was found that the transition to turbulence takes place in two steps [Bohr et al. 1990a]. At some value $\alpha = \alpha_{trans}(\beta)$ transient turbulent states appear. As in the inset of Fig. 5.4a the turbulence lasts only for a finite time T. Then at $\alpha = \alpha_{turb}(\beta)$ the transient time T seems to diverge and the system goes into a sustained turbulent state. Fig. 5.4b shows the maximum Lyapunov exponent as a function of α. For $\alpha < \alpha_{turb}$ the true Lyapunov exponent is of course zero – what is plotted is the best estimate of the maximum Lyapunov exponent in

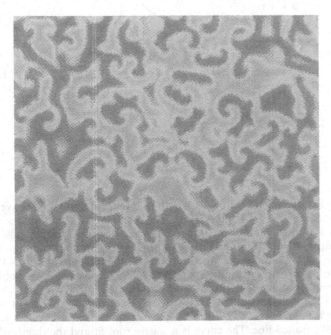

Figure 5.3 Turbulent state of the complex Ginzburg–Landau equation with $\alpha = 1.9$ and $\beta = 0$. The real part of the complex field is shown.

the transient turbulent state. It is seen that the Lyapunov exponent grows like $\lambda \sim \sqrt{\alpha_{\text{trans}} - \alpha}$.

What determines α_{trans} and α_{turb}? At first one might expect that they are related in some way to the BFN instability. But this is not the case: in particular the transitions remain at $\beta = 0$, where the BFN instability disappears. Moreover, we know that the spirals select a particular wave number $q = q(\alpha, \beta)$ and thus it might be more relevant to look at the stability of a plane wave with the

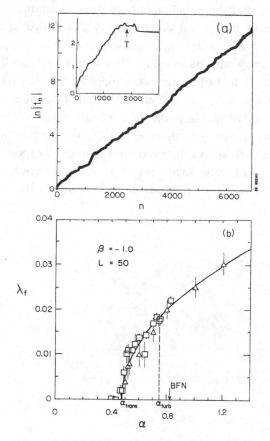

Figure 5.4 (a) The logarithm of the length of a tangent vector ($\ln |t_n|$) versus discrete time n for $\alpha = 0.74$ and $\beta = -1$. The inset has the same β but $\alpha = 0.71$ and shows a transient turbulent state. It only has a positive Lyapunov exponent up to $n = T$ [Bohr et al. 1990a]. (b) The maximum Lyapunov exponent λ_f at finite time as a function of the parameter α (with $\beta = -1$). The squares are obtained by iterating a tangent vector. The triangles were obtained by measuring the distance between two nearby states. For α less that $\alpha_{\text{turb}} \approx 0.75$ the turbulent state is transient and the Lyapunov exponent is only defined for a finite time T. The true Lyapunov exponent is zero below α_{turb} as shown by the dotted line. The curve is a square root fit and the threshold of the BFN instability is marked by the arrow. The lattice is 50×50 [Bohr et al. 1990a].

selected wavelength. We do not have an analytical expression for $q(\alpha, \beta)$ in a two-dimensional system, but numerically it can be found from simulations. Figure 5.5 shows, for two different values of β, the wavelength $\lambda_q = 2\pi/q(\alpha, \beta)$ versus α. It also shows the analytically computed value of the smallest unstable wavelength (computed for the map version (5.52)). The crossing between these two curves determines the α at which the selected plane waves become unstable. These values are very close to the values of α_{trans} found from computation of the Lyapunov exponents. We therefore expect the first transition to be related to the emerging instability at the wavelength selected by the spirals [Bohr et al. 1990b].

What about the transition to sustained turbulence, α_{turb}? Is that also related to a linear instability in the system? This question was answered by Aranson et al. [1992a], who noted that α_{turb} is very close to the onset of absolute instability for the selected wavelength. Thus, in the transient turbulent state, the spirals are able to convect away the disturbance, whereas the onset of absolute instability makes this impossible and the system remains turbulent. The phase diagram showing the various transition lines is shown in Fig. 5.6.

5.6.1 Transient turbulence and nucleation

What is the structure of the asymptotic state in the transiently turbulent regime, i.e. between the two curves (SB) and (T) in Fig. 5.6? This is clearly seen on the sequence of pictures in Fig. 5.7 on which, not the phase, but the modulus of the field is shown. Spirals are, far from their cores, locally almost plane waves, and thus characterized by an almost constant modulus. At the core of a spiral the modulus vanishes and thus the spiral cores are easily seen in a plot of the modulus. As seen from the figures, the final state has rather few vortices and

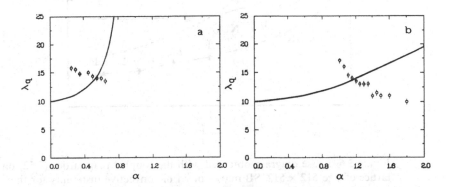

Figure 5.5 Linear stability crossing for transient turbulence. The full curve shows the wavelength λ_q and the corresponding critical α determined from (5.41–42) (slightly corrected because of the coupled map approach) and the circles represent the characteristic wavelength in the final frozen states determined from simulations. (a) $\beta = -1$ (b) $\beta = 0$ [Bohr et al. 1990b].

they do not move – they seem to have formed a disordered frozen state. When this was discovered [Bohr et al. 1989] it came as a surprise, since the analogous variational system (the real Ginzburg–Landau equation, which describes the so-called XY model of statistical mechanics or vortices in superfluid helium, see for example, Chaikin and Lubensky [1995]) has a simple logarithmic interaction, which can not give bound states.

In Bohr et al. [1989] these bound states were called 'entangled'. From Fig. 3.7 it is clear that they are not really entangled. In fact the large vortices have divided the space between them in a very clear way. There is a line of increased modulus separating the vortices and these lines form an interesting network reminiscent of soap froths. These lines or walls are almost shocks at which the transverse gradient of the phase would become discontinuous. We shall return to their structure and importance in the next section. It can also be seen that a number of very small vortices (we shall call them *edge vortices*) remain in the corners formed by the walls. Some of the edge vortices remain slightly mobile even at very large times, but the large vortices seem completely fixed. Whether they will remain so ad infinitum is an important open question. Due to their disordered and metastable appearance we shall refer to such states as *vortex glasses*.

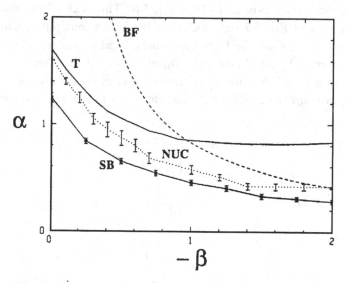

Figure 5.6 Phase diagram obtained with the coupled map lattice (5.52) on a lattice of size 512×512. SB marks onset of convective instability for the selected wavelength, which corresponds to $\alpha = \alpha_{\text{trans}}(\beta)$. T marks the transition to sustained turbulence $\alpha = \alpha_{\text{turb}}(\beta)$ coinciding with the onset of absolute instability, and the dashed line marked BF is the BFN line $1 + \alpha\beta = 0$. The region between the lines SB and T is where nucleation of vortex glass structures is expected. In fact, it is not seen below the line marked NUC [Huber et al. 1992].

In the turbulent state, the spirals are small and their number fluctuates strongly. After a characteristic transient time T the system settles in a vortex glass. T increases from a very small value at α around α_{trans} and becomes infinite at $\alpha = \alpha_{turb}$. This variation was studied by Huber et al. [1992] and the result is shown in Fig. 5.8. The transient time was computed by determining, when the vortex density had decayed by two standard deviations from the average value in the metastable turbulent state. As one can see, the divergence of T is fitted well by the form

$$\log\left(\frac{T}{T_L}\right) \sim (\alpha_{turb} - \alpha)^{-2}. \tag{5.58}$$

We can understand this result in terms of nucleation [Huber et al. 1992]. Around the time T, one or several vortices start growing and push away smaller vortices. We now assume that the winners are the ones that, by chance, have

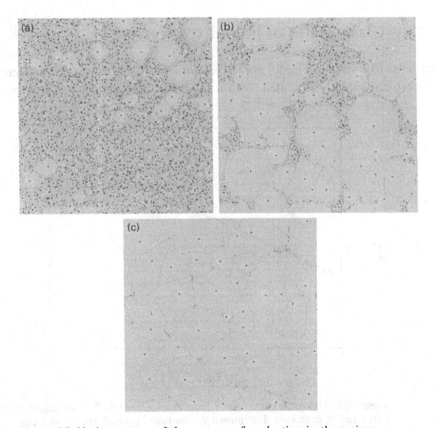

Figure 5.7 Various stages of the process of nucleation in the regime $\alpha_{trans} < \alpha < \alpha_{turb}$. The parameters are: $a = 0.8$ and $\beta = -1$. A 760×760 part of a 1024×1024 lattice is shown. Note that α_{turb} is slightly higher than in Fig. 5.4b which was computed for a small lattice of 50×50 [Huber 1993].

grown larger than some *critical* radius R_*. Since we are *below* the absolute instability, the exponent for stability at a fixed point in space, $\lambda_0 = \lambda(v = 0)$, is negative. Correspondingly $\tau = 1/|\lambda_0|$ is a characteristic 'healing time'. Imagine the large, growing vortex immersed in a swarm of small perturbing vortices with characteristic velocity v. If the distance $v\tau$ that they move during the healing time is much smaller than R_*, it is reasonable to assume that the vortex can recover and keep growing. Approaching the absolute instability where $\lambda_0 \simeq \text{const.}(\alpha_{\text{turb}} - \alpha)$ we thus assume that R_* scales as $1/|\lambda_0| \sim 1/(\alpha_{\text{turb}} - \alpha)$. If we further assume that the motion, until a 'critical droplet' is formed, is completely random (i.e. Poissonian), the time T it takes for this to happen is exponentially large in the area, i.e.

$$T \sim e^{\rho \pi R_*^2}.$$ (5.59)

Inserting the scaling assumption for R_* then gives the form (5.58).

5.7   Glassy states of bound vortices

Patterns like those in Fig. 5.9 appear to consist very accurately of a collection of unperturbed spirals separated by thin shocks and the structure of such patterns

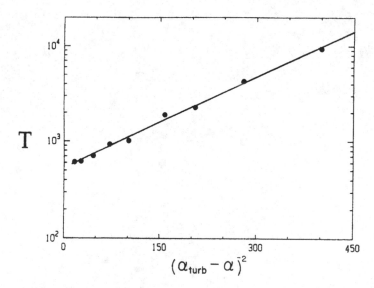

Figure 5.8 The nucleation time T versus $(\alpha_{\text{turb}} - \alpha)^{-2}$. T is averaged over five runs, and in each case determined as the time at which the density had decreased by two standard deviations (towards the frozen state) from its average value in the metastable turbulent state (subtracting the brief initial time after which the system has reached the turbulent state) [Huber et al. 1992].

was studied in [Bohr et al. 1996, 1997]. Each vortex is thus described by a phase ϕ such that the ith vortex has the phase of a simple Archimedian spiral

$$\phi_i = \sigma_i \theta_i - qr_i + C_i \qquad (5.60)$$

where the direction of spiraling is specified by the charge σ_i ($= \pm 1$) of the ith spiral (which seems, on the figure, to be randomly distributed). θ_i and r_i are the polar coordinates measured from the centre of the ith spiral (with respect to a fixed direction), and C_i is a phase constant for the spiral. The wave number q is the same for all spirals; it is the *selected* wave number for a given set of parameters α and β. On the shock lines, the phases of two vortices are equal, and the domain of each vortex is simply the region where its phase is larger than the phase of any others. This phase matching strongly constrains the geometry of the pattern. If, say, three shocks meet at a corner, the location of the corner as well as the locations and charges of the three surrounding vortices uniquely determine the three shock curves emanating from that corner.

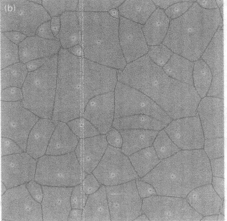

Figure 5.9 Glassy or frozen state. (a) Phase plot. (b) Amplitude plot. Note that (a) and (b) can be overlaid. The parameter values are $\alpha = 0.879$ and $\beta = -1$ [Bohr et al. 1997].

In particular, since the three phases ϕ_i match at the corner, the three constants C_i are determined only up to an additive constant. Also it is possible to see why corners predominantly have three, and not four or more shocks emanating from them: two phases become equal on a line, three phases are typically equal only at single points and four phases only become equal under special conditions.

To a good approximation, the shocks are segments of hyperbolae with the two nearest vortices as foci. This has been noted from numerical computation [Huber 1994] and can be easily understood. The shocks follow lines where the phases of two vortices, ϕ_1 and ϕ_2 are equal. Assuming that the distance from each vortex to the shock is much larger than the characteristic wavelength $\lambda = 2\pi/q$, one finds simply

$$r_1 - r_2 = (C_1 - C_2)/q \tag{5.61}$$

and the locus of points having constant difference in distance from two centres is a hyperbola with focal points at those centres.

Within this approximation there is a simple rule giving the local direction of a shock line: for any point P on a shock line, the direction of the line is such that it bisects the angle from P to the two neighbouring vortices. This follows directly from the fact that the difference between the distances has to remain constant along the shock. The distances to the two vortex centres changes proportionally to $\cos\theta_i$, where θ_i is the angle between the tangent of the shock line at P and the ray from the ith vortex centre to P ($i = 1$ or 2). Thus $\cos\theta_1 = \cos\theta_2$ and $\theta_1 = \theta_2$.

In Fig. 5.10a we display a close-up of the wall pattern. In Fig. 5.10b we have reproduced almost the same pattern (aside from the small edge-vortices) by feeding the position of the vortex centres and the constants C_i, estimated from the

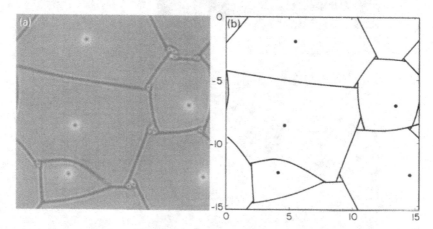

Figure 5.10 (a) Close up of wall structure. (b) Reconstruction of the wall structure using the hyperbolic approximation (5.61) [Bohr et al. 1997].

distances to the walls, into (5.60). In contrast, the so-called Voronoi construction [Voronoi 1908] finds the region nearest to any vortex by constructing the perpendicular bisectors of all connecting lines. Obviously the vortex domains are *non-Voronoi*, but at the same time the construction represents a generalization of the Voronoi construction, to which it corresponds when all $r_i \gg \lambda$ and all the C_i are identical.

5.8   Vortex interactions

The existence of 'glassy' vortex states implies that the interactions between vortices are rather complex. In two recent works [Rica and Tirapegui 1990, Elphick and Meron 1991] this interaction is computed for a dilute gas of spirals. One finds [Rica and Tirapegui 1991] that the motion of the kth spiral centre satisfies

$$\frac{\mathrm{d}\mathbf{r}_k}{\mathrm{d}t} = -2\sigma_k \hat{z} \times \partial_{\mathbf{r}_k} \Phi^{(k)}(\mathbf{r}_k) + 2\beta \partial_{\mathbf{r}_k} \Phi^{(k)}(\mathbf{r}_k) \tag{5.62}$$

where $\sigma_k = \pm 1$ is the topological charge of the kth spiral and $\Phi^{(k)}$ is the phase seen by that spiral, when its own field is subtracted. The first term corresponds to the usual result obtained in the variational limit ($\alpha = \beta = 0$) i.e. the term responsible for the logarithmic interaction leading to the Kosterlitz–Thouless transition in planar equilibrium systems, see, for example, Chaikin and Lubensky [1995].

The difficulty comes when one has to determine $\Phi^{(k)}$: the problem is how strongly the vortices influence each other. Here it becomes very important to take into account the shocks that form between the vortices and which seem to divide the space into compartments where a single vortex dominates. A theory of vortex interactions taking into account the shocks has been developed by Aranson et al. [1993a]. They have studied two spirals separated by a straight wall. From the outset it is assumed that the spirals move with some velocity and this velocity has to be determined from the entire solution including boundary conditions at the wall. It turns out that only particular velocities are compatible with given parameters and separations between the spirals. In particular, a bound state must correspond to a solution with $v = 0$ and this can happen in a regime of parameters where the interaction is oscillatory and allows for zeros as function of distance.

In the following we shall take a simpler approach [Bohr et al. 1997] and represent the spirals as plane waves. We thus assume that the centres are well separated and the walls are thin compared to this distance. In that way we shall be able to recover some of the results of Aranson et al. [1993a] and in addition understand the structure of the walls between the spirals.

5.8.1   Microscopic theory of shocks

Consider a straight wall (segment) running along the x_2-axis as shown in Fig. 5.11. To the left and right of it we assume plane wave states with $\mathbf{k}_l = (-k_1, k_2)$ and $\mathbf{k}_r = (k_1, k_2)$, respectively, since, as seen in the simple approximation for the walls of section 5.7, the wall must locally bisect the angle from any point on it to the two neighbouring vortices. From section 5.3 we know that the plane wave states have the form $A(x_1, x_2, t) = R_k e^{i(\mathbf{k}\cdot\mathbf{r} - \omega t)}$ where $k = \sqrt{k_1^2 + k_2^2}$, $R_k^2 = 1 - k^2$ and $\omega = \alpha R_k^2 + \beta k^2 = \alpha + (\beta - \alpha)k^2$. Let $A(x_1, x_2, t) = e^{ik_2 x_2 - i\omega t} G(x_1)$. The equation for G is

$$G''(x) = (k_2^2 - \frac{1 + i\omega}{1 + i\beta})G + \frac{1 + i\alpha}{1 + i\beta}|G|^2 G, \tag{5.63}$$

where $x = x_1$ or

$$G'' = \left(k_2^2 - \frac{1 + \omega\beta}{1 + \beta^2} + i\frac{\beta - \omega}{1 + \beta^2}\right)G + \left(\frac{1 + \alpha\beta}{1 + \beta^2} + i\frac{\alpha - \beta}{1 + \beta^2}\right)|G|^2 G. \tag{5.64}$$

Now let $G = Re^{i\phi}$. Use the relations between ω, k and R_k (5.34), and separate (5.64) into real and imaginary parts to obtain

$$R'' - R((\phi')^2 - k_1^2) = \frac{1 + \alpha\beta}{1 + \beta^2} R(R^2 - R_k^2), \tag{5.65}$$

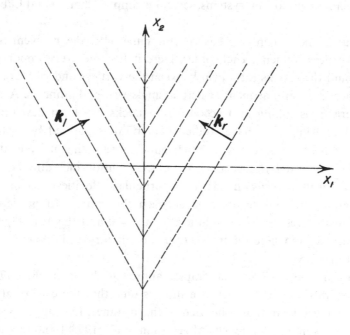

Figure 5.11 A shock separating two plane wave states.

$$R\phi'' + 2R'\phi' = \frac{\alpha - \beta}{1 + \beta^2} R(R^2 - R_k^2). \tag{5.66}$$

We now scale by introducing $\xi = xk_1$ and let y be the (normalized) 'local wave number' $y = \phi'(\xi)$ and $r = R/R_k$. Then

$$r'' - r(y^2 - 1) = Ar(r^2 - 1) \tag{5.67}$$

$$ry' + 2r'y = Br(r^2 - 1) \tag{5.68}$$

where

$$A = \frac{1 + \alpha\beta}{1 + \beta^2} \frac{R_k^2}{k_1^2} \quad \text{and} \quad B = \frac{\alpha - \beta}{1 + \beta^2} \frac{R_k^2}{k_1^2}.$$

One should note that we are interested in the structure of the wall, when the angle between the plane waves and the wall is varied, but the wavelength $\lambda = 2\pi/k$ is fixed. If the final wave number $\mathbf{k}_r = (k\cos\theta, k\sin\theta)$, the k_1 appearing in the scaling and in A and B is $k_1 = k\cos\theta$ and the angle θ is the angle between the phase contours and the wall.

To find the wall structure, we should solve (5.67–68) on $-\infty < \xi < \infty$ such that $r \to 1$ and $y \to \pm 1$ as $\xi \to \pm\infty$. But then the position of the wall would not be given. Therefore it is easier to fix the wall centre at 0 by solving on $0 < \xi < \infty$ with the boundary condition $y(0) = r'(0) = 0$ and $r, y \to 1$ as $\xi \to \infty$. That leaves as unknown the wall height $r(0)$, which has to be found by solving the equations. These boundary conditions at $\xi = 0$ correspond to the condition $\partial_x A = 0$ used by Aranson et al. [1993a] on the wall to calculate the spiral interactions.

The result of a numerical solution of (5.67–68) is shown in Figs. 5.12–13 for two different incident angles. The calculated variation of r is compared to the one found in shocks in the simulations of the full complex Ginzburg–Landau equation. The equations (5.67–68) can be simplified by solving (5.68) for y as

$$y(\xi) = \frac{B}{r^2(\xi)} \int_0^\xi r^2(\xi') \left(r(\xi')^2 - 1 \right) d\xi'. \tag{5.69}$$

Finally it is convenient, especially for taking the limit $\theta \to \pi/2$, to use the function $w(x) = r(x)^2 - 1$, whereby (5.67–68) take the form

$$\frac{1}{2}w'' - \frac{(w')^2}{4(1 + w)} = (1 + w)(Aw + y^2 - 1) \tag{5.70}$$

$$[(1 + w)y]' = Bw(1 + w). \tag{5.71}$$

5.8.2 Asymptotic properties

For $x \to \pm\infty$, $w \to 0$ and $y \to y_0 = \pm 1$. Letting $y = y_0 - \delta$, (5.70) and (5.71) can be linearized as

$$w'' = 2Aw - 4y_0\delta,$$

$$\delta' = y_0 w' - Bw, \tag{5.72}$$

and by differentiation and substitution of δ' we get

$$w''' + 2(2 - A)w' - 4y_0 Bw = 0. \tag{5.73}$$

Assuming solutions of the form $e^{\lambda x}$, we find the characteristic equation

$$\lambda^3 + 2(2 - A)\lambda - 4y_0 B = 0. \tag{5.74}$$

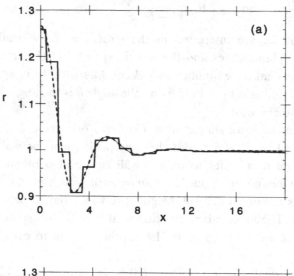

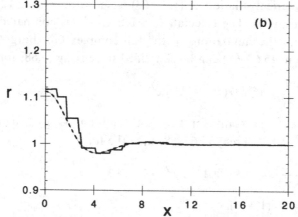

Figure 5.12 Transverse structure of the walls or shocks separating the spiral waves. The dashed line is the solution for the (normalized) modulus r of equations (5.67) and (5.68) for a shock between two plane waves and the full lines are the result of a lattice simulation. The parameters were $\alpha = 0.792$ and $\beta = -1$. (a) Shows the *head − on* case where the angle between the phase contours and the shock is 0. (b) The case, where the angle is $\pi/3$. Note that the structure of the wall is oscillatory for the cases shown [Bohr et al. 1997].

The discriminant is $D = Q^3 + R^2$, where $Q = \frac{2}{3}(2 - A)$ and $R = 2y_0 B$, and the solutions can be expressed in terms of $S_\pm = (R \pm \sqrt{D})^{1/3}$ as

$$\lambda_1 = S_+ + S_-, \qquad (5.75)$$

$$\lambda_2 = -\frac{1}{2}(S_+ + S_-) + \frac{\sqrt{3}}{2}i(S_+ - S_-), \qquad (5.76)$$

$$\lambda_3 = -\frac{1}{2}(S_+ + S_-) - \frac{\sqrt{3}}{2}i(S_+ - S_-). \qquad (5.77)$$

If $D > 0$ one solution is real and two are complex conjugates. Moreover, the sign of the real solution (5.75) is the same as that of R, whereas the signs of the

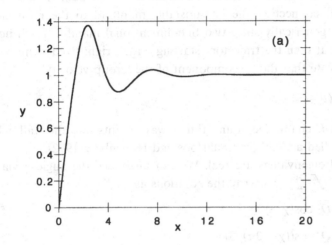

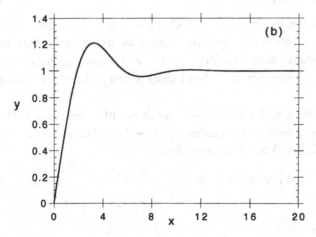

Figure 5.13 The solution for y of equations (5.67) and (5.68) for a shock between two plane waves. The parameters were $\alpha = 0.792$ and $\beta = -1$. (a) Shows the head-on case where the angle between the phase contours and the shock is 0. (b) Shows the case where the angle is $\pi/3$ [Bohr et al. 1997].

real parts of the two others are opposite. Thus the signs of the real parts are opposite in the two sides of the shock, $x \to \pm\infty$ where y_0 has opposite signs. If $B > 0$, $R < 0$ at $x \to -\infty$ where $y_0 = -1$. Therefore the stable manifold of the left fixed point is one-dimensional and the unstable one is two dimensional with outward spiraling trajectories. At the right fixed point the situation is opposite. Here the stable manifold has inward spiralling trajectories.

The wall solution, that we are looking at is a heteroclinic connection, where the unstable manifold of the left fixed point has a trajectory that lies in the stable manifold of the right one. For $D > 0$ we thus have an oscillating wall. If $B < 0$ (which doesn't change the sign of D) the wall would, superficially, be monotonic since the unstable manifold of the left fixed point is now 'real', but, in fact a wall connecting the left unstable manifold to the right stable one does not exist generically since two one-dimensional manifolds will not in general intersect. But then the trajectory starting on the right fixed point would generically exist. Note that the x_1-component of the 'group velocity'

$$\frac{\partial \omega}{\partial k_1} = 2k_1(\beta - \alpha) \tag{5.78}$$

changes sign when B changes sign and thus always points into the wall, which can therefore be called a 'sink' [van Saarloos and Hohenberg 1992].

When $D < 0$ all eigenvalues are real. We can introduce the angle χ via the relation $\cos \chi = R/\sqrt{-Q^3}$ and write the solutions as

$$\lambda_1 = 2\sqrt{-Q^3} \cos(\chi/3) \tag{5.79}$$

$$\lambda_2 = 2\sqrt{-Q^3} \cos((\chi + 2\pi)/3) \tag{5.80}$$

$$\lambda_3 = 2\sqrt{-Q^3} \cos((\chi + 4\pi)/3). \tag{5.81}$$

In this case a change in the sign of R changes χ to $\pi - \chi$ which produces a set of three eigenvalues with opposite signs and, again, we either have a two-dimensional unstable manifold for $y_0 = -1$, and a two-dimensional stable manifold for $y_0 = 1$ or vice versa. In both cases a heteroclinic connection will generically exist.

To summarize, we have wall solutions for any choice of parameters. If $D > 0$ it oscillates with an asymptotic wave number $\frac{\sqrt{3}}{2}(S_+ - S_-)$, whereas it is monotonic for $D < 0$. The condition $D > 0$ can be written

$$A^3 - \left(\frac{27}{2}C^2 + 6\right) A^2 + 12A - 8 < 0, \tag{5.82}$$

where

$$C = \frac{B}{A} = \frac{\alpha - \beta}{1 + \alpha\beta}.$$

In any case the disturbance of each plane wave state is exponentially small away from the wall. We thus expect that the same holds for the interaction between vortices. We also expect that true bound states should occur only where the interaction is oscillatory as determined by the above asymptotic analysis. These criteria exactly correspond to the results of Aranson et al. [1993a]. Equation (5.82) is identical to their equation (15), with $\beta = 0$ and their $p = k\lambda$.

5.8.3 Weak shocks

The wall equations (5.67–68) depend on the incidence angle θ of the wave vector $\mathbf{k}$ through the constants A and B which have $k_1^2 = k^2 \cos^2 \theta$ in the denominator, as well as in the scaling variable $\xi = k_1 x_1$. If we let $t = 1/\cos\theta$, we can write $A = A_0 t^2$ and $B = B_0 t^2$ where

$$A_0 = \frac{1 + \alpha\beta}{1 + \beta^2} \frac{R_k^2}{k^2} \tag{5.83}$$

and

$$B_0 = \frac{\alpha - \beta}{1 + \beta^2} \frac{R_k^2}{k^2}. \tag{5.84}$$

In the limit $\theta \to \pi/2$, $t \to \infty$, and in this limit the wall disappears, since the wave vector becomes parallel to the wall and thus there is no change in wave vector across the wall. This can be seen clearly in Fig. 5.10. Approaching this limit we can solve the equations in the form (5.70–71) order by order in t^{-1}.

We thus assume the following expansion for w and y:

$$w = w_0 t^{-2} + w_1 t^{-3} + w_2 t^{-4} + \dots, \tag{5.85}$$
$$y = y_0 + y_1 t^{-1} + y_2 t^{-2} + \dots, \tag{5.86}$$

where $y_0 = \pm 1$. We shall here only solve for the lowest order: y_0 and w_0. Further terms have been calculated in [Bohr et al. 1997]. To lowest order we find, by equating coefficients of order t^0:

$$A_0 w_0 - 1 + y_0^2 = 0 \tag{5.87}$$

and

$$y_0' = B_0 w_0, \tag{5.88}$$

or

$$y_0' = \frac{B_0}{A_0}(1 - y_0^2). \tag{5.89}$$

The solution with boundary conditions $y_0(0) = w_0'(0) = 0$ is

$$y_0 = \tanh \frac{B_0}{A_0} \xi \tag{5.90}$$

and

$$w_0 = \frac{1}{A_0 \cosh^2 \frac{B_0}{A_0} \xi} = \frac{1}{A_0} \left(1 - \tanh^2 \frac{B_0}{A_0} \xi \right). \tag{5.91}$$

Obviously, in this limit, the main variation is in the *phase*, whereas the modulus of the field changes very little. Let us try to use the nonlinear phase equation to describe the shocks. As we found in section 5.2 the phase equation has the form

$$\partial_t \phi = \nu \nabla^2 \phi + \lambda (\nabla \phi)^2 \tag{5.92}$$

where $\nu = 1 + \alpha\beta$ and $\lambda = \alpha - \beta$ and where ϕ is the deviation of the phase from the uniformly rotating value.

We now look for a shock solution connecting two plane waves. Thus ϕ must have the form

$$\phi = \Omega t + \Phi(x) = \Omega t + f(x) \tag{5.93}$$

where $\Omega = -(\beta - \alpha)k^2$. Thus

$$\Omega = \nu \nabla^2 f + \lambda (\nabla f)^2. \tag{5.94}$$

Introducing again $y = f'(x)$ we get the 'eikonal equation':

$$y' = \frac{\Omega}{\nu} - \frac{\lambda}{\nu} y^2 = \frac{\alpha - \beta}{1 + \alpha\beta} (k^2 - y^2). \tag{5.95}$$

In terms of $\xi = kx$ (and rescaling y to $f'(\xi)$) we get

$$y'(\xi) = \frac{\alpha - \beta}{1 + \alpha\beta} (1 - y^2) \tag{5.96}$$

which is exactly equivalent to (5.89).

5.9   Phase turbulence and the Kuramoto–Sivashinsky equation

Above the BFN instability, the phase diffusion coefficient ν in (5.92) becomes negative, and, as discussed in section 5.2 the phase diffusion equation turns into the Kuramoto–Sivashinsky equation

$$\partial_t \phi = -|\nu| \nabla^2 \phi - \mu \nabla^4 \phi + \lambda (\nabla \phi)^2, \tag{5.97}$$

where $\mu = \frac{1}{2}\beta^2(1 + \alpha^2)$. This equation has two important symmetries. One is the translational invariance $\phi = \phi' + \text{const}$. The other one is 'Galilean invariance', i.e. the simultaneous transformation of field and coordinates to the primed variables

$$x' = x - 2\theta\lambda t,$$
$$t' = t',$$
$$\phi = \phi' - \theta x' - \lambda\theta^2 t'. \qquad (5.98)$$

The name *Galilean invariance* become more obvious by taking the gradient on both sides of (5.97) and writing an equation for the field $\mathbf{u} = \nabla\phi$. The resulting equation bears a close resemblance to the Navier–Stokes equations for a velocity field $\mathbf{u}$, as we shall see later.

One should note that, by rescaling the phase field, the coordinates and time, all coefficients in (5.97) can be removed. Thus the scaling

$$x = ax',$$
$$t = bt',$$
$$\phi = c\phi',$$

gives, upon dropping the primes, the standard form

$$\partial_t \phi = -\nabla^2\phi - \nabla^4\phi + (\nabla\phi)^2 \qquad (5.99)$$

if

$$a = \sqrt{\mu/|v|},$$
$$b = \mu/|v|^2,$$
$$c = |v|/\lambda, \qquad (5.100)$$

where the only left over parameter is the system size.

As pointed out in section 5.2, (5.99) is perhaps the simplest example of a turbulent nonlinear field theory. It will be discussed further in the context of 'rough interfaces' in chapter 7. The turbulent state has interesting long range correlations as well as short range order. Both stem from the linear instabilities present in (5.97). Thus, by Fourier transformation, and neglecting the nonlinear terms, (5.97) becomes

$$\frac{d\phi_k}{dt} = (|v|k^2 - \mu k^4)\phi_k \qquad (5.101)$$

which gives exponential instability in the range $[0, \sqrt{|v|/\mu}]$ for k. The maximally unstable wave numbers have the length $k_c = \sqrt{|v|/2\mu}$ and the cellular but disordered pattern selected by the system has a characteristic length $l_c \approx 2\pi/k_c$. The dynamics in the turbulent state is dominated by singular events in which cells that have become too large split in two (due to the influence of the nonlinear term) or a small cell between two larger cells is squeezed out.

If (5.97) is a good approximation to the complete complex Ginzburg–Landau equation even in the BFN unstable regime, it would therefore imply the existence of a new kind of turbulent state in which phase fluctuations are large while fluctuations in the modulus are small and vortices unimportant. This possibility has been studied recently in several works [Sakagutchi 1990a,b, Shraiman et al. 1992, Chaté and Manneville 1995, Egolf and Greenside 1994, 1995, Bohr et al. 1994, Grinstein et al. 1996] but it is not quite clear at present under what circumstances the formation of vortices is suppressed. In the following we shall give a short survey of the properties of the phase turbulent state and the transition between phase turbulence and 'vortex turbulence'.

5.9.1   Correlations in the Kuramoto–Sivashinsky equation

The statistical properties of the turbulent state of the Kuramoto–Sivashinsky equation in one spatial dimension have been studied in many recent works. It is known [Pomeau et al. 1984, Manneville 1988] that the number of positive Lyapunov exponents grows linearly with the system size L for a one-dimensional system. This feature is, as we pointed out in chapter 1, typical of turbulent systems. Using the so-called Kaplan–Yorke conjecture (see section 1.5.2 and appendix C) we can calculate the dimension of the chaotic attractor in phase space and thus the finite density of Lyapunov exponents translates to a finite dimension density δ_λ. It has been calculated accurately by Manneville [1988] as

$$\delta_\lambda \approx 0.23\sqrt{2}k_c. \tag{5.102}$$

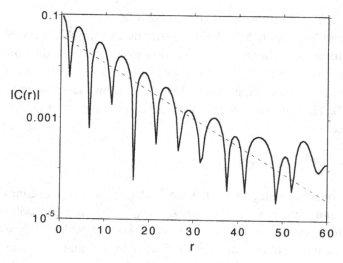

Figure 5.14 Correlation function for the phase gradients of the Kuramoto–Sivashinsky equation.

The long-range fluctuations of the phase are very large. Thus, in analogy with fully developed hydrodynamical turbulence, the structure function

$$S_2(r) = \langle |\phi(x+r,t) - \phi(x,t)|^2 \rangle \tag{5.103}$$

(where the average is over space and/or time) scales as a power of r. In this case, it scales linearly with r in the regime $l_c \ll r \ll L$. One thus has

$$S_2(r) \approx c_\phi \, r \tag{5.104}$$

and distant points can have widely different phases. Going back to (5.100) we see that the constant c_ϕ varies with the parameters as

$$c_\phi \sim c^2/a. \tag{5.105}$$

At the same time the correlations in the phase field are short ranged. Figure 5.14 shows the correlation function for the phase gradients:

$$C(r) = \langle \partial_x \phi(x+r,t) \partial_x \phi(x,t) \rangle - \langle \partial_x \phi(x,t) \rangle^2. \tag{5.106}$$

It clearly has an exponentially decaying envelope.

5.9.2 Dimension densities and correlations in phase turbulence

To find phase turbulence one has to move rather far out to the right in the phase diagram Fig. 5.6. For a detailed phase diagram, the reader is referred to Shraiman et al. [1992]. If we fix $\beta = -3.5$ and vary α we find the phase turbulent state above the BFN point $\alpha = 1/3.5 = 0.286$. At $\alpha \approx 0.7$ vortices start nucleating and we get vortex turbulence. The corresponding dimension densities have been computed by Egolf and Greenside [1995] and are shown on Fig. 5.15, where one can also see the result (5.102) for the Kuramoto–Sivashinsky equation. Close to the BFN instability, the Kuramoto–Sivashinsky equation should be a good approximation and indeed the curve nicely agrees with the numerical results.

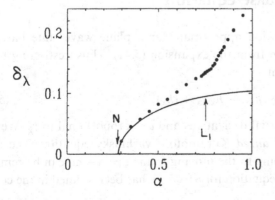

Figure 5.15 Dimension density as function of α, with $\beta = -3.5$. The dots are numerical results for the complex Ginzburg–Landau equation, while the solid line is the value (5.102) for the Kuramoto–Sivashinsky equation. L_1 denotes the transition between phase and vortex turbulence and N denotes the BFN transition [Egolf and Greenside 1995].

Further away there are stronger deviations, which means that the parameters v, μ and λ are renormalized. Finally, at the transition to vortex turbulence, the slope of the dimension density curve changes by approximately a factor of two and is markedly different from the one for the Kuramoto–Sivashinsky equation. As we saw earlier the correlation length for the phase in the Kuramoto–Sivashinsky equation is of the order of the cell size, l_c. Thus the chaotic fluctuations have a dimension per correlation volume of order $\delta_\lambda l_c \approx 2\pi 0.23\sqrt{2}$, independent of the parameters.

Egolf and Greenside [1994] computed the correlation function for the full, complex A-field and found strong violations of this simple relation between dimension density and correlation length. Near the transition to phase turbulence, they found that the correlation length determined in this way becomes very large without similar variations in the dimension density. This is quite understandable, as pointed out in Bohr et al. [1994]. Thus, assuming that only the phase varies, i.e. that $A \approx \text{const } e^{i\phi(x,t)}$ on can use a gaussian approximation for the phase field [Shraiman et al. 1992, Grinstein et al. 1996]:

$$\langle e^{i(\phi(x+r,t)-\phi(x,t))} \rangle \approx e^{-\frac{1}{2}\langle (\phi(x+r,t)-\phi(x,t))^2 \rangle}. \tag{5.107}$$

The argument of the exponent is the structure function (5.103) so that

$$\langle e^{i(\phi(x+r,t)-\phi(x,t))} \rangle \sim e^{-c_\phi r}, \tag{5.108}$$

with the corresponding correlation length $\xi = 1/c_\phi = a/c^2$ (see (5.100)). Close to the BFN instability, $v \to 0$. Since $l_c \sim |v|^{-1/2}$ and $\xi \sim |v|^{-5/2}$, the ratio of ξ and l_c will increase as $|v|^{-2}$ and diverge at the BFN transition as will the dimension-content of a ξ-volume, $\delta_\lambda \xi$. The length scale of relevance for the understanding of the chaotic fluctuations is thus not ξ but rather l_c. Analogous length scales can be found from the decay of the information function for A [Bohr et al. 1994] or from the decay of fluctuation in the modulus $|A|$ of the order parameter [Egolf and Greenside 1995].

5.10 Anisotropic phase equation

In general, the phase equation appropriate for a plane wave state has additional terms, as can be seen from the expansion (5.41). Thus, restricting to one dimension, we find the form

$$\partial_t h = -ah_x - h_{xx} + ch_{xxx} - h_{xxxx} + (h_x)^2, \tag{5.109}$$

where subscripts denote partial derivatives and a is proportional to v_g. We shall call this equation the *generalized Kuramoto–Sivashinsky equation*. The linear term can be removed by going to the moving frame $\xi = x - at$, can be removed. This equation (actually the equation for $u = \partial_x h$), has been studied in the context of surface waves [Kawara 1983, Elphick et al. 1991, Alfaro et al. 1992]. Even small values of c reduce the maximum Lyapunov exponent considerably and for large c the turbulence disappears altogether and the state turns into a train of solitons.

To conclude this chapter we shall describe the the spreading of a localized disturbance of the strongly unstable state $h =$ const. It will be shown that it is possible to calculate the asymptotic form of such a turbulent spot exactly (including, of course, the nonlinear terms) [Conrado and Bohr 1994, 1995]. Further, by increasing c, a 'non-linearization' [van Saarloos 1988, 1989] of one of the fronts occur in which pulses (localized travelling solutions) escape through the linear front [Chang et al. 1995].

5.10.1 The shape of a spreading spot

On Figs. 5.16–17, one sees the results of numerical integration of (5.109) with initial conditions given by a narrow gaussian centred at $x = 0$. First, in Fig. 5.16, we show the standard Kuramoto–Sivashinsky equation, i.e. $c = 0$. In Fig. 5.17 we choose $c = 0.1$, and the $x \to -x$ symmetry is clearly broken. The system size is $L = 9000$ and (5.109) is solved by explicit discretization with grid spacing $= 0.5$. Fig. 5.16a shows three snapshots at different times and as can be seen explicitly on Fig. 5.16b, the initial bump clearly approaches a *growth shape* [Krug and Spohn 1991], i.e. asymptotically the overall shape of $h(x, t)$ is given by

$$h(x, t) = tH(x/t). \tag{5.110}$$

The bump grows sideways with velocity $v_f \approx 1.60$ which can be found by linear analysis as described in appendix D. This is done under the assumption that the transition from the flat state in front of the spot to the expanding spot is smooth. Further, the bump grows upwards with a velocity $v_0 \approx 0.43$ in agreement with the numerically estimated mean velocity for a large, homogenous system: $v_0 = \langle (\partial_x h)^2 \rangle$. From Figs. 5.16–17 it is seen that the growth shapes are well approximated by a central parabolic part joined (with continuous h_x, but discontinuous h_{xx}) to linear segments nearer to the fronts. When the overall shape (5.110) is subtracted, the field $h(x, t)$ looks like the solution of (5.109) grown from uniformly random initial conditions except for a short region around the front, see the inset in Fig. 5.16b. Note in passing that the width of the subtracted field increases approximately as $t^{0.25}$ in accordance with the fact that the effective length of the system increases linearly, never allowing the width to saturate. This will be much clearer later in the context of the 'rough interfaces' of chapter 7. If we add a small amount of noise to the initial gaussian, the macroscopic shape is unaltered. For both types of initial conditions the h-field is clearly chaotic. When computing Lyapunov exponents one has to keep in mind the strong linear divergence of the flat state. To avoid this, tangent vectors

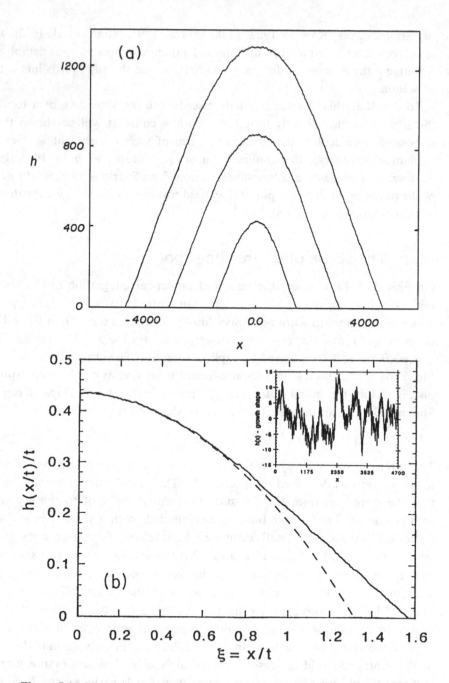

Figure 5.16 Numerical solution of the Kuramoto–Sivashinsky equation. The initial state was flat except for a localized disturbance in the centre. The grid spacing was 0.5. (a) The solution is shown at times $t_1 = 1000$, $t_2 = 2000$ and $t_3 = 3000$. (b) The solution at the three times of (a) are superimposed by the scaling $h \to h/t$ and $x \to x/t$. Due to the mirror symmetry only positive x are shown. The dashed curve is the parabola $y(\xi) = 0.434 - 0.25\xi^2$. The inset shows the remaining field (after the growth shape is subtracted from $h(x, t)$) at $t = 3000$ [Conrado and Bohr 1994].

with support inside the growth region should be used exclusively, and this is accomplished most easily by taking initial conditions for the tangent vector with support only in the initial seed. In this way the largest Lyapunov exponent for the growing shape is (within the accuracy of the simulations) identical to that of a large homogenous system – regardless of whether the initial conditions are noisy. It is interesting to note that in many other cases, notably the 1D complex Ginzburg–Landau equation [Nozakki and Bekki 1983], coupled logistic maps [Pikovsky 1991] and reaction-diffusion equations [Sherratt 1994] the front of an expanding spot goes from laminar to turbulent via a periodic state.

In Fig. 5.18 we show the result for larger c ($= 0.2$). The right-hand part of the growth shape still seems to follow the predictions, but the left-hand part now consists of a series of 'steps'. Let us now determine the growth shapes and their possible limitations.

To find the structure of H, assume that

$$h(x, t) = tH(x/t) + g(x, t), \tag{5.111}$$

where we demand that the spatial average g has no growth rate of its own, i.e. $\langle dg/dt \rangle = 0$. Inserting this into (5.109) gives (for $t \to \infty$)

$$\partial_t g \approx A(\xi) - g_{xx} - g_{xxxx} + c\, g_{xxx} + (g_x)^2 + 2H'(\xi)g_x, \tag{5.112}$$

where $\xi = x/t$ and $A(\xi) = (H'(\xi))^2 + \xi H'(\xi) - H(\xi)$ and where the neglected terms are $O(1/t)$. For very large t we can think of ξ as a slowly varying variable,

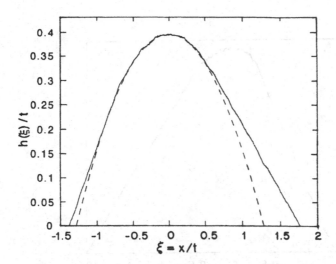

Figure 5.17 Numerical solution of equation (5.109) with $c = 0.1$. The field and variable have been scaled as $h \to h/t$ and $x \to x/t$. The dashed curve is the parabola $y(\xi) = 0.396 - 0.24\xi^2$. The difference between the coefficient 0.24 and the predicted value 0.25 (from (5.114)) is presumably due to the discretization of (5.109) [Conrado and Bohr 1994].

essentially constant over a wide range of x. Thus (5.112) differs from (5.109) only by a constant $A(\xi)$ and a linear term in $\partial_x g$. But such linear terms can be removed by a Galilean transformation, i.e by going into a moving frame, which should not alter the growth rate of g. Now, equation (5.109) with homogenous initial conditions in a sufficiently large system has growth rate $v_0(c)$. Thus, for g to have zero growth rate, we must demand

$$A(\xi) = (H'(\xi))^2 + \xi H'(\xi) - H(\xi) = -v_0(c). \tag{5.113}$$

The boundary conditions are $H(0) = v_0(c)$, and $H(\xi) \to 0$ at the edges $\xi = v_f{}^{\pm}$. Together with the condition that H be monotonic (this is a consequence of the fact that H is the Legendre transform of the angle-dependent growth velocity [Krug and Spohn 1991]), the boundary conditions determine H as

$$H(\xi) = v_0(c) - \frac{1}{4}\xi^2 \tag{5.114}$$

for small ξ. This parabolic shape is valid in the interval $\xi_- < \xi < \xi_+$, where $\xi_- = v_f{}^- + \sqrt{(v_f{}^-)^2 - 4v_0(c)}$ and $\xi_+ = v_f{}^+ - \sqrt{(v_f{}^+)^2 - 4v_0(c)}$. At $\xi = \xi_\pm$ the curvature of the shape is discontinuous and the growth shape becomes linear all the way out to $\xi = v_f{}^{\pm}$ as shown in Fig. 5.19a. In other words the shape is found by joining the inner parabola to its tangents through $(v_f{}^{\pm}, 0)$. Note that the velocity of the moving frame $|v| = 2|H'(\xi)| \leq |\xi_\pm|$, is never outside of the growth region $[v_f{}^-, v_f{}^+]$, where $A(\xi)$ would be zero instead of $-v_0(c)$. Thus the solution is self-consistent.

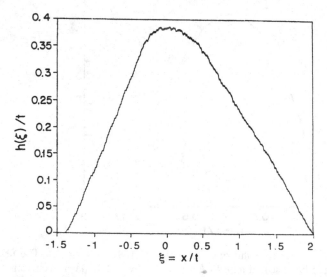

Figure 5.18 Numerical solution of (5.109) with $c = 0.2$. The field and variable have been rescaled by $h \to h/t$ and $x \to x/t$. Note the step structure on the left side [Conrado and Bohr 1994].

One can formulate the results in terms of average growth velocities. Here one assumes that, effectively, $\dot{h} = v(\nabla h)$, where $v(\nabla h)$ is the slope dependent growth velocity (see, for example, Krug and Spohn [1991]). Again Galilean invariance (5.98) can be used to relate growth rates for systems with different mean slopes. Indeed (5.98) implies (with $\lambda = 1$) that $v(u) = v_0(c) + u^2$, where $u = \nabla h$, and by the standard Vul'f construction (Legendre transformation) (see, for example, Krug and Spohn [1991]) we then find $H(\xi) = v_0(c) - \frac{1}{4}\xi^2$. In performing the Legendre transformation one minimizes $v(u) + u\xi$ over u, but effectively the linear theory restricts u to some interval $u_- < u < u_+$ and gives us again the straight edges of Fig. 5.19.

However this construction is not always possible. If c is large enough none of the tangents on the left side will lie within the allowed growth region. Presumably the growth shape should then be parabolic all the way down to

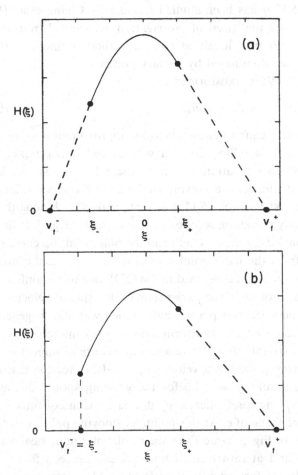

Figure 5.19 Schematic drawing of growth shapes: (a) for $c < c_r$ and (b) for $c > c_r$ [Conrado and Bohr 1994].

$\xi = v_f^-$, where it should, discontinuously, jump to zero, as shown in Fig. 5.19b. The meaning of such a growth shape is not clear and presumably gives rise to an instability in the system, although it is not easy to simulate numerically. Within the numerical accuracy, we find instead, as seen on the left-hand side of Fig. 5.18, a step-like structure. This increases the left-moving front velocity and thus allows the system to avoid the catastrophe. Thus the left-propagating edge is clearly nonlinear [van Saarloos and Hohenberg 1992] and moves with a speed $v_f^- \approx -1.3$ overshooting the velocity -1.23 predicted by linear theory. The origin of this behaviour is the existence of *pulses*, which we shall discuss in the next section.

5.10.2 Turbulent spots and pulses

The spreading of localized disturbances in the generalized Kuramoto–Sivashinsky equation (5.109) has been studied recently by Chang et al. [1995]. In particular, they compare the speed of propagation determined from the linearized equations with that of a localized pulse and thereby find a transition from a 'linear front' to one dominated by solitary pulses.

Introducing $u = h_x$ (5.109) is transformed to

$$\partial_t u = -u_{xx} + cu_{xxx} - u_{xxxx} + 2uu_x. \tag{5.115}$$

As mentioned earlier this equation appears as an approximation to the equation for the height of a fluid falling down a vertical wall [Sivashinsky 1977, Chang 1994] and a *pulse* is a solution $h = h(\xi)$, where $\xi = x - vt$ and $h \to 0$ for $x \to \pm\infty$ [Pumir et al. 1983], i.e. a moving analogue of the shock solution of section 5.9. Writing the last term on (5.115) as $(u^2)_x$ makes it obvious that the equation for the uniformly travelling solutions admits a first integral. It thus becomes a three-dimensional dynamical system and the pulse solution corresponds to a homoclinic orbit from the fixed point $u = u_x = u_{xx} = 0$, i.e. the situation where the (1D) stable manifold is contained in the (2D) unstable manifold. This happens only for special choices of the parameters i.e. for special velocities v, in contrast to the shocks between two plane waves, which was shown generically to exist in section 5.9. Note that the transformation $x \to -x$ interchanges stable and unstable manifolds, so only the relative assignments are of importance.

The simplest left-moving pulse has a velocity $v_p^- = -1.216$ for the case $c = 0$ compared to the linear result $v_f^\pm = \pm1.6$ for the growing spot. Now, as c is increased both v_f^+ and v_f^- increase, whereas v_p^- decreases. Consequently v_p^- and v_f^- become equal at some value of c. At this point, left-moving pulses can escape out of the growing spot. In Fig. 5.20 we show the results of a numerical solution of (5.115) for $c = 0.15$ and at rather small times. It can be seen that a quite regular sequence of pulses is formed near the left edge. Also note that near the right-hand edge more complicated pulse patterns are formed, and they appear inside the 'linear' front stationary with respect to the coordinate system (the 'lab frame') in contrast to the pulses on the left side, which are stationary relative to the front. We know of no theoretical explanation for this striking fact although something similar has been observed before for periodic structures (see, for example, Dee and Langer [1983]).

To understand the relation to the shape transition discussed in the last section, various propagation velocities were computed in [Lundbek Hansen and Bohr 1996] and the results are shown as function of c in Fig. 5.21. Shown are the velocity of the (left) front v_f^- (computed by following a spreading spot for a long time) and the velocity of the left-moving pulse v_p^-. Further, we show the limiting velocity $v^*(c) = 2\sqrt{v_0(c)}$ (where $v_0(c)$ is the velocity with which the turbulent solution of (5.109) translates upward). It marks the breakdown of the growth shape of the last section. Finally, the linear front velocity v_l^- is shown. It is seen that the linear velocity v_l^- is a good approximation to v_f^- as long as c is small. In this regime $v_0(c)$ (and thus $v^*(c)$) is almost constant. If the curves for v_f^- and $v^*(c)$ crossed, the growth shape would become discontinuous. But this never happens: instead v_p^- and v_l^- cross, which turns the front into a train of pulses and from that point on, the front velocity v_f^- follows v_p^- (with a small correction due to the weak interaction between the pulses). This happens at $c \approx 0.14$.

Thus the destruction of the growth shape *almost* occurs. But the system manages to avoid it by sending out pulses. And, for very large system sizes, it

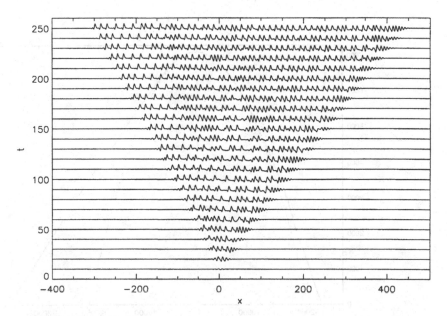

Figure 5.20 Numerical solution of (5.115) with $c = 0.15$. It can be seen that the left-hand edge is dominated by a series of pulses. [Lundbek Hansen and Bohr 1996].

seems that one recovers the growth shapes of the last section even in the pulse-dominated regime – except for the fact that the left edge cannot be calculated by the saddle point method. In Fig. 5.22 we thus show the growth shape for $c = 0.4$.

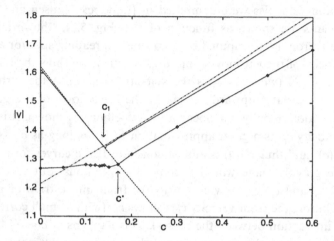

Figure 5.21 Various velocities of relevance for the motion of fronts in (5.109), as defined in the text. v_f^- (full line), v_p^- (dash-dotted line), $v^*(c)$ (diamonds connected by full line), v_l^- (dashed line) [Lundbek Hansen and Bohr 1996].

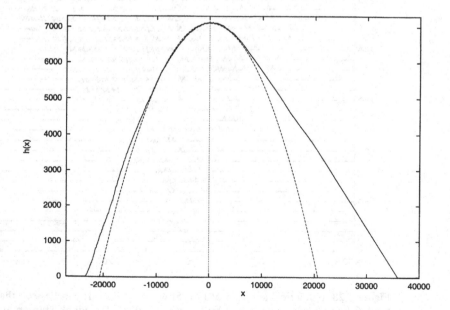

Figure 5.22 Growth shape for $c = 0.4$ [courtesy of J. Lundbek Hansen].

5.10.3 Anisotropic turbulent spots in two dimensions

The spreading of turbulent spots in higher-dimensional systems has been studied within the linear theory in several works, especially in connection with the so-called Emmons spots, which can form when a fluid boundary layer becomes turbulent [Criminale and Kovasznay 1962, Gaster 1962]. In particular, Gaster [1962] gives detailed predictions of the shape of the Emmons spot, based on the spectrum of the Orr–Sommerfeld equation controlling small deviations of the laminar state to lowest order. In this work, he makes use of the so-called Squire's theorem, which allows the two-dimensional problem to be transformed into a one-dimensional one.

In Conrado and Bohr [1995] similar problems were considered in the context of amplitude equations. Thus, simple turbulent field theories, like the ones considered in this chapter, were studied in two dimensions including spatially anisotropic terms. Like the case of the Emmons spot, they allowed for the breaking of the $x \rightarrow -x$ symmetry, while maintaining the $y \rightarrow -y$ symmetry. As an example, consider

$$\partial_t h = \epsilon h + a_1 \partial_x^2 h + a_2 \partial_y^2 h - \nabla^4 h + c_1 \partial_x^3 h + c_2 \partial_x \partial_y^2 h + (\nabla h)^2 \qquad (5.116)$$

where $\nabla^4 h = \nabla^2(\nabla^2 h)$. Figure 5.23 gives an example of the spreading of a localized disturbance in such a field theory. It is clearly seen that the non-isotropic linear terms in (5.116) can generate anisotropic spreading spots and, for example, the boomerang-shapes characteristic of the Emmons spots.

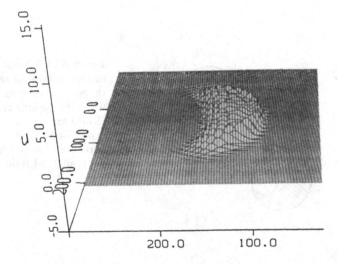

Figure 5.23 A turbulent spot in (5.116) with $\epsilon = -0.15$, $a_1 = -0.75$, $a_2 = -1$ and $c_1 = c_2 = 1$. Note the similarity with the Emmons spot, which forms in a fluid boundary layer near a wall (shown in Fig. 1.2) [Conrado and Bohr 1995].

The computation of shapes in two dimensions is analogous to the one-dimensional problem described in appendix D. The difference is that one now has to compute saddle points in a two-dimensional complex space, i.e. a four-dimensional real space. This makes the calculation a lot more difficult – in particular, the question of which saddle point is the 'right' one. We shall not enter into these issues here, but refer the reader to Conrado and Bohr [1995]. An interesting outcome is a prediction of the typical morphological transitions of the spot shapes which occur when the coefficients of equations like (5.116) are varied. The important point is that, just as in one dimension, the shapes are zeros of the scalar function $z = \mathrm{Re}[\lambda(\mathbf{v})]$, with the difference that $\mathbf{v}$ is now a two-dimensional vector. Thus the shape is the intersection between a surface (z) and a plane ($z = 0$) and typical shape variations correspond to moving the plane for a fixed shape of the function $\lambda(\mathbf{v})$. Figure 5.24 shows a typical sequence of growth shapes obtained in this way.

For the particular choice of nonlinear terms represented by (5.116), the asymptotic (nonlinear) growth shapes can be computed as in the one-dimensional case. One finds the same problems as discussed in the preceding sections: breakdown of the linear front description and pulse formation [Conrado and Bohr 1995] although the transitions involved have not been explored in detail.

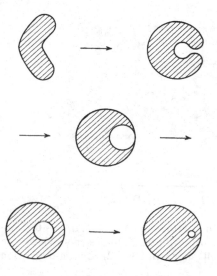

Figure 5.24 Typical metamorphosis of the shapes of spreading spots, when the parameters of (5.116) are varied. The exponentially growing parts are cross-hatched [Conrado and Bohr 1995].

Chapter 6

Predictability in high-dimensional systems

Prediction of the future state of a chaotic system is a problem with an obvious interest in many fields such as geophysics or astrophysics. Based on the classical deterministic point of view of Laplace [1814], it is in principle possible to predict the state of a system, at any time, once the evolution laws and the initial conditions are known. Now, it is evident that predictability has severe limitations in the presence of deterministic chaos because of the exponential divergence of the distance between two initially close trajectories. Typically an uncertainty $\delta \mathbf{x}(0)$ on the state of the system at time $t = 0$ increases as

$$|\delta \mathbf{x}(t)| \simeq |\delta \mathbf{x}(0)| \, e^{\lambda t}, \tag{6.1}$$

where λ is the maximum Lyapunov exponent (see section 1.5). An external observer will thus consider a system predictable up to the time at which the uncertainty reaches a threshold value δ_{max}, determined by the particular needs. As a consequence, starting with $|\delta \mathbf{x}(0)| = \delta_0$, the *predictability time* is

$$T \sim \frac{1}{\lambda} \ln \frac{\delta_{max}}{\delta_0}. \tag{6.2}$$

This relation tells us that the predictability time is proportional to the inverse Lyapunov exponent, since the dependence on the precision of the measure and the threshold is very weak and, in a first approximation, can be neglected. However (6.2) is a too naive answer to the predictability problem since it does not take into account some important features of chaotic systems. Let us briefly list the different reasons for a failure of a simple link between predictability and maximum Lyapunov exponent in a generic dynamical system;

(a) The Lyapunov exponent λ is a global quantity: it measures the average exponential rate of divergence of nearby trajectories. In general there exist finite-time fluctuations of this rate and it is possible to define an 'instantaneous' rate, called effective Lyapunov exponent $\gamma_t(\tau)$ [Paladin and Vulpiani 1987a]. For finite time delay τ, $\gamma_t(\tau)$ depends on the particular point of the trajectory $\mathbf{x}(t)$ where the perturbation is performed. In the same way, the predictability time T fluctuates, following the γ-variations.

(b) In dynamical systems with many degrees of freedom, the interactions among different degrees of freedom play an important role in the growth of the perturbation. If one is interested in the case of a perturbation concentrated on certain degrees of freedom (e.g. small length scales in weather forecasting), and of a prediction on the evolution of other degrees of freedom (e.g. large length scales), the knowledge of the statistics of the effective Lyapunov exponent is not sufficient. One has to analyse the behaviour of the tangent vector $\mathbf{z}(t)$, i.e. of the direction along which an infinitesimal perturbation grows, see, for example, Pikovsky [1993]. In such a situation a relevant quantity is the time T_R, necessary to the tangent vector to relax on the time dependent eigenvector $\mathbf{e}_1(t)$ of the stability matrix, corresponding to the maximum Lyapunov exponent λ_1. It is worth stressing that a generic tangent vector $\mathbf{z}(t)$ relaxes exponentially fast to $\mathbf{e}_1(t)$ [Orszag et al. 1987]. When the perturbations on a system are small enough, and a linear approach can be used, nonlinear terms of type $\delta x_i \delta x_j$ are negligible in the evolution equation of $\delta \mathbf{x}$, and one has that

$$T \sim T_R + \frac{1}{\lambda} \ln \frac{\delta_{max}}{\delta_0}. \tag{6.3}$$

Note that the mechanism of transfer of the error $\delta \mathbf{x}$ through the degrees of freedom of the system, which determines T_R, could be more important than the rate of divergence of nearby trajectories.

(c) In systems with many characteristic times, such as the eddy turn-over times in fully developed turbulence, T is determined by the detailed mechanism due to the nonlinear effects in the evolution equation for $\delta \mathbf{x}$, if the perturbations are not infinitesimal or if the threshold of accepted error is not small. In this case, the predictability time could be unrelated to the maximum Lyapunov exponent and T might depend, in a non-trivial way, on the details of the system [Aurell et al. 1996, Bofetta et al. 1997].

(d) In the presence of noise, or in general of a non-deterministic part in the evolution laws, there are two different ways to define the predictability, by considering either two trajectories of the system with the same noise or two trajectories of the same system evolving with different realizations of the noise. Both these definitions are physically relevant in different contexts but the results can be very different in the presence of a strong dynamical intermittency [Paladin et al. 1995, Loreto et al. 1996].

6.1   Predictability in turbulence

The sensitive dependence on initial conditions renders long term forecasting impossible. For instance, Ruelle [1979] remarked that thermal fluctuations in the atmosphere produce observable changes on a scale of centimetres after only few minutes. One thus expects that after one or two weeks, the earth's atmospheric circulation would be unpredictable, even if the exact evolution equations were known. This is the so-called *butterfly effect*, in the words of Lorenz: 'A butterfly moving its wings over Brazil might cause the formation of a tornado over Texas.'

In fully developed turbulence the maximum Lyapunov exponent diverges with the Reynolds number. However, the predictability time for the large length scale variables seems to be independent of Reynolds number. Both these results are derived on the basis of phenomenological arguments and are well reproduced by numerical simulations in the GOY shell model. The paradox of the presence of strong chaos with a rather weak butterfly effect can be explained by a careful stability analysis of systems with many characteristic time scales. This is one of the major applications of the theory of chaotic dynamical systems to the phenomenology of turbulent flows and will be discussed in some detail in the following.

6.1.1   Maximum Lyapunov exponent of a turbulent flow

In three-dimensional fully developed turbulence, the maximum Lyapunov exponent should be roughly proportional to the inverse of the smallest characteristic time of the system, the turn-over time τ of eddies of the size of the Kolmogorov length ℓ_D. Let us introduce, in terms of the spatial average of the energy dissipation $\bar{\epsilon}$ and of the typical large length scale of the system L, the corresponding velocity $V = (\bar{\epsilon} L)^{1/3}$ and time $T_0 = L/V = (L^2/\bar{\epsilon})^{1/3}$. The turn-over time of an eddy of size ℓ is, by dimensional counting,

$$\tau(\ell) \sim \frac{\ell}{v(\ell)} \sim T_0 \left(\frac{\ell}{L}\right)^{1-h}, \tag{6.4}$$

where h is the Hölder exponent of the velocity field:

$$v(\ell) \equiv |\mathbf{v}(\mathbf{x}+\mathbf{r}) - \mathbf{v}(\mathbf{x})| \sim V \left(\frac{\ell}{L}\right)^h \qquad |\mathbf{r}| = \ell.$$

As discussed in chapter 2 the nonlinear transfer of energy is stopped at the Kolmogorov scale ℓ_D:

$$\ell_D(h) \sim L \, Re^{-1/(1+h)}. \tag{6.5}$$

So that the maximum Lyapunov exponent scales with Re like

$$\lambda \sim \frac{1}{\tau(\ell_D)} \sim \frac{1}{T_0} Re^\alpha \qquad \text{with } \alpha = \frac{1-h}{1+h}, \tag{6.6}$$

since it should be proportional to the inverse of the shortest characteristic time of the system, i.e. the turn-over time of the smallest eddy of size ℓ_D. In the Kolmogorov theory $h = 1/3$ for all space points, and $\alpha = 1/2$, as first pointed out by Ruelle [1979].

The presence of quiescent quasi-laminar periods changes the chaotic features of the flow and the intermittency of energy dissipation must be described by introducing a spectrum of singularities h. In the multifractal approach, see section 2.3, the singularities $h \in [h, h + dh]$ are concentrated on a fractal set of dimension $D(h)$ so that the probability that the velocity difference scales with an exponent h is

$$P_\ell(h) \sim \ell^{3-D(h)}. \tag{6.7}$$

Therefore, there exists a spectrum of viscous cut-offs, since each h selects a different damping scale, and hence a spectrum of turn-over times. To obtain the maximum Lyapunov exponent, we have to integrate $\tau(h)^{-1}$, given by (6.4) at the scale $\ell = \ell_D(h)$, over the h-distribution $P_\ell(h)$:

$$\lambda \sim \int \tau(h)^{-1} P_\ell(h) \, dh \sim \frac{1}{T_0} \int \left(\frac{\ell_D}{L} \right)^{h-D(h)+2} dh. \tag{6.8}$$

From (6.5) the viscous cut-off vanishes in the limit $Re \to \infty$ and the integral can be estimated by the saddle point method,

$$\lambda \sim \frac{1}{T_0} Re^\alpha \qquad \text{with } \alpha = \max_h \left[\frac{D(h) - 2 - h}{1 + h} \right]. \tag{6.9}$$

The value of α depends on $D(h)$. By using the function $D(h)$ obtained by fitting the exponents ζ_q with the random beta model, see equation (2.63), we find $\alpha = 0.459\ldots$, slightly smaller than the Ruelle prediction $\alpha = 0.5$.

6.1.2   The classical theory of predictability in turbulence

The classical theory of predictability has been developed by Lorenz [1969], see also Lilly [1973], using physical arguments, and by Leith and Kraichnan [1972], on the basis of closure approximations. The fundamental ingredients of the Lorenz approach stem from dimensional arguments on the time evolution of a perturbation in an energy cascade picture. In that framework, it is rather natural to assume that the time $\tau(k)$ for a perturbation at wave number say $2k$ to induce a complete uncertainty on the velocity field on the wave number k, is proportional to the typical eddy turn-over time at scale k:

$$\tau(k) \sim \frac{1}{kv_k},\tag{6.10}$$

where v_k is the typical velocity difference at scale $1/k$ determined by

$$v_k^2 \sim \int_k^{2k} E(k)\,\mathrm{d}k.$$

In the Kolmogorov theory $\tau(k) \sim k^{-2/3}$. The predictability time to propagate an uncertainty $O(v(\ell_D))$ from the Kolmogorov scale $\ell_D \sim k_K^{-1}$ up to the scale of the energy containing eddies $L_0 \sim k_0^{-1}$, is given by:

$$T \simeq \sum_{n=0}^{N} \tau(2^n k_K)\tag{6.11}$$

where $k_K \sim Re^{3/4} k_0$ and $N = \log_2(k_K/k_0) \sim \ln Re$. The geometrical series (6.11) is dominated by the last term ($n = N$) so that

$$T \sim T_0 \sim L_0/V_0.\tag{6.12}$$

It is easy to compute the distance δU between two velocity fields at time $t_j = \sum_{n=0}^{j} \tau(2^n k_K)$. Noting that $t_j \sim k_j^{-2/3}$ and $\delta U^2(t_j) \sim v_{k_j}^2 \sim k_j^{-2/3}$ one obtains

$$\delta U^2(t) \sim t,\tag{6.13}$$

i.e. a power law. Closure approximations, where one still uses dimensional arguments, confirm the results (6.12) and (6.13).

It is important to stress that, in the Lorenz approach, the predictability time T is independent of the Reynolds number. In fact, many characteristic times are involved so that (6.12) and (6.13) depend strongly on the physical mechanism for the inverse cascade of the perturbation.

From the point of view of an observer interested on the forecasting on large length scales for not infinitesimal perturbation the Lyapunov exponent is not physically relevant. We thus find the amazing situation that a large maximum Lyapunov exponent coexists with a long predictability time [Boffetta et al. 1997]. In the next section, we provide a mathematical justification of this feature in a simplified dynamical model of the energy cascade.

6.2 Predictability in systems with many characteristic times

The presence of a long predictability time in strongly chaotic systems can be understood in a model of coupled maps with different time scales. This is an idealized situation where each degree of freedom has a chaotic dynamic with its own characteristic time and the different degrees of freedom are locally coupled by a small interaction. The model can be regarded as a simple prototype for several physical phenomena where one has many different scales in the time evolution, such as the eddy turn-over times in the energy cascade. Despite its simplicity, it displays non-trivial properties which may be close to the behaviour of more realistic systems.

The system is given by a chain of chaotic coupled maps

$$\left.\begin{aligned}
&x_1(n+1) = (1-\epsilon)f_1(x_1(n)) + \epsilon g_1(x_2(n)),\\
&x_2(n+1) = (1-\epsilon)f_2(x_2(n)) + \epsilon g_2(x_3(n)),\\
&\qquad\vdots\\
&x_{N-1}(n+1) = (1-\epsilon)f_{N-1}(x_{N-1}(n)) + \epsilon g_{N-1}(x_N(n)),\\
&x_N(n+1) = f_N(x_N(n)),
\end{aligned}\right\} \qquad (6.14)$$

where the N variables x_k are defined on the interval $I = [0,1]$, ϵ is the local coupling constant and $f_k, g_k : I \to I$ are generic functions. In order to mimic the interactions among the eddies at different scales we consider f_k such that the Lyapunov exponent λ_k of the map

$$x(n+1) = f_k(x(n)) \qquad (6.15)$$

increases with k. A simple choice is

$$f_k(x) = e^{\lambda_k}x \mod 1, \quad g_k(x) = x. \qquad (6.16)$$

In order to have a dynamical model related to the energy cascade in turbulence, x_k has to be considered as a typical representative variable (e.g. velocity difference) of the eddies on the octave of length scales $l_k = L_0 2^{-k}$. From the Kolmogorov arguments, defining the Lyapunov exponent λ_k as the inverse characteristic time and using (6.10),

$$\lambda_k = \frac{1}{\tau_0} 2^{2/3(k-1)}.$$

However, different choices, e.g. $\lambda_k = k\lambda_0$, do not change the main features. The coupling parameter ϵ is fixed to the same (small) value for each length scale, although more realistic models should consider a scale-dependent coupling.

Let us consider the effects of a perturbation δ_0 at the initial time on the smallest scale N on the growth of the perturbation δx_1 at the largest length scale. After N steps the perturbation will affect the largest scale x_1 which then grows with exponential rate corresponding to the maximal Lyapunov exponent λ_N:

$$\delta x_1(n) = \epsilon^{N-1} e^{n\lambda_N} \delta_0.$$

This is true only for infinitesimal perturbations, i.e. up to the time $M_N \simeq -\ln(\delta_0)/\lambda_N$ at which the most unstable variable saturates i.e. $\delta x_N(M_N) = O(1)$. On times longer than M_N the growth of δx_1 is governed by x_{N-1}:

$$\delta x_1(M_N + n) = \epsilon^{N-2} e^{n\lambda_{N-1}} \delta x_{N-1}(M_N),$$

up to the time $M_N + M_{N-1}$ where also δx_{N-1} reaches its saturation threshold, and so on.

The evolution of the system (6.14) leads to a linear evolution for the uncertainty according to the law

$$\Delta(t+1) = \mathbf{A}\, \Delta(t)$$

where $\Delta = (\delta x_1, \delta x_2, \cdots, \delta x_N)$ and $\mathbf{A}$ is the $N \times N$ matrix

$$\mathbf{A} = \begin{pmatrix} l_1 & \epsilon & 0 & \dots & 0 & 0 \\ 0 & l_2 & \epsilon & \dots & 0 & 0 \\ .. & .. & .. & .. & .. & .. \\ 0 & 0 & 0 & \dots & 0 & l_N \end{pmatrix} \qquad l_k = \exp(\lambda_k).$$

Let us denote with T_k the time at which scale k saturates, i.e. $\delta x_k(T_k) = 1$, and with M_k the interval (number of steps) during which scale k dominates the dynamics, i.e. $M_k = T_k - T_{k+1}$. If we suppose that the main contribution to the growth of uncertainty during period M_k is given by the faster scale δx_k (the hypothesis is correct whenever ϵ is small and the λ_k are not too large) we can write, for any $i < k$

$$\delta x_i(T_k) = \delta x_i(T_{k+1}) + (\mathbf{A}^{M_k})_{i,k} \delta x_k(T_{k+1}) \tag{6.17}$$

while $\delta x_k(T_k) = \Delta_{sat}$ is the saturation level. Matrix $\mathbf{A}^M$ can be easily computed and at leading order is given by

$$(\mathbf{A}^M)_{i,k} = \epsilon^{k-i} \frac{e^{M\lambda_k} - e^{M\lambda_i}}{e^{\lambda_k} - e^{\lambda_i}}. \tag{6.18}$$

The saturation intervals are, by definition, given as

$$M_k = \frac{1}{\lambda_k} \log \frac{\Delta_{sat}}{\delta x_k(T_{k+1})} \tag{6.19}$$

so that at the leading order

$$M_k = \frac{1}{\lambda_k} \log \frac{(\alpha - 1)\lambda_k}{\epsilon}, \tag{6.20}$$

with $\alpha = 2^{2/3}$.

For a numerical check of these arguments, let us integrate the system (6.14) with two different initial conditions $\mathbf{x}(0)$, $\mathbf{x}'(0)$ where $\delta x_k(0) \equiv x'_k(0) - x_k(0) = \delta_{k,N}\delta_0$. The initial uncertainty is thus confined to the smaller and faster scale x_N while we are interested in the knowledge of the system at the largest scale x_1. The perturbation on the largest scale δx_1 follows different exponential laws at different times, so that the global envelope $\delta x_1(t)$ is very complex. Figure 6.1 shows the uncertainty δx_1 on the large scale as a function of time in a numerical experiment.

Some exponential regimes are still recognizable as straight lines with different slopes in a linear–logarithmic plot. Figure 6.2 shows (in log–log scale) the result obtained averaging over a large number of initial conditions. We see that the agreement with the direct simulation of (6.14) is quite good.

The global behaviour can be fitted by a power law, whose slope can be roughly estimated by the following argument. Apart from logarithmic corrections one has $M_k \sim \alpha^{-k}$. From (6.17–20) we estimate the growth of δx_1 during period M_k to be $\sim \epsilon^k$. Then we expect to have $\delta x_1(t) \sim t^\gamma$ with $\gamma = -\log \epsilon / \log \alpha$.

From the model of coupled maps with different characteristic times one learns that it is possible to have non-trivial behaviours for a *non-infinitesimal* uncertainty at the slowest scale which can be wrongly interpreted as power laws.

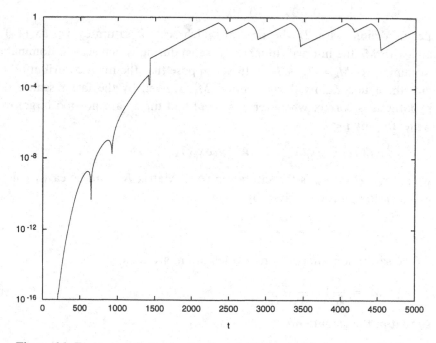

Figure 6.1 Error growth δx_1 for the largest (slowest) scale in system (6.14) with $N = 10$, $\tau_0 = 200$ and $\epsilon = 10^{-4}$. Straight segments correspond to exponential growth driven by different dominant scales. The initial condition is chosen at random and the initial uncertainty on the small scale is $\delta_0 = 10^{-8}$ [Aurell et al. 1996].

The complex envelope of $\delta x_1(t)$ is actually generated by a sequence of saturation processes. For this kind of system, if we are interested in long-time predictions, the maximum Lyapunov exponent is of little meaning and one should consider all the time scales present in the system.

6.3   Chaos and butterfly effect in the GOY model

6.3.1   Growth of infinitesimal perturbations and dynamical intermittency

It is easy to analyse the time evolution of an infinitesimal perturbation in the GOY model. Let us indicate the state of the shell model at time t by $\mathbf{u}(t) \equiv \{u_1(t), \ldots, u_n(t)\}$. At a certain time t, a perturbed state $\mathbf{u}'(t)$ is produced by adding a small increment δ_0 to the velocity component in some of the shells, and the distance between the two trajectories is defined as

$$D(\tau) = |\mathbf{u}(t + \tau) - \mathbf{u}'(t + \tau)|. \tag{6.21}$$

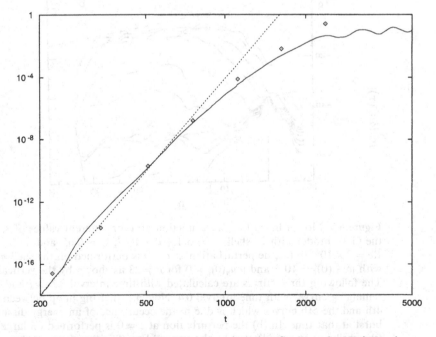

Figure 6.2 Full line: as in Fig. 6.1 in log–log plot. Scatter plot: estimated growth of the largest scale δx_1 during different periods M_k according to the quasi-linear approximation given by equations (6.17) and (6.20). Dashed line: power law expected by the argument described in the text [Aurell et al. 1996].

If the evolution is chaotic $D(\tau)$ grows exponentially in time, i.e.

$$\langle \ln D(\tau) \rangle \approx \lambda \tau, \qquad \text{for } \tau \gg 1$$

where the average is over several perturbations.

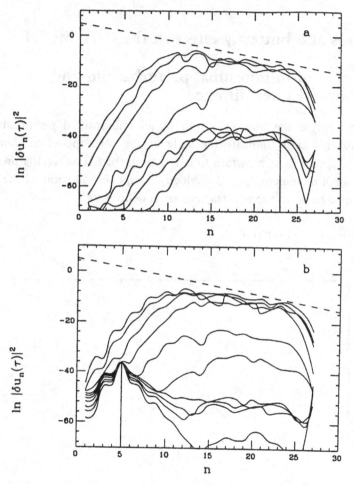

Figure 6.3 Plot of $\ln |\delta u_n(\tau)|^2$ as a function of n for different values of τ, for the GOY model with 27 shells, $f = 5(1 + i) \times 10^{-3}$, $k_0 = 0.05$, and $Re = 2 \times 10^9$. In (a) the perturbation at $\tau = 0$ is performed on the shell $n = 23$, with $|\delta u_{23}(0)| = 10^{-8}$ and $|\delta u_n(0)| = 0$ for $n \neq 23$ as shown by the vertical line. The following three curves are calculated with time interval 0.2 in τ and the remaining curves with time interval 0.4. Note a rather big jump between the 4th and the 5th curves which is due to the occurrence of an energy dissipation burst at that time. In (b) the perturbation at $\tau = 0$ is performed on large scale (the shell $n = 5$) as indicated by the vertical line. The time intervals between the curves are the same as in (a). The disturbance moves to small scales before the exponential growth starts. The dashed lines show the scaling law $|\delta u_n|^2 \simeq k_n^{-2/3}$ of the Kolmogorov theory [Crisanti et al. 1993b].

To study the butterfly effect the model is perturbed at high wave numbers, close to the dissipative cut-off given by the Kolmogorov length. To get a deeper insight on the growth of the disturbance, we introduce the difference on the nth shell at a time τ of the two fields

$$|\delta u_n(\tau)|^2 = |u_n(t + \tau) - u'_n(t + \tau)|^2.$$

The exponential growth of $|\delta u_n(\tau)|^2$ is triggered by a large energy burst localized at the small length scales and associated with a sharp increase of the instantaneous Lyapunov exponent. After such a burst, the value of $|\delta u_n(\tau)|^2$ increases with τ at smaller and smaller k_n, so that the initial disturbance localized on small scales propagates towards lower k_n by a sort of inverse cascade, as shown in Fig. 6.3a. The disturbance eventually reaches the beginning of the inertial range affecting the flow at large scales: the 'butterfly' disturbance has grown to macroscopic scales.

If the disturbance is not initially localized on the Kolmogorov scale, we do not find an exponential growth of $|\delta u_n(\tau)|^2$ for any shell k_n, until the disturbance has spread all the way down to the small (dissipative) scales. The time needed by this 'precursor' disturbance to move from small to large wave numbers is very fast, of order $\sim \lambda^{-1}$. After this time interval, the exponential growth is again triggered by a chaotic burst localized on large k_n, as shown in Fig. 6.3b, leading to the inverse cascade previously discussed for Fig. 6.3a. In other words

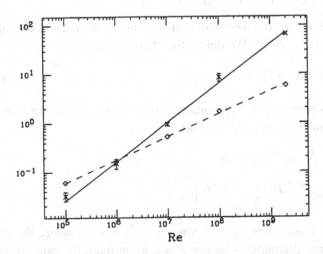

Figure 6.4 The Lyapunov exponent, λ (diamonds) and μ (crosses), as a function of the Reynolds number from a shell model calculation with $N = 27$ shells, $f = 5(1 + i) \times 10^{-3}$ and $k_0 = 0.05$. The dashed line is the multifractal prediction $\lambda \sim Re^\alpha$ with $\alpha = 0.459$, where the function $D(h)$ is given by the random beta model fit of the ζ_p exponents. The full line indicates $\mu \sim Re^w$ with $w = 0.8$ [Crisanti et al. 1993a].

the butterfly effect always stems from the small scales close to the dissipative cut-off even when the perturbation is performed on the large scales.

We can verify the multifractal prediction for the dependence of the maximum Lyapunov exponent on the Reynolds number (6.9) in the GOY model as shown in Fig. 6.4 using $N = 27$ shells. The different values of Re are obtained by changing the value of the viscosity v. The correction to the Ruelle prediction $\lambda \sim Re^{1/2}$ is well evident and agrees with (6.9). The same figure also shows the variance μ of the finite-time fluctuations. Note that λ and μ give the main characterization of the distribution of the effective Lyapunov exponent, and that the value $\mu/\lambda = 1$ separates weak from strong intermittency [Benzi et al. 1985].

The numerical data show that the variance diverges as

$$\mu(Re) \sim Re^w \qquad \text{with } w \simeq 0.8. \tag{6.22}$$

Noting that

$$\gamma_t(\tau) = \frac{1}{\tau} \int_t^{t+\tau} \gamma_{t'}(0) \, dt' \tag{6.23}$$

an explicit calculation leads to

$$\mu \sim \int_0^\infty \langle (\gamma_{t+t'}(0) - \lambda)(\gamma_t(0) - \lambda) \rangle \, dt' = \langle (\gamma(0) - \lambda)^2 \rangle \int_0^\infty C(t') \, dt' \tag{6.24}$$

where $C(t')$ is the normalized correlation function of the effective Lyapunov exponent

$$C(t') = \langle (\gamma_t(0) - \lambda)(\gamma_{t+t'}(0) - \lambda) \rangle / \langle (\gamma(0) - \lambda)^2 \rangle, \tag{6.25}$$

which has the same qualitative behaviour of the energy dissipation correlation function [Crisanti et al. 1993b]. We define the characteristic time

$$t_c = \int_0^\infty C(t') \, dt' \sim T_0 \, Re^{-z}, \tag{6.26}$$

which vanishes as a power of Re. The quantity $\langle (\gamma(0) - \lambda)^2 \rangle$ can be estimated by repeating the arguments used for λ, so that

$$\langle \gamma(0)^2 \rangle \sim \int \tau(h)^{-2} P_{\ell_D}(h) \, dh \sim \frac{1}{T_0^2} Re^y$$

$$\text{with } y = \max_h \left[\frac{D(h) - 1 - 2h}{1 + h} \right] = 1.$$

The result $y = 1$ is model independent, since $\langle \gamma(0)^2 \rangle \sim Re \, \bar{\epsilon}$, where the spatial average of the energy dissipation density $\bar{\epsilon}$ is a finite quantity independent of Re. The fact that $\langle \gamma(0)^2 \rangle \gg \lambda^2$ at high Re implies

$$\mu \sim \langle \gamma(0)^2 \rangle \, t_c \simeq \frac{1}{T_0} Re^w \qquad \text{with } w = 1 - z.$$

In the shell model this relation is satisfied with $z \simeq 0.2$.

We stress that in the absence of intermittency one would expect that $t_c \sim \lambda^{-1}$, and thus $z \simeq 1/2$. As a consequence, $z \simeq 0.2$ indicates that the presence of quiescent periods in the turbulent activity is much more relevant for the decay rate of time correlations than for the Lyapunov exponent.

Although it is sensible to expect $w > 1/2$ in real turbulent fluids, we cannot exclude that $w \simeq 0.8$ is due to the particular form of the time correlations in the shell model.

The basic qualitative feature of these results is just the dynamical counterpart of the multifractality of energy dissipation in three-dimensional space. In generic chaotic systems a lower bound of t_c is given by λ^{-1}. It follows $w \geq 1/2$ and $w > \alpha$, implying that μ/λ diverges as $Re \to \infty$, and so the dynamical intermittency.

6.3.2 Statistics of the predictability time and its relation with intermittency

Let us define the predictability time for the shell model as follows. Consider two initial realizations $\mathbf{u}$ and $\mathbf{u}'$, at time $t = 0$

$$u'_n(0) = u_n(0) + \delta u_n(0)$$

with $\delta u_n(0) \neq 0$ only for $n = n^*$, $n^* + 1$, where n^* corresponds to the Kolmogorov length. The predictability time T is defined as the maximum time t such that

$$|\delta u_4(t)|^2 + |\delta u_5(t)|^2 < \delta_{\max}^2. \tag{6.27}$$

By changing the initial condition, e.g. by taking the same trajectory at different times, the above computation can be repeated many times and one can obtain the probability distribution of T. The time T is not constant but strongly dependent on the degree of chaos: if the system undergoes an energy burst, T is very short. On the other hand, if the system is in a laminar period, T can be very large. Figure 6.5 shows the probability distribution function (PDF) of T for two different values of Re. At $Re \approx 10^6$ we observe a rather peaked PDF of almost gaussian type. For larger values of Re ($Re \approx 2 \times 10^9$) the distribution gets an exponential tail, indicating the possibility of large excursions in the value of T, depending on whether the system is in a turbulent or in a purely laminar period. Moreover, the typical predictability time T_t, i.e. the value of T where the PDF reaches it maximum, is very dependent on the Lyapunov exponent and hence on the Reynolds number. In the shell model, the typical predictability time decreases as a power of Re, and at increasing Re the occurrence of large values of $(T - T_t)/T_t$ is more and more likely.

We want to stress that many features of the above scenario do not depend on the values of the threshold $\delta_{\max}$, if $\delta_{\max}$ is small enough, i.e. we can neglect the nonlinear terms in the evolution of $\delta\mathbf{u}$. The gross features of the probability distributions shown in Fig. 6.5 do not depend on the particular dynamical system considered but only on the degree of intermittency measured by μ/λ: when $\mu/\lambda \gg 1$ the probability distribution of the predictability time has long exponential tails, while for $\mu/\lambda \leq 1$ it is very peaked. Long exponential tails also appear in the Lorenz model with r slightly larger than $r_c \simeq 166.07$, near the intermittent transition [Pomeau and Manneville 1980], as μ/λ increases, see Crisanti et al. [1993b].

The mechanism for the occurrence of exponential tails is not an artifact of the shell model, but a rather robust feature of highly intermittent systems. A

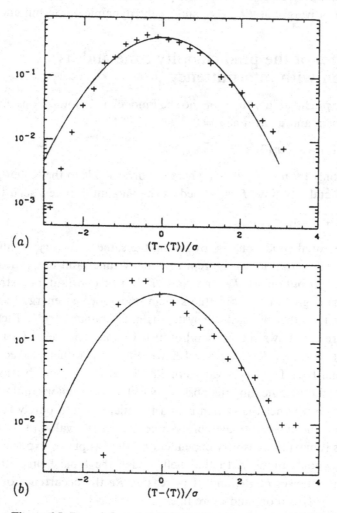

Figure 6.5 Rescaled probability distribution functions PDF of the predictability time T for the shell model: $\sigma P(T)$ versus $(T - \langle T \rangle)/\sigma$ for (a): $Re = 10^6$ and (b): $Re = 2 \times 10^9$. The respective average values are $\langle T \rangle = 84.0$ and 6.32 and the standard deviations $\sigma = [\langle (T - \langle T \rangle)^2 \rangle]^{1/2}$ are 22.2 and 3.16, respectively. The full line is the standard gaussian [Crisanti et al. 1993b].

simple argument shows the relation between the PDF of T and the fluctuations of the effective Lyapunov exponent γ. As a first rough approximation, we can assume that [Fujisaka 1983, Benzi et al. 1985]

$$\ln \frac{|\delta \mathbf{u}(t)|}{|\delta \mathbf{u}(0)|} = \lambda t + \sqrt{\mu}\, w(t) \tag{6.28}$$

where $w(t)$ is a Wiener process with $w(0) = 0$ and

$$\langle w(t) \rangle = 0, \qquad \langle w(t)\, w(t') \rangle = \min\,[t, t'].$$

By this approximation the predictability problem is reduced to a first exit problem: T is the largest time such that

$$\sqrt{\mu}\, w(t) < \ln(\delta_{\max}/\delta_0) - \lambda t. \tag{6.29}$$

This is a standard problem in stochastic process theory, whose solution [Burgers 1974] gives the PDF of T:

$$P(T) = \frac{|\ln(\delta_{\max}/\delta_0)|}{\sqrt{2\pi\mu T^3}} \exp\left[-\frac{(\lambda T - \ln^2(\delta_{\max}/\delta_0))^2}{2\mu T}\right]. \tag{6.30}$$

For small values of μ/λ the PDF is almost gaussian and the mean value of T is close to the most probable value T_t given by the maximum of (6.30)

$$T_t \simeq \frac{1}{\lambda} \ln(\delta_{\max}/\delta_0). \tag{6.31}$$

On the contrary for $\mu/\lambda \gg 1$ the PDF exhibits an asymmetric 'triangular shape' and

$$T_t \simeq \frac{1}{3\mu} \ln^2(\delta_{\max}/\delta_0). \tag{6.32}$$

In the approximation (6.28) one neglects correlations between $\gamma_t(0) - \lambda$ and $\gamma_{t'}(0) - \lambda$ for $t \neq t'$. This is reasonable only for times much larger than λ^{-1}. Since $\ln(\delta_{\max}/\delta_0)$ is not large, T/λ cannot be too large and the previous argument is not very accurate. The qualitative behaviour predicted by the approximation (6.28) is, however, confirmed both in the shell model and the Lorenz system.

6.3.3 Growth of non-infinitesimal perturbations

In a physical approach to the problem of predictability in turbulence one has to consider the amplification in the time of physically realistic errors on the initial conditions. This is one of the central problems in meteorology, where the initial inaccuracy comes from the measurement and, obviously, cannot in general be considered as infinitesimal.

Let us discuss the problem of predictability in the GOY model comparing the temporal evolution of pairs of different realizations of the velocity field, say u_n and u'_n. Both fields evolve according to (3.10) from initial conditions such that:

i) the energy spectra of u_n and u'_n at the initial time are equal

ii) u_n and u'_n at time zero differ only on small scales, corresponding to wave numbers $k_n \geq k_{n^*}$:

$$u'_n = \begin{cases} u_n & \text{for } n < n^*, \\ e^{iw_n} u_n & \text{for } n \geq n^*, \end{cases} \qquad (6.33)$$

where k_{n^*} is the inverse of the Kolmogorov length and w_n is a random number uniformly distributed in the range $[0, \theta]$.

By changing the value of θ we can modify the correlation between the two fields. The extreme case is $\theta = 2\pi$, which corresponds to completely uncorrelated fields, meaning no knowledge for $n \geq n^*$, i.e. one has the Lorenz prescription discussed in section 6.1.2.

In Figs. 6.6 and 6.7 we show the behaviour of the growth of moments of the difference defined as

$$\langle |\mathbf{u} - \mathbf{u}'|^q \rangle = \left\langle \left[\sum_n |u_n - u'_n|^2 \right]^{q/2} \right\rangle, \qquad (6.34)$$

for the cases $\theta = 2\pi$ and 0.001.

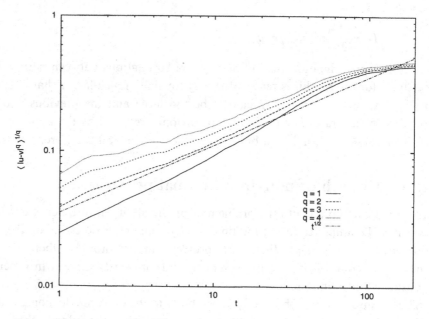

Figure 6.6 Moments $\langle |\delta u|^q \rangle^{1/q}$ versus t for $\theta = 2\pi$, the straight line indicates $t^{1/2}$ [Aurell et al. 1996].

The numerical simulations confirm the validity of the Lorenz argument in the shell model ($\theta = 2\pi$). The unique effect of the intermittency is the fact that one has an anomalous scaling i.e.

$$\langle |u - u'|^q \rangle \sim t^{q\alpha(q)} \tag{6.35}$$

where $\alpha(q)$ is a decreasing function of q while in the Lorenz argument one has $\alpha(q) = 1/2$. The value of $\alpha(2)$ is close to $1/2$ as predicted by (6.13) which is obtained by neglecting intermittency. For the case of small θ, see Fig. 6.7, one has a scenario rather close to that observed in section 6.2 for the simple model (6.14), i.e. the perturbation grows exponentially at very short times and then turns into a sort of power law.

In conclusion, the predictability time in a system with many characteristic times is not related to the maximum Lyapunov exponent. In a phenomenological theory of fully developed turbulence, a maximum Lyapunov exponent growing with a power of the Reynolds number can coexist with a predictability time given by the lifetime of the largest structures at the integral length scales, therefore independent of the Reynolds number as predicted by the Lorenz theory and confirmed by numerical computations on the model (6.14) and the GOY model.

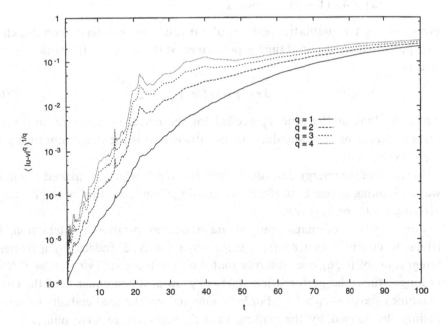

Figure 6.7 Moment $\langle |\delta u|^q \rangle^{1/q}$ versus t for $\theta = 0.001$ [Aurell et al. 1996].

6.4   **Predictability in extended systems**

Let us study the predictability in extended systems with particular emphasis on the relevance of the local interactions, considering a system of coupled maps on a one-dimensional lattice. The equations of the model are

$$x_i(t+1) = (1 - \epsilon_0)f(x_i(t)) + \frac{1}{2}\sum_{j=1}^{N/2-2} \epsilon_j \{f(x_{i+j}(t)) + f(x_{i-j}(t))\}, \quad (6.36)$$

where f is a chaotic map of the interval $[0, 1]$ into itself, $\epsilon_0 = \sum_{j=1}^{N/2-2}\epsilon_j$, and periodic boundary conditions $x_i = x_{i\pm N}$ are assumed.

The local case, $\epsilon_j = 0$ for $j \geq 2$, is equivalent to CML discussed in chapter 4. In addition we consider the non-local couplings

$$\epsilon_1 = C_1 \quad \text{and} \quad \epsilon_j = \frac{C_2}{j^\alpha} \quad \text{for } j \geq 2, \quad\quad (6.37)$$

where the power α measures the strength of non-locality.

There is no direct relation between the evolution laws (6.36) and those of a fluid. However, the system of coupled maps can be regarded as a discrete (time–space) version of an integro-differential equation. Let us recall that the Navier–Stokes equations for an incompressible fluid can be written as an integro-differential equation with non-local interactions among the fluid points.

Here we discuss the predictability in equation (6.36) with the logistic map

$$f(x) = 4x(1-x) \quad\quad \text{mod } 1.$$

Nevertheless, the qualitative features of our results do not depend on this choice. The perturbation at initial time is performed at the centre of the lattice $i = N/2$, that is

$$|\delta x_{N/2}(0)| = \delta_0 \quad ; \quad \delta x_i(0) = 0 \text{ for } i \neq N/2. \quad\quad (6.38)$$

Then, we look at the time T_p needed for the perturbation to reach a certain threshold $\delta_{\max}$ on the boundary of the lattice, that is the maximum time t such that $|\delta x_1(t)| \leq \delta_{\max}$.

In terms of the energy cascade in turbulent fluids which motivated our model, we are looking at the 'butterfly effect' starting from the centre of the lattice and arriving up to the first site.

For a system of maps coupled via a nearest neighbour interaction, it is trivial to conclude that $\delta x_1(t) = 0$ for times $t < N/2$. Indeed, by a numerical integration of (6.36) one observes that $\delta x_1(t) = 0$ for times $t < t^* = CN$ and at longer times the perturbation starts to grow as a consequence of the chaotic dynamics, $\delta x_1(t) \sim \delta_0 e^{\lambda(t-t^*)}$. For local interactions, the predictability is therefore mainly determined by the waiting time t^*, necessary to have $\min_i |\delta x_i| > \delta_0$, which is roughly proportional to the system size N, as shown in Fig. 6.8. The mean predictability time $\langle T \rangle$ is shown, averaged over a large number of initial conditions. One has the linear law

$$\langle T \rangle = t_1 + C N, \tag{6.39}$$

where the time $t_1 \sim \lambda^{-1}$ is due to the exponential error growth after the waiting time and can be neglected when the lattice is large enough. This agrees with the existence of a finite speed spreading as given by (4.19).

The scenario is very different in the case of non-local interactions, since the perturbation on the centre of the lattice may propagate toward the boundaries without any time delay, due to the system size. The numerical integration of (6.36) shows that even for weak non-locality (e.g. $C_1 << C_2$ and rather large α-values), the waiting time t^* does not increase, or increases very slowly, with the system size N and

$$\langle T \rangle \sim t_1 \sim \lambda^{-1}.$$

As shown in Fig. 6.8, the same qualitative behaviour is given by weakly non-local couplings, and mean field interactions ($\epsilon_j = C_2/N$).

In conclusion, the predictability time in spatially extended systems is given by two contributions: the waiting time t^* and the characteristic time $t_1 \sim \lambda^{-1}$ associated with chaos. For non-local interactions, the waiting time is constant with respect to the size of the system N while for local interactions it is

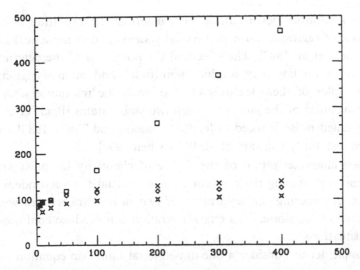

Figure 6.8 Averaged predictability time $\langle T \rangle$ versus N for: local coupling $C_1 = 0.3$ (squares); non-local coupling $C_1 = 0.3$, $C_2 = 0.01$ with $\alpha = 2$ (crosses) and $\alpha = 3$ (diamonds); mean field coupling $\epsilon_i = C_2/N$ with $C_2 = 0.3$ (crossed squares). The initial perturbation is performed at the centre of the lattice (site $i = N/2$) and has an amplitude 10^{-14}; the maximum admitted error is $\delta_{\max} = 0.1$ [Paladin and Vulpiani 1994].

proportional to N. Let us stress that for these results the nonlinear terms in the evolution of a small perturbation $\delta x(t)$ are quite important. One numerically observes that the waiting time t^* is not just the relaxation time T_R of δx on the tangent eigenvector. In fact, we find that T_R is much larger than t^*.

Basically the mechanism of the perturbation growth is the following. At the beginning the instability on the centre of the lattice δx_j with $j \approx N/2$ increases exponentially with a rate given by the maximum Lyapunov exponent. It thus attains the saturation level in a time $\sim \lambda^{-1}$. Only after the saturation at the lattice centre does the perturbation start to spread (either in a slow or in fast way, according the nature of the coupling). In a time t^* it reaches the boundary of the system.

If one assumes that the perturbation is infinitesimal and considers the linearized evolution of δx, i.e of the tangent vector, one gets a relaxation toward the typical stability eigenvector (corresponding to the maximum Lyapunov exponent) whose components are of the order of $N^{-1/2} \exp(\lambda t)$ for times $t = O(T_R)$ with $T_R \gg t^*$. The spatial spreading is thus accelerated by the nonlinear terms of the evolution law of δx and as a consequence the predictability time decreases substantially.

6.5   Predictability in noisy systems

The effect of a random perturbation on a deterministic evolution law is an important issue of relevance both in physical systems and in numerical simulations [Crutchfield et al. 1982]. The effects of the noise and of the deterministic part of the evolution law may produce non-trivial, and often intriguing, behaviours. Examples of these features are the stochastic resonance where one has a synchronization of the jumps between two stable states [Benzi et al. 1981, 1982], the so-called noise-induced order [Matsumodo and Tsuda 1983] and the noise-induced instability [Bulsara et al. 1990, Chen 1990].

The simplest characterization of the degree of chaoticity for noisy systems can be obtained by treating the random term as a usual time-dependent term, and therefore, considering the separation of two nearby trajectories with the same realization of the noise. This characterization can produce confusion and paradoxical situations.

As an example, let us consider a one-dimensional Langevin equation

$$\frac{dx}{dt} = -\frac{\partial V(x)}{\partial x} + \sqrt{2\sigma}\, \eta, \tag{6.40}$$

where $V(x)$ diverges for $|x| \to \infty$ and has more than one minimum, e.g. the usual double well $V = -x^2/2 + x^4/4$, and η is a white noise. We can define the Lyapunov exponent λ_σ associated with the separation rate of two nearby trajectories with the same realization of the stochastic term η, i.e.

$$\lambda_\sigma = \lim_{t\to\infty} \frac{1}{t} \ln|z(t)| \qquad (6.41)$$

where the evolution of the tangent vector is:

$$\frac{dz}{dt} = -\frac{\partial^2 V(x(t))}{\partial x^2} z(t).$$

Since the system is ergodic and the invariant probability distribution is $P(x) = Ce^{-V(x)/\sigma}$ one has:

$$\lambda_\sigma = \lim_{t\to\infty} \frac{1}{t} \ln|z(t)| = -\lim_{t\to\infty} \frac{1}{t} \int_0^t \partial_{xx}^2 V(x(t')) dt'$$

$$= -C \int \partial_{xx}^2 V(x) e^{-V(x)/\sigma} \, dx = -\frac{C}{\sigma} \int (\partial_x V(x))^2 e^{-V(x)/\sigma} \, dx < 0.$$

This result is rather intuitive: the trajectory $x(t)$ spends most of the time in one of the 'valleys' where $-\partial_{xx}^2 V(x) < 0$ and only short intervals on the 'hills' where $-\partial_{xx}^2 V(x) > 0$ so that the average of the logarithm of the distance between two trajectories computed with the same realization of the noise decreases.

The fact that $\lambda_\sigma < 0$ is relevant for the predictability problem only if the realization of the noise is known. In the more physical case of two initially close trajectories with two different noise realizations, the two trajectories are very distant after a certain time T_σ, i.e. they are in two different 'valleys'. For $\sigma \to 0$, by the Kramers formula [Chandrasekhar 1943], one has $\langle T_\sigma \rangle \sim \exp \Delta V/\sigma$ where ΔV is the difference between the values of V on the top of the hill and at the bottom of the valley.

This example shows the limitation of the Lyapunov exponent which is computed by treating the noise term as a usual time-dependent term for the characterization of the 'complexity' of noisy systems. In addition, it is easy to realize that it is very difficult to extract λ_σ from experimental data. In order to introduce a more natural indicator of complexity it is convenient to follow a quite different approach, where two realizations of the noise, instead of only one as in (6.41), are used [Paladin et al. 1995]. This is exactly what happens when experimental data are analysed by the Wolf et al. [1985] algorithm.

For the sake of simplicity let us discuss a 1D map (the generalization to N-dimensional maps or ordinary differential equations is obvious):

$$x(t+1) = f[x(t), t] + \sigma w(t) \qquad (6.42)$$

where t is an integer and $w(t)$ is an uncorrelated random process, e.g. w are independent random variables uniformly distributed in $[-1, 1]$. The Lyapunov exponent λ_σ associated with the separation rate of two nearby trajectories with the same realization of the stochastic term $w(t)$ is given by (6.41) where now

$$z(t+1) = f'[x(t), t] z(t)$$

and $f' = df/dx$. Of course, for $\sigma = 0$, one gets the Lyapunov exponent λ_0 of the unperturbed dynamical system. Some authors argue for the existence of the so-called noise-induced order [Matsumoto and Tsuda 1983] if λ_σ passes from positive to negative values when the strength of the fluctuation σ increases.

Let us define a new indicator of complexity in the framework of a deterministic system (i.e. equation (6.42) with $\sigma = 0$) where it coincides with the unperturbed maximum Lyapunov exponent λ_0. Let $x(t)$ be the trajectory starting at $x(0)$ and $x'(t)$ be the trajectory starting at $x'(0) = x(0) + \delta x(0)$ with $\delta_0 = |\delta x(0)|$ and indicate by τ_1 the maximum time such that $|x'(t) - x(t)| < \Delta$. Then, we put $x'(\tau_1 + 1) = x(\tau_1 + 1) + \delta x(0)$ and define τ_2 as the maximum τ such that $|x'(\tau_1 + \tau) - x(\tau_1 + \tau)| < \Delta$ and so on.

We can define the effective Lyapunov exponent as

$$\gamma_i = \frac{1}{\tau_i} \ln \frac{\Delta}{\delta_0}. \tag{6.43}$$

However, we sample the expansion rate in a non-uniform way, at time intervals τ_i. As a consequence the probability of picking γ_i is $p_i = \tau_i / \sum_j \tau_j$ so that

$$\lambda_0 = \langle \gamma_i \rangle = \frac{\sum_i \tau_i \gamma_i}{\sum_i \tau_i} = \frac{1}{\bar{\tau}} \ln \left(\frac{\Delta}{\delta_0} \right) \tag{6.44}$$

where

$$\bar{\tau} = \lim_{N \to \infty} \frac{1}{N} \sum_{i=1}^{N} \tau_i. \tag{6.45}$$

This definition without any modification can be extended to noisy systems by introducing the rate

$$K_\sigma = \frac{1}{\bar{\tau}} \ln \left(\frac{\Delta}{\delta_0} \right) \tag{6.46}$$

which coincides with λ_0 for a deterministic system ($\sigma = 0$). When $\sigma = 0$, there is no reason to determine the Lyapunov exponent in this apparently odd way, of course. However, the introduction of K_σ is rather natural in the framework of information theory [Ford 1983]. Considering again the noiseless situation, if one wants to transmit the sequence $x(t)$ ($t = 1, 2, \ldots, T_{max}$) accepting only errors smaller than a tolerance threshold Δ, one can use the following strategy:

(1) Transmit the rules which specify the dynamical system (6.42), using a finite number of bits which does not depend on the length T_{max}.

(2) Specify the initial condition with precision δ_0 using a number of bits $n = \log_2(\Delta/\delta_0)$ which is necessary to reach the time τ_1 where the error equals Δ. Then specify again the new initial condition $x(\tau_1 + 1)$ with a precision δ_0 and so on. The number of bits necessary to specify the sequence with a tolerance Δ up to $T_{max} = \sum_{i=1}^{N} \tau_i$ is $\simeq Nn$ and the mean information for time step is $\simeq Nn/T_{max} = K_\sigma / \ln 2$ bits.

In the presence of noise, the strategy of the transmission is unchanged but since it is not possible to transmit the realization of the noise $w(t)$, one has to estimate the growth of the error $\delta x(t) = x'(t) - x(t)$, where $x(t)$ and $x'(t)$ evolve in two different noise realizations $w(t)$ and $w'(t)$, and $|\delta x(0)| = \delta_0$.

The resulting equation for the evolution of $\delta x(t)$ is:

$$\delta x(t+1) \simeq f'[x(t), t]\, \delta x(t) + \sigma \tilde{w}(t) \qquad \tilde{w}(t) = w'(t) - w(t). \tag{6.47}$$

For the sake of simplicity we discuss the case $|f'[x(i), i]| = \text{const} = \exp \lambda_0$, where (6.47) gives the bound on the error:

$$|\delta x(t)| < e^{\lambda_0 t}(\delta_0 + \tilde{\sigma}) \qquad \text{with} \qquad \tilde{\sigma} = \frac{2\sigma}{e^{\lambda_0} - 1}. \tag{6.48}$$

This formula shows that δ_0 and $\tau = \bar{\tau}$ are not independent variables but they are linked by the relation

$$e^{\lambda_0 \tau}(\delta_0 + \tilde{\sigma}) \simeq \Delta. \tag{6.49}$$

As a consequence, we have only one free parameter, say τ, to optimize the information entropy K_σ in (6.46), so that the complexity of the noisy system can be estimated by

$$G_\sigma = \min_\tau K_\sigma = \lambda_0 + O(\sigma/\Delta) \tag{6.50}$$

where the minimum is reached at an optimal time $\tau = \tau_{\text{opt}}$ from the transmitter point of view.

In the case of a deterministic system K_σ does not depend on the value of τ (i.e. it is equivalent to using a long $\bar{\tau}$ and to transmitting many bits few times or to using a short $\bar{\tau}$ and to transmitting few bits many times). On the contrary, in noisy systems there exists an optimal time τ_{opt} which minimizes K_σ: using relation (6.48) one sees that $\Delta = \exp(\lambda_0\bar{\tau})(\delta_0 + \tilde{\sigma})$ and K_σ have a minimum for $\tau_{\text{opt}} \simeq 1/\lambda_\sigma$. This result might appear trivial but has a relevant consequence from a theoretical point of view in the presence of noise, even if the value of the entropy G_σ changes only $O(\sigma/\Delta)$, there exists an optimal time for the transmission.

In the case of weak intermittency (i.e. slight fluctuations of the effective Lyapunov exponent γ around λ_0) there are no substantial modifications of the previous discussion. The interesting situation occurs for strong intermittency when there is an alternation of positive and negative γ during long time intervals [Paladin et al. 1995]. In this case the existence of an optimal time for the transmission induces a dramatic change for the value of G_σ. This is particularly clear when considering the limiting case of positive γ_1 in an interval $T_1 \gg 1/\gamma_1$ followed by a negative γ_2 in an interval $T_2 \gg 1/|\gamma_2|$, and again a positive effective Lyapunov exponent. In the expanding intervals, one can transmit the sequence using $\simeq T_1/(\gamma_1 \ln 2)$ bits, while during the contracting interval one can use only few bits. Since, in the expanding intervals, the transmission has to be repeated rather often and moreover $|\delta x|$ cannot be lower than the noise amplitude σ, in contrast to the noiseless case, it is impossible to use the contracting intervals to compensate the expanding ones. This implies that in the limit of very large T_i only the expanding intervals contribute to the evolution of the error $\delta x(t)$ and the information entropy is given by an average of the positive effective Lyapunov exponents:

$$G_\sigma \simeq \langle \gamma\, \theta(\gamma) \rangle. \tag{6.51}$$

For the approximation considered above, $G_\sigma \geq \lambda_\sigma = \langle \gamma \rangle$. Note that by definition $G_\sigma \geq 0$ while λ_σ can be negative. The estimate (6.51) stems from the fact that δ_0 cannot be smaller than σ so the typical value of τ_i is $O(1/\gamma_i)$ if γ_i is positive. We stress again that (6.51) holds only for strong intermittency, while for uniformly expanding systems or rapid alternations of contracting and expanding behaviours $G_\sigma \simeq \lambda_\sigma$.

It is not difficult to estimate the range of validity of the two limiting cases $G_\sigma \simeq \lambda_\sigma$ and $G_\sigma \simeq \langle \gamma\, \theta(\gamma) \rangle$. Denoting by $\gamma_+ > 0$ and $\gamma_- < 0$ the typical values of the effective Lyapunov exponent in the expanding and contracting time intervals of length T_+ and T_- respectively, (6.51) holds if during the expanding intervals there are at least two repetitions of the transmission and the duration of the contracting interval is long enough to allow the noise to dominate over the contracting deterministic effects. In practice, one should require

$$\exp(\gamma_+ T_+) \gg \frac{\Delta}{\sigma} \qquad \exp(|\gamma_-| T_-) \gg \frac{\Delta}{\sigma}. \tag{6.52}$$

In a similar way, $K \simeq \lambda_0$ holds if:

$$\exp(|\gamma_-| T_-) \ll \frac{\Delta}{\sigma}. \tag{6.53}$$

The above results have been checked by some simple numerical computations [Paladin et al. 1995]. The random perturbation $w(t)$ is an independent variable uniformly distributed in the interval $[-1/2, 1/2]$.

The first system considered is given by periodic alternation of two piecewise linear maps of the interval $[0, 1]$ onto itself:

$$f[x, t] = \begin{cases} a x \quad \mathrm{mod}\ 1 & \text{if } (2n - 1)T \leq t < 2nT; \\ b x & \text{if } 2nT \leq t < (2n + 1)T \end{cases} \tag{6.54}$$

where $a > 1$ and $b < 1$. Note that in the limit of small T, $G_\sigma \to \max[\lambda_\sigma, 0]$ since it is a non-negative quantity as shown in Fig. 6.9 where for $b = 1/4$, λ_σ is negative.

The second system is strongly intermittent without an external forcing. It is the Beluzov–Zhabotinsky map [Matsumoto and Tsuda 1983, Herzel and Pompe 1987] related to the famous chemical reaction, see chapter 5:

$$f(x) = \begin{cases} [(1/8 - x)^{1/3} + a]e^{-x} + b & \text{if } 0 \leq x < 1/8; \\ [(x - 1/8)^{1/3} + a]e^{-x} + b & \text{if } 1/8 \leq x < 3/10; \\ c(10\,x\,e^{-10x/3})^{19} + b & \text{if } 3/10 \leq x \end{cases} \quad (6.55)$$

with $a = 0.506\,073\,57, b = 0.023\,288\,527\,9, c = 0.121\,205\,692$. The map exhibits a chaotic alternation of expanding and very contracting time intervals. Although the value of T_- is very small because $|\gamma_-| >> 1$, the first inequality (6.52) is not satisfied because the expanding time intervals are rather short. As a consequence the asymptotic estimate $G_\sigma \simeq \langle \gamma\,\theta(\gamma) \rangle$ cannot be observed. In Fig. 6.10, one sees that while λ_σ passes from negative to positive values at decreasing σ, G_σ is not sensitive to this transition to 'chaos'. Another important remark is that in the usual treatment of the experimental data, if some noise is present, one practically computes G_σ and the result can be completely different from λ_σ. Let us mention for example Chen [1990] where the author studies a one-dimensional nonlinear time-dependent Langevin equation. A simple numerical computation shows that λ_σ is negative while the author claims to find, using the Wolf method, a positive 'Lyapunov exponent'.

The above results show that the same system can be regarded either as regular (i.e. $\lambda_\sigma < 0$) when the same noise realization is considered for two nearby

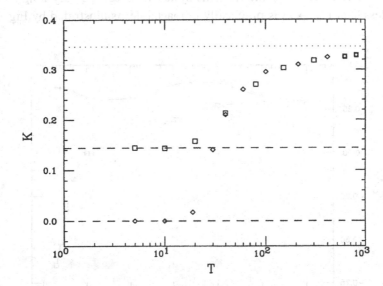

Figure 6.9 K_σ versus T with $\sigma = 10^{-7}$ for the map (6.54). The parameters of the map are $a = 2$ and $b = 2/3$ (squares) or $b = 1/4$ diamonds. The dotted line indicates the Pesin-like relation (6.51) while the dashed lines are the noiseless limit of K_σ. Note that for $b = 1/4$ the Lyapunov exponent λ_σ is negative [Paladin et al. 1995].

trajectories or as chaotic (i.e. $G_\sigma > 0$) when two different noise realizations are considered, see e.g. Fig. 6.9.

The situation is similar to what is observed in fluids with Lagrangian chaos, see chapter 8. There, a pair of particles passively advected by a chaotic velocity field might remain close following together a 'complex' trajectory. The lagrangian Lyapunov exponent is thus zero. However a data analysis gives a positive Lyapunov exponent because of the 'eulerian' chaos. We can say that λ_σ and G_σ correspond to the lagrangian Lyapunov exponent and to the exponential rate of separation of a particle pair in two slightly different velocity fields, respectively.

The indicator of complexity G_σ for a chaotic system perturbed by noise is very similar to the Kolmogorov–Sinai entropy h for deterministic systems, since it is given by a sort of minimization of Shannon entropies over time intervals of transmission instead of space partitions. The relation $G_\sigma \simeq \langle \gamma \, \theta(\gamma) \rangle$ is, in some sense, the analogue in the time domain of the Pesin relation $h = \sum_i \lambda_i \, \theta(\lambda_i)$ between h and the Lyapunov spectrum [Eckmann and Ruelle 1985] where the negative Lyapunov exponents do not decrease the value of h since the contraction along the corresponding directions cannot be observed for any finite space partition. In the same way the contracting time intervals, if long enough, do not decrease G_σ.

It is important to note that the limit $\sigma \to 0$ is very delicate. Indeed for small σ, say $\sigma < \sigma_c$, the inequality (6.53) will be fulfilled and $G_\sigma \simeq \lambda_\sigma \to \lambda_0$ for $\sigma \to 0$. However in strongly intermittent systems T_- can be very long so that the noiseless limit $G_\sigma \to \lambda_0$ is practically unreachable, as illustrated by Fig. 6.10.

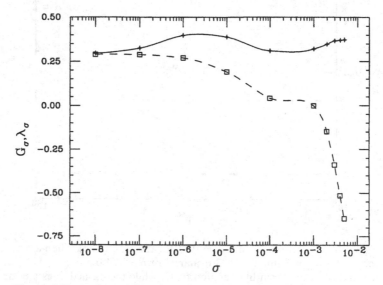

Figure 6.10 λ_σ (squares) and K_σ (crosses) versus σ for the map (6.55) [Paladin et al. 1995].

From the previous considerations one sees that the predictability problem in systems containing some randomness has to be treated in a careful way. In particular the 'usual' Lyapunov exponent may have no relevance. Without entering into details we just mention the sand-piles models which are considered as a paradigm for the self-organized criticality [Chen et al. 1990]. It is possible to show that the Lyapunov exponent computed with the same realization of randomness, i.e. considering two trajectories initially very close and adding the sand in the same places for the two systems, is negative $O(1/L^2)$ where L is the linear size of the system [Caglioti and Loreto 1996].

Of course this result does not imply that the system is 'predictable'. From a physical point of view, the growth of the distance $d(t)$ between two initially close trajectories computed by adding sand in different sites is much more interesting. Let us consider a perturbed realization obtained by 'minimal errors' with respect to a reference realization. In the reference realization, to initialize an avalanche, one adds sand on a site i chosen at random. In the perturbed realization one adds a sand grain at one of the nearest neighbour sites to i instead. Even in this case the distance, $d(t)$, between the two realizations increases up to a maximal distance after a few avalanches [Loreto et al. 1996]. Practically, the same situation has been observed for the Langevin equation (6.40) discussed at the beginning of this section.

6.6 **Final remarks**

Let us close this chapter with some remarks.

(I) The predictability problem can be formulated in terms of the Lyapunov exponent, or effective Lyapunov exponent, only for infinitesimal perturbations.

(II) In systems with many degrees of freedom and many characteristic times one has to take some subtle points into account, particularly if the interest is focused on the physical aspects of the problem such as small but non-infinitesimal perturbations. For these cases the Lyapunov exponent, and also the effective Lyapunov exponent, is not able to characterize the predictability in a complete way and one has to consider at least two other aspects:

  (a) A generic tangent vector relaxes only after a 'waiting time' T_R on the time-dependent first eigenvector e_1 of the stability matrix. Sometimes T_R can be much longer than the time given by the inverse Lyapunov exponent.

  (b) For perturbations which cannot be considered infinitesimal, nonlinear effects are very important and the time growth of the perturbation is not exponential but closer to a power law, actually a sequence of exponentials with different rates.

(III) Even at very high Reynolds number, there exist well-defined coherent structures, such as vortex tubes, which move roughly maintaining their shape. In this case, one should reformulate the predictability problem, if one is interested only in qualitative behaviours. For instance, a reasonable question is the prediction of the centre and the orientation of the vortex tubes. In this case one could hope to have a long predictability time. Of course, this problem cannot be studied within the shell model, where all the spatial structures are lost. For example in 2D turbulence the L_2 distance $d(t)$ between two velocity fields:

$$d(t)^2 = \int |\mathbf{u}(\mathbf{x}, t) - \mathbf{u}'(\mathbf{x}, t)|^2 d\mathbf{x}$$

is not a good indicator of the decorrelation at large scales between $\mathbf{u}(\mathbf{x}, t)$ and $\mathbf{u}'(\mathbf{x}, t)$. It is possible to have large values of $d(t)$, e.g. $d(t) \simeq 0.3\sqrt{\int |\mathbf{u}(\mathbf{x}, t)|^2 d\mathbf{x}}$, for two fields which appear rather similar, as it happens when one considers the same vortices slightly deformed and shifted [Boffetta et al. 1997]. It is thus necessary to introduce an appropriate norm which is able to define the physically relevant notion of distance taking into account the existence of large scale structures.

(IV) In systems containing some randomness, such as Langevin equations or sand-piles models, there are two different ways to define the predictability: considering two trajectories of the system with the same noise or two trajectories of the system with two different realizations of the noise. Both these definitions can be appropriate according to the particular context although they generally give different results. The usual definition of complexity via the Lyapunov exponent can lead to paradoxical conclusions so that the noisy systems must be analysed in a careful way.

Chapter 7

Dynamics of interfaces

7.1 Turbulence and interfaces

In this chapter we describe the static and dynamical properties of rough interfaces and surfaces. One may naturally ask: what has the dynamics of interfaces to do with turbulence in dynamical systems? In fact there are quite a lot of similarities shared by the two topics. The dynamics of some of the interfaces investigated in this chapter can be chaotic; a good example is the front motion in the Kuramoto–Sivashinsky equation introduced in (5.32); other examples are various forms of coupled map lattices for interface growth. Furthermore, a large class of interfaces exhibit scale invariance, meaning that the typical fluctuations of the interface scale as a power law with the system size; such interfaces are called rough. Various other forms of invariances are also relevant for interface growth: an example is Galilean invariance which is a property shared with the Navier–Stokes equations. This is related to the fact that the standard non-linearity in many interface equations, of the form of a gradient squared, is of the same type as in hydrodynamic equations: by a simple transformation as used in (5.115) one can bring this non-linearity onto a convective form like in the Navier–Stokes equations. There are therefore analogous features between the dynamics of rough interfaces and turbulence. Fronts and interfaces are mostly (although not only) studied in one dimension. The prototype model for rough fronts, the Kardar–Parisi–Zhang [1986] equation, is equivalent to the Burgers equation with additive noise, and the Burgers equation is considered as the one-dimensional analogue of the Navier–Stokes equations. For a recent review on interface motion, see [Halpin-Healey and Zhang 1995].

7.2   The Burgers equation

An interesting simplification of the Navier–Stokes equations is obtained neglecting the pressure term, thereby relaxing the incompressibility condition. One then obtains the following partial differential equation

$$\partial_t \mathbf{v} + (\mathbf{v} \cdot \nabla)\mathbf{v} = \nu \Delta \mathbf{v} + \mathbf{f}(\mathbf{x}, t). \tag{7.1}$$

In fact, the corresponding one-dimensional equation,

$$\partial_t v + v \partial_x v = \nu \partial_{xx}^2 v, \tag{7.2}$$

was introduced by Burgers [1940] as a model of turbulence; for a general discussion on the Burgers equation see Platzman [1964], Gurbatov et al. [1991]. There are many differences between Burgers turbulence, or 'Burgulence', as it is often called, and hydrodynamical turbulence [Kida 1979, Fournier and Frisch 1983]. In our context the most important one is that (7.2) is integrable and so does not display any chaos, even when a forcing is added to the right-hand side. In addition, the Navier–Stokes equations, because of incompressibility, have a non-local behaviour in real space while the Burgers equation is local.

In order to solve (7.2), we first introduce a potential function h, defined by $u = -\partial_x h$. Now the ingenious Hopf–Cole transformation [Hopf 1950, Cole 1951] is applied

$$h = 2\nu \ln \theta.$$

Noting that $v \partial_x v = \partial_x(v^2/2)$, we obtain the linear heat equation for θ:

$$\partial_t \theta = \nu \partial_{xx}^2 \theta. \tag{7.3}$$

In the absence of boundaries, the solution of (7.3), for $t > 0$, is

$$\theta(x, t) = \frac{1}{\sqrt{4\pi\nu t}} \int_{-\infty}^{\infty} e^{-\frac{(x-a)^2}{4\nu t}} \theta_0(a)da, \tag{7.4}$$

where

$$\theta_0(a) = \exp\left[\frac{1}{2\nu}h_0(a)\right], \tag{7.5}$$

and $h_0(a)$ is the initial potential function such that $u_0(a) = -\partial_x h_0(x = a)$. One can apply the saddle point estimate of the integral in (7.4) when $\nu \to 0$ (limit of infinite 'Reynolds' number), so that one obtains

$$h(x, t) = \max_a [\psi_0(a) - \frac{(x-a)^2}{2t}]. \tag{7.6}$$

The solutions of the Burgers equation in the inviscid limit for smooth initial data and smooth forcing, develops structures with a vague flavour of the intermittent bursts of 3D turbulence. After some time, there appear isolated discontinuities (shocks), separated by smooth regions. As a consequence low order moments of the velocity difference are dominated by smooth behaviour while high order moments are dominated by shocks. In other words, one has a bifractal model: in the limit $v \to 0$, the velocity field has singularities of exponent $h = 0$ on a set of dimension $D(h = 0) = 0$ (for the isolated points of the shocks), and singularities of exponent $h = 1$ on a set of dimension $D(h = 1) = 1$, corresponding to the regular regions. Using the multifractal formalism of section 2.3, adapted to the one-dimensional case, the exponents of the structure functions are given by the relation

$$\zeta_p = \min_h (h\,p - D(h) + 1). \qquad (7.7)$$

This allows one to determine the scaling in the inertial range as

$$\zeta_p = p \quad \text{for } p \leq 1,$$
$$\zeta_p = 1 \quad \text{for } p \geq 1. \qquad (7.8)$$

It is worth stressing that the Burgers equation arises naturally in a number of different problems [Halpin-Healey and Zhang 1995], such as the formation of large-scale structures in the Universe [Vergassola et al. 1994], phase diffusion equations [Kuramoto 1984] (see (5.31)), and the physics of disordered systems. However, the Burgers equation, being equivalent to the linear heat equation, exhibits a non-trivial behaviour similar to those of chaotic dynamical systems only if one considers either random initial conditions or random forcing [Bertini et al. 1994]. Both these cases have been considered. In the former case, Sinai [1992] and She et al. [1992] have investigated the solutions of the unforced Burgers equation with random non-smooth self-similar initial velocities. Such initial conditions arise naturally in the formation of the large scale structures in the Universe [Gurbatov et al. 1989] and in experiments of neural signal transmission. In this context, the main problem is to determine the scaling which is dynamically generated through the formation of shocks and its relation with the imposed scaling on the initial velocity profile. Using the Legendre transform (7.6) one can show that the location x at time t of a fluid particle starting at a and moving with velocity $v(a, 0)$ is, before arriving at a shock:

$$x = a + t\,v(x = a, t = 0). \qquad (7.9)$$

One can thus analyse the solution in terms of the function $a(x)$ or equivalently of $x(a)$, the value of x at time $t = 1$. For initial conditions given by fractional

Brownian profiles such that $|v(x+r,0) - v(x,0)| \sim r^H$, the Lagrangian function $a(x)$ is found to be a devil's staircase, following the scaling law

$$|a - a'| \sim |x - x'|^{1/H} \tag{7.10}$$

for small $a - a'$. Moreover the cumulative probability of Lagrangian shock intervals Δa scales as $(\Delta a)^{-H}$ for small Δa. The remaining (regular) Lagrangian locations form a Cantor set of dimension H. In Eulerian coordinates, the shock locations are everywhere dense. Similar results hold in the multidimensional case [Vergassola et al. 1994]. One thus has strong numerical evidence, supported by the rigorous results of Sinai [1991] for $H = 1/2$ in one dimension, that in the Burgers equation, non-trivial forms of scaling invariance appear in the shock structure which is determined by the scaling properties of the initial data.

A famous variant of the Burgers equation, in this context is known as the KPZ equation [Kardar et al. 1986, Schwarz and Edwards 1992]. It has become a paradigm in the field of growing interfaces which is the main issue in this chapter. The KPZ equation is obtained by integrating (7.1), using a velocity field which can be expressed in term of the potential of a function $h(x,t)$, say $\mathbf{v} = -\nabla h$, and a random potential forcing, $\mathbf{f} = -\nabla \eta(\mathbf{x}, t)$,

$$\frac{\partial h(\mathbf{x},t)}{\partial t} = v\nabla^2 h(\mathbf{x},t) + \frac{\lambda}{2}(\nabla h(\mathbf{x},t))^2 + \eta . \tag{7.11}$$

One is interested in the solution obtained after an appropriate average on the random forcing. The KPZ equation is, as we shall discuss at length in this chapter, used in particular to describe the growth of a surface under a random 'rain' of particles, the flux of which is given by η. In general one would therefore assume that η is a gaussian white noise which satisfies $\langle \eta(\mathbf{x},t)\eta(\mathbf{x}',t')\rangle = D\delta(\mathbf{x} - \mathbf{x}')\delta(t - t')$.

It is interesting that the KPZ equation opens a bridge between hydrodynamics and the theory of disordered systems. Indeed, via the Cole–Hopf transformation, $h = \frac{2v}{\lambda} \log Z$, (7.11) is transformed into the so-called directed polymer problem

$$\frac{\partial Z(\mathbf{x},t)}{\partial t} = v\nabla^2 Z(\mathbf{x},t) + V(\mathbf{x},t)Z(\mathbf{x},t) \tag{7.12}$$

where $V(\mathbf{x},t) = \lambda\eta(\mathbf{x},t)/(2v)$ plays the role of a random potential. With the constraint that the end point of the directed polymer is fixed at $(\mathbf{x},t)$, one obtains the following path integral solution for the partition function Z

$$Z(\mathbf{x},t) = \int \rho(\mathbf{x}_0)d\mathbf{x}_0 \int_{\mathbf{x}(t=0)=\mathbf{x}_0}^{\mathbf{x}(t)=\mathbf{x}} d[\mathbf{x}(\tau)] \exp(-\mathcal{H}) \tag{7.13}$$

where the probability distribution of the initial point of the polymer is related to the initial velocity by $\mathbf{v}(\mathbf{x}, t = 0) = -2v\nabla\rho(\mathbf{x}_0)$ and the Hamiltonian $\mathcal{H}$ is:

$$\mathcal{H} = \int_0^t d\tau \left(\frac{1}{4v}\left(\frac{d\mathbf{x}}{d\tau}\right)^2 + V(\mathbf{x}(\tau),\tau) \right) . \tag{7.14}$$

The difficulty of explicitly calculating the statistical properties of the velocity field stems from the fact that the various moments of the velocity field are related to averages of the logarithm of the partition function Z (the so-called quenched average). That is one of the central problems in the physics of disordered systems, solved only in very particular cases such as the celebrated Parisi [1979] solution of the infinite range Sherrington–Kirkpatrick model. The existence of a connection between fluid dynamic turbulence and spin glasses is fascinating, although it is still too early to decide whether it can pave the way to a deeper understanding of intermittency in fully developed turbulence. From (7.13), it is indeed possible to use the standard tool of spin glasses, the replica trick together with a variational method which becomes exact in the limit $d \to \infty$, to provide an analytic solution given by (7.8) for the exponents ζ_p of the d-dimensional forced Burgers equation with a finite correlation length of the forcing. In fact it is possible to compute the full probability distribution of the velocity difference between points separated by a distance r much smaller than the correlation length of the forcing [Bouchaud et al. 1995] which is the regime where the intermittent exponents (7.8) are found.

7.3 The Langevin approach to dynamical interfaces: the KPZ equation

The growth of interfaces is studied in the context of a variety of physical systems, most notably in fluid displacements in porous media [Rubio et al. 1989, Kahanda et al. 1992, He et al. 1992, Kahanda and Wong 1993, Horvárth et al. 1991], growth of solids from the vapor, stochastic ballistic depositions [Family and Vicsek 1985, Meakin et al. 1986, Family 1986, Jullien and Botet 1985, Wolf and Kertész 1987, Kertész and Wolf 1988, Kim and Kosterlitz 1989], paper burning [Zhang et al. 1992] and even in the growth of bacterial colonies [Vicsek et al. 1990]. The experiments on the porous medium uses a cell constrained by two glass plates, which is called a Hele-Shaw cell [Hele-Shaw 1898, Bensimon et al. 1986]. The cell is filled with mono-size spheres and randomly packed; in some experiments the spheres are glued to the glass plates. The fluid is pushed from one side of the cell at constant pressure. For fluids of low capillary number, the interfaces can often have a fractal-like structure with overhangs, in particular if the pressure is high. For higher capillary numbers and 'intermediate' pressures, the interfaces can become self-affine with no overhangs, and this is the situation we restrict ourselves to when considering rough interfaces. Figure 7.1 shows the experimental results of Rubio et al. [1989] for water penetrating a porous medium. The interface is digitized at various time intervals.

In many experimental situations, the restriction to self-affine interfaces is not natural. In the porous flow experiment, for instance it is possible to perform controlled experiments in which overhangs do not appear (as shown in Fig.

7.1), but in most situations overhangs will be present. Robbins and co-workers [Cieplak and Robbins 1988, 1990] did pioneering work by developing a dynamical model for the invasion of a fluid in a porous medium. Using a 2D array of disks with random radii, they studied the structures of an invading fluid as a function of its wetting properties. Starting from a non-wetting fluid an interface with fractal properties appears in which the dynamics is similar to that of invasion percolation [Wilkinson and Willemsen 1983, Lenormand and

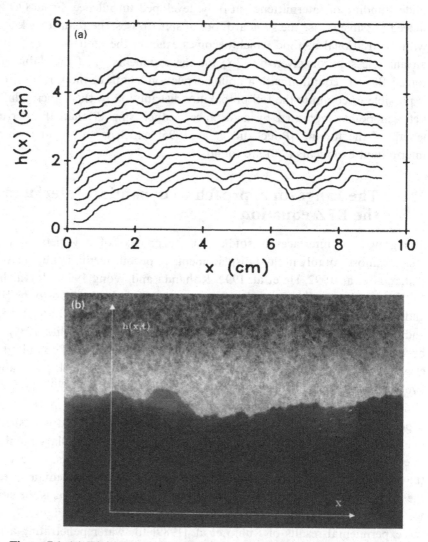

Figure 7.1 (a) Digitized interfaces from an experiment where water is penetrating a porous medium. The time interval between the interfaces is 30 s. The capillary number is $Ca = 4.93 \times 10^{-3}$ [Rubio et al. 1989]. (b) Photograph of slowly burning paper, clearly showing the advancing front [Zhang et al. 1992].

Zarcone 1985]. For an invasion by a wetting fluid, the geometrical properties of the interfaces change and in fact, the structures becomes closer and closer to a flat interface when the wetting of the medium becomes stronger. The invading clusters are then almost compact and the front of the cluster is nearly self-affine.

This chapter concentrates on the self-affine interfaces and most of the theoretical effort here has been devoted to the characterization of the interface roughening and to the understanding of the exponents appearing in the relation between the width of the interface and the linear size of the system or the time. There has been two main theoretical approaches to the subject: in approach A, one relies on a stochastic differential equation description of the interface growth, whereas in approach B, one employs discrete deposition models on a lattice substrate. Two typical representatives of these two approaches are the KPZ equation and the Kim–Kosterlitz (KK) model [Kim and Kosterlitz 1989]. The KK model randomly selects a site on a cubic d-dimensional lattice and permits growth in the height of the interface by one unit whenever the local gradient does not exceed a chosen threshold. It has been found by extensive numerical simulations that both KK and KPZ models predict the same exponents for the width of the interface, defined by $\tilde{h}(\mathbf{x}, t) = h(\mathbf{x}, t) - \langle h(\mathbf{x}, t) \rangle$ and

$$W^2(t) = \langle |\tilde{h}(\mathbf{x}, t)|^2 \rangle , \tag{7.15}$$

where $\langle ... \rangle$ stands for an average over $\mathbf{x}$. For short periods of time, the width was found to scale like

$$\frac{W}{W_0} \sim \left(\frac{t}{t_0} \right)^\beta \tag{7.16}$$

whereas for long periods of time, the width is saturated to an L-dependent limit which satisfies

$$\frac{W}{W_0} \sim \left(\frac{L}{L_0} \right)^\chi . \tag{7.17}$$

The static exponent χ also characterizes the structure function of the interface

$$S_1(\ell) = \langle |h(\mathbf{x} + \ell, t) - h(\mathbf{x}, t)| \rangle \sim \ell^\chi \tag{7.18}$$

for times long enough such that stationary growth has been achieved. Family and Vicsek [1985] showed that these two scalings can be combined via a scaling function $g(y)$ by

$$W(t, L) \sim L^\chi g(t/L^z) , \tag{7.19}$$

where we have introduced the dynamical exponent $z = \chi/\beta$. The values of χ and β found by simulations seem to agree approximately with the ansatz [Kim and Kosterlitz 1989]

$$\chi(d) = \frac{2}{(d+3)} , \qquad \beta(d) = \frac{1}{(d+2)}. \tag{7.20}$$

There is really no analytic theory to account for these values except in $d = 1$, where the exponent $\chi = 1/2$ is dictated by standard random walk theories, see Halpin-Healey and Zhang [1995]. When the KPZ equation is rescaled $x \rightarrow bx$, $h \rightarrow b^{\chi}h$, $t \rightarrow b^{z}t$, then, if the nonlinear term dominates over the linear diffusive term, one obtains the 'Galilean' relation between the exponents $\chi + z = 2$. A renormalization group analysis indicates that the long wavelength limit of (7.11) may be governed by a strong coupling fixed point which may or may not account for the exponents found numerically [Kardar et al. 1986].

These theoretical models seem however to overlook something fundamental in the physics of fluid displacement in porous media, or in general for experiments in quenched random systems. Experiments in this field discovered that the values of χ and β differ significantly from those quoted above [Rubio et al. 1989, Kahanda and Wong 1993, Horvárth et al. 1991]. One way to account theoretically for the differences in the exponents within approach A is to change the properties of the random noise term in (7.11). Indeed, such a task was undertaken by Zhang [1990] and Parisi [1992]. Zhang introduced a non-gaussian, power law noise distribution $p(\eta) \sim \eta^{-\alpha-1}$ which is uncorrelated in space. Using this distribution in numerical simulations, he was able to obtain a continuum of values of χ depending on the values of α. Parisi introduced a quenched gaussian noise which depends on h and x, but not explicitly on t. He observed that there is a substantial transient behaviour (both in space and time) where the exponents are different than proposed by KPZ, and some agreement with experimental results was found. This, however, leaves open the question of what is the physical reason for the different exponents. Jensen and Procaccia [1991] and Tang et al. [1991] modified the KK model by introducing pinning centres according to the distribution of voids found in numerical simulations of randomly packed spheres. Indeed, a numerical simulation of this model shows a slow cross-over phenomenon, where the exponent χ is very close to the experimentally determined value of ~ 0.72 and then, after some time, it again becomes like KPZ, with $\chi = 0.5$. This is a resonable explanation, but not quite satisfactory because one has to introduce the pinning structure from 'outside'.

In contrast to this, there has been an interesting development [Tang and Leschhorn 1992, Buldyrev et al. 1992, Sneppen 1992] of using the idea to relate interface growth through a quenched random environment to directed perco-lation [Schögl 1972, Grassberger and Sundermeyer 1978]. Employing this idea in various models the experimental findings are understood more satisfacto-rily. Sneppen [1992] defined a model where the dynamics by itself induces a pinning mechanism on the quenched randomness, and at the same time brings the motion of the interface to a critical state, i.e. the interface becomes self-organized-critical [Bak et al. 1987]. Before describing these models in detail, we shall present some general results for deterministic systems, which are not driven by additive noise but only by internal, chaotic motion.

7.4   Deterministic interface dynamics: the Kuramoto–Sivashinsky equation

The Kuramoto–Sivashinsky (KS) equation is of the form [Nepomnyashchy 1974, Kuramoto 1984, Sivashinsky 1977]

$$\frac{\partial h(\mathbf{x},t)}{\partial t} = -\nabla^2 h(\mathbf{x},t) - \nabla^4 h(\mathbf{x},t) + (\nabla h(\mathbf{x},t))^2 \,, \tag{7.21}$$

where $h(\mathbf{x},t)$ is the scalar field defining the interface. It was introduced in chapter 5 as the phase equation for the complex Ginzburg–Landau equation (see (5.32) and (5.99)) and studied further in sections 5.11 and 5.12. This equation applies in certain cases (Sivashinsky [1977]) to the propagation of flame fronts and the *negative surface tension* term is due to the fact that a small instability, in a burning liquid, will grow simply because there is more liquid to burn around it. With only the negative surface tension term present, the equation would be unstable and it is therefore necessary to balance this term by a fourth-order term. The linear terms cause a flat interface to be unstable, with perturbations of wave vector $\mathbf{k}$ growing as $\omega(\mathbf{k}) = \mathbf{k}^2 - \mathbf{k}^4$ (as also discussed in (5.101)). This leads to a locally cellular structure with a wavelength $\ell \simeq 2\pi\sqrt{2}$ given by the maximum $k_0 = 1/\sqrt{2}$ of $\omega(\mathbf{k})$. In fact, burning fronts are observed to exhibit such cellular structure. The nonlinear term, however, couples the modes (allowing for lateral growth) and this keeps the moving interface from diverging. The motion is chaotic; as shown by Pomeau et al. [1984], it has a finite density of positive Lyapunov exponents. The instabilities generated by the chaotic dynamics are transferred by the nonlinear term to smaller length scales, where they are dissipated by the fourth-order derivative. A snapshot of the interface is shown in Fig. 7.2.

The Kuramoto–Sivashinsky equation seems as far as statistical properties are concerned, to be describable by the KPZ equation (7.11) on long scales (at least in one dimension). This was conjectured by Yakhot [1981] and later Zaleski [1989] attempted to give a kind of 'computer assisted proof' of this correspondence. If the KS equation on a scale s, much larger than the cell structure is equivalent to the KPZ equation we have effectively, in one dimension,

$$\frac{\partial h}{\partial t} = v(s)\nabla^2 h + \frac{\lambda(s)}{2}(\nabla h)^2 + f_s(x,t) \,.$$

In fact, we must have $\frac{\lambda(s)}{2} = 1$. This follows from the Galilean invariance (5.98) of the equation, which we have already used several times. Since the value of $\lambda(s)$ enters into this transformation, a renormalization of it would imply that the result depends on the frame of reference in which it is carried out. Zaleski fitted $v(s)$ such that the last term $f(x,t)$ would behave like a noise term, without any correlations to itself or to h. This criterion determines an effective $v(s)$, which, surprisingly enough, is rather large and becomes basically independent of the scale s, when s is large enough. On the other hand, direct simulations of the Kuramoto–Sivashinsky equation, measuring, for example, the width of the interface or the correlations [Zaleski 1989, Fujisaka and Yamada 1977, Manneville 1981, Zaleski and Lallemand 1985, Hyman et al. 1986], never indicated any clear deviations from the behaviour of the linear noisy Edwards–Wilkinson (EW) equation and this was attributed to an unusually slow crossover between the linear and nonlinear theory.

To settle this issue, Sneppen et al. [1992] simulated the Kuramoto–Sivashinsky equation for very large systems. They studied mainly the transient behaviour, where the interface is growing up to its saturated width. In order to be able to

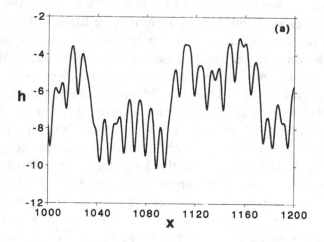

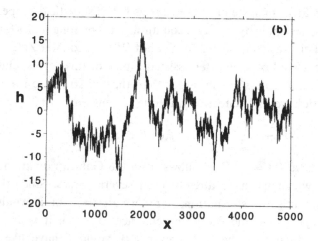

Figure 7.2 The interface obtained from integration of the Kuramoto–Sivashinsky equation. Note the cell structure (a) and the roughening of the interface at large length scales (b).

handle very large systems, they used a simple finite difference technique together with an explicit Euler time step. As we shall see below, the error introduced by this numerical scheme gives rise to a 'renormalization' of the coupling constant λ.

Numerically, it is observed that the system for a long time follows the linear KPZ equation, also called the Edwards–Wilkinson (EW) [1982] equation where the dynamical exponent is $\beta = \frac{1}{4}$. After a cross-over time, the dynamical behaviour turns away from linear scaling. Figure 7.3a shows the squared width

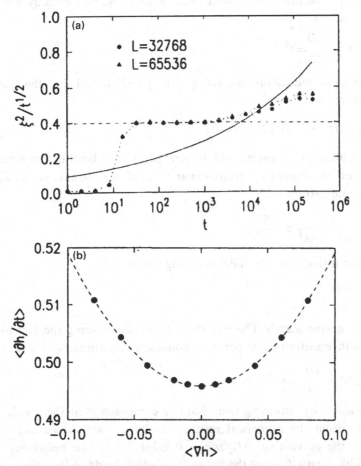

Figure 7.3 (a) Ensemble averaged time dependence of the interface width as a function of time for the Kuramoto–Sivashinsky equation. The dashed horizontal line corresponds to a scaling $W(t) \sim t^{1/4}$, as predicted by the Edwards–Wilkinson [1982] theory. At the cross-over time t_c, the scaling bends away from linear behaviour. The solid line corresponds to the asymptotic KPZ behaviour. (b) Tilt dependence of the velocity of a Kuramoto–Sivashinsky interface. From the parabolic fit, one can obtain the strength λ of the nonlinear term [Sneppen et al. 1992].

divided by $t^{1/2}$: a horizontal line corresponds to the behaviour of the linear theory. It is not completely clear from the numerical data whether the scaling indeed bends over to a KPZ behaviour, as indicated by the solid line. The χ exponent is equal to $1/2$, as in KPZ.

7.4.1 Cross-over to KPZ behaviour

Since the transient behaviour appears to cross-over to KPZ behaviour, one can investigate the scaling, by assuming that it does indeed behave as KPZ in the long time limit. The linear EW equation can be integrated directly giving the width as a function of time

$$W_{\lambda=0} = \frac{D}{\sqrt{2\pi v}} t^{1/2} . \tag{7.22}$$

From dimensional considerations, Krug et al. [1992] found that the width for the KPZ case must follow

$$W_{KPZ}(t)^2 = c_2((D/2v)^2 \lambda t)^{2/3} , \tag{7.23}$$

where c_2 is a universal constant which numerically has been determined to be 0.40. To obtain the cross-over from linear to nonlinear behaviour, (7.22) and (7.23) are equated so that

$$t_c = \frac{2^5 v^5}{\pi^3 c_2^6 D^2 \lambda^4} \simeq \frac{252}{vg^2} , \tag{7.24}$$

where we have introduced the KPZ coupling constant

$$g = \frac{\lambda^2 D}{v^3} , \tag{7.25}$$

with units of inverse length. The argument proceeds by using the fact that the saturated width, constrained to periodic boundary conditions, scales as

$$W_\infty^2(L) = \frac{DL}{24v} , \tag{7.26}$$

which is obtained by summing over Fourier components of the width in the limit $t \rightarrow \infty$. From the numerical results for short times shown in Fig. 7.3a, one reads on the curve that $D/\sqrt{2\pi v} = 0.397 \pm 0.001$. The behaviour of the saturated width versus L gives the value D/v, which leads to the effective KPZ parameters $D = 3.2 \pm 0.1$ and $v = 10.5 \pm 0.6$. The high value of the viscosity, which is in accordance with Zaleski [1989], is the core of the physics of the Kuramoto–Sivashinsky equation and is responsible for the the long intermediate scaling regime which has hampered previous numerical work, see Manneville [1981], Zaleski [1989]. One can give a very simple argument for this large value of the viscosity: taking the cells as the basic unit, $l \simeq 8$ and the time as the inverse of the fastest growing mode $\tau \simeq 1/\omega(k_0) \simeq 4$, an estimate of the effective viscosity is $v_{\text{eff}} \simeq l^2/\tau \simeq 16$ [Sneppen et al. 1992].

To complete the comparison with the KPZ equation, an estimate of the effective value of λ is needed – especially in view of the rather crude approximation to the Kuramoto–Sivashinsky equation used in the simulations. As noted earlier (section 5.9) Galilean invariance implies that the growth rate for a tilted interface, depends on the angle θ as $v(\theta) = v_0 + (\lambda/2)\theta^2$, where the coefficient $(\lambda/2)$ must be the same as the one multiplying $(\nabla h)^2$ in (7.21), i.e. 1. In the simulations Galilean invariance is broken, since the discretization of $(\nabla h)^2$ introduces higher order non-linearities. We can, however, compute an effective λ_{eff} by tilting the interface a small angle θ (through the boundary conditions) and measuring $v(\theta)$. Using the expansion $v(\theta) \simeq (\lambda_{\text{eff}}/2)\theta^2 + \ldots$ for the ensemble averaged velocity $v = \langle \partial h/\partial t \rangle$, one obtains an equation for λ_{eff}. Figure 7.3b shows the numerical results for various tilted Kuramoto–Sivashinsky interfaces. From the curvature of the graph, the value $\lambda_{\text{eff}} = 4.65 \pm 0.15$ is found.

If these estimates of the numerical constants are inserted into (7.24), the cross-over time becomes $t_c \simeq 7000$ and the corresponding cross-over size $L_c \simeq 2500$, meaning that L has to exceed L_c in order to observe the cross-over to nonlinear behaviour. Note that the estimated cross-over time, (7.24), is very close to the site of bending to the nonlinear behaviour in the numerical data shown in Fig. 7.3a. This provides evidence for the correctness of the conjecture by Yakhot [1981] that the Kuramoto–Sivashinsky equation behaves as the KPZ equation in very large systems and in the long time limit. The same conclusion has been reached from considerations of the saturated state by L'vov et al. [1993].

In recent years, there have been several attempts to get a simpler understanding of the turbulent states of the Kuramoto–Sivashinsky equation. Frisch et al. [1986] showed that the equation actually possesses a continuum of stable stationary 'cellular' states, that these states are elastic and can even generate viscosity. Shraiman [1986] used those states as a starting point for an investigation of the statistical properties of the turbulent states. He stressed the fact that, although the system is often very near (at least locally) to one of the stationary states, it is never able to settle into a particular one of them and remains in a state of frustration, if the system size is large enough. Thus, presumably, many turbulent states are, at least in a technical sense, transient (just like the coupled map lattices discussed in chapter 4), but the transient times grow e.g. exponentially with system size.

Chow and Hwa [1995] provide an effective hydrodynamic theory for the Kuramoto–Sivashinsky equation in terms of the long wave components of the fields $h(x, t)$ and the *local drift rate fluctuation* $\pi(x, t)$. The drift rate is the average growth velocity $\langle \partial_t h \rangle$ of the interface. The effective viscosity is related to the response of the local drift rate to the curvature of the h-field, and this quantity can be computed directly by imposing a slow curvature on the Kuramoto–

Sivashinsky state. For these computations it suffices to use small systems. Chow and Hwa [1995] note that the effective viscosities of the turbulent state and of the cellular states have very different origins. The latter follows again from Galilean invariance: a tilted interface leads to a drift (i.e. a Galilean boost). Thus a 'valley' leads to a compression of the cells and to a larger drift rate, and vice versa for a 'hilltop'. In the turbulent state there is not a unique selected wavelength and the local wave number is no longer a conserved quantity, that is, creation and annihilation of cells can take place. They call this type of viscosity 'defect mediated' since it is mediated by space-time defects in the cell structure. They also note that the effective coupling constant λ_{eff} discussed above for the numerical simulations by Sneppen et al. [1992] does not replace the bare coupling constant in all cases. For example for the *advection rate* as function of tilt, the bare λ rather than λ_{eff} is relevant.

In a recent work by Rost and Krug [1995a], a completely different approach is used for the description of the turbulent states of the Kuramoto–Sivashinsky equation. They attempt to create a 'minimal' dynamics by replacing the cells with interacting particles which can coalesce or be created. This shows clearly the connection to the Burgers equation, which has a non-interacting particle representation, given by (7.9). Within this model the statistical properties are very accurately given by the KPZ equation and thus, by not resolving the details within each cell, the slow cross-over from the KS equation to the KPZ equation is avoided.

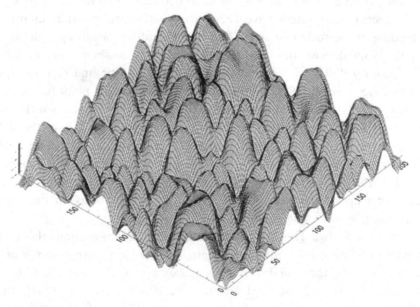

Figure 7.4 A surface generated by the Kuramoto–Sivashinsky in 2+1 dimensions. Note the intricate structure of the cells.

7.4.2   The Kuramoto–Sivashinsky equation in 2+1 dimensions

The dynamical behaviour of the Kuramoto–Sivashinsky equation in 2+1 dimensions is a very delicate problem and it is probably fair to say that it is still unresolved. Nevertheless, there have been several intriguing attempts in recent years which we now discuss. Figure 7.4 shows a typical example of the two-dimensional surface. Note the cellular structure of the surface, similar to what is observed in 2D flame fronts. Figure 7.5 shows numerical data from simulations of the KS equation in $d = 2 + 1$. Figure 7.5a is the Fourier transform the spatial h-autocorrelation function, that is $\langle h_k h_{-k} \rangle$, which shows a bump at a well-defined wave vector corresponding to the fastest growing mode of the linear theory. This bump relates to the cellular structure on the surface as shown in Fig. 7.4. The dotted line in Fig. 7.5a has slope 2, indicating that, away from the fastest growing mode, the correlation function scales as $1/k^2$.

Naturally, one would expect that the behaviour of the KS equation in $d = 2 + 1$ follows that of the KPZ equations, like in $d = 1 + 1$. The scaling exponents for the KPZ equation in $d = 2 + 1$ are from numerical simulations known to be $\chi \simeq 0.4$ and $\beta \simeq 0.25$. Whether the Kuramoto–Sivashinsky equation is also described by

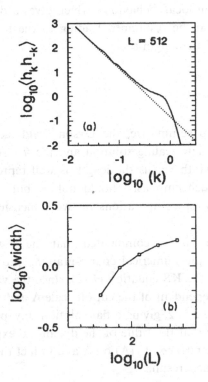

Figure 7.5 Numerical results for the Kuramoto–Sivashinsky equation in 2D. (a) Fourier transform of the correlation function in a system of size 512×512. (b) The width W versus length L [Procaccia et al. 1992].

these exponents or will behave in a completely different way is unclear. In fact, Jayaprakash et al. [1993] used the method of Zaleski [1989], described in 7.4, to numerically estimate the effective equation by a coarse-graining method. This equation is in the KPZ-form (7.11), now in 2+1 dimensions, so the procedure leads to an effective stochastic noise term without any correlations to itself or the field of the interface. It was again found that the effective diffusion constant saturates at a rather high value $v(s) \rightarrow v_B \sim 11 \pm 1.5$ as the coarse-graining proceeds, and this value appeared to be independent of the cut-off scale s. The conclusions of Jayaprakash et al. [1993] were therefore that the KS equation indeed behaves like the KPZ in $d = 2 + 1$.

Procaccia et al. [1992] presented a completely different viewpoint. They used a diagrammatic approach based on the elegant Wyld [1961] technique. This technique is particularly useful in renormalization approaches where the linear part of the system is unstable, which is the case here. Any perturbation theory that expands around the linear propagator will therefore diverge. Instead, using the Wyld approach, one may expand around the renormalized propagator. The main result of this approach is that in $d = 2 + 1$, it is possible to find two solutions. One is a 'local' solution, where wave vectors in the vertex are of the same order. This is consistent with the KPZ solution in 2+1 dimensions. The other solution is obtained from the Dyson–Wyld equations by assuming that one of the wave vectors in the vertex is of the order of the most unstable mode, i.e. $k_1 \sim k_{max}$. This can be called 'non-local' behaviour, which gives a different solution. The diagrammatic series can be resummed leading to the following results for the width as a function of the length scale L

$$W \sim \ln \ln(L/L^*), \tag{7.27}$$

where L^* is of the order $1/k_{max}$. To summarize, the Dyson–Wyld technique gives two solutions: the KPZ power law scaling solution and the *flat* solution (7.27). The numerical data for the width versus the length is well represented by the expression (7.27) but too much emphasis should not be put on that since the restrictions on the lengths for computations in two dimensions are severe.

L'vov and Procaccia [1994] have further commented that the results of Jayaprakash et al. [1993], based on the numerical coarse-graining procedure, precisely support the flat solution to the KS equation (7.27), since the value of the effective viscosity exponent is independent of the cut-off scale Λ and the dynamical exponent should therefore be $z = 2$, giving a flat solution. Jayaprakash et al. [1994] on the other hand argue that the value of the dynamical exponent $z = 2$ does not follow from the independence of $v(\Lambda)$ on Λ and in fact the KPZ value $z \sim 1.6$ is in accordance with their results.

7.4.3 Interfaces in coupled map lattices

This section introduces an alternative way of generating a rough interface, namely by means of coupled map lattices (CML), which are discussed in detail in chapter 4. A CML is very different from the equations studied so far in this chapter. The KPZ equation is a PDE with a random forcing. The Kuramoto–Sivashinsky equation is a PDE where the instabilities and the nonlinearities interact in an intricate way such that the motion in the full phase space is chaotic. The rationale behind a CML is to insert chaotic dynamics locally, i.e. on each spatial point, in a controlled way such that the noise is generated deterministically. The dynamics of the full system is also chaotic. Furthermore, a CML is in many respects much simpler to study than a PDE simply because time as well as space is discrete. The CML described in chapter 4 do not generate a rough interface with scale invariances. The main point of this section is therefore to modify the models already discussed in such a way that they generate interfaces. A main point is to break the symmetry so the field will only grow in one direction. For the KPZ and Kuramoto–Sivashinsky equations this was insured by the presence of the nonlinear term in the gradient. Here it is achieved by defining a nonlinear, chaotic map on the entire real axes.

Consider a 1D chain indexed by $i = 1, 2, ..., L$. On each site, there is a height or 'velocity' variable $u_i(n)$, where n is the discrete time. The dynamics is given by

$$u_i(n+1) = u_i(n) + f(u_i(n) - [u_i(n)]) + \epsilon(u_{i+1}(n) + u_{i-1}(n) - 2u_i(n)), \quad (7.28)$$

where [ ] indicates the integer part and periodic boundary conditions are assumed [Bohr et al. 1992]. The last term is, as in other CML, a discrete form of the Laplacian, whereas the two first terms provide the forcing. The function f is a nonlinear map of the unit interval, typical the logistic map, $f(x) = Rx(1-x)$. The argument of the function f is therefore the fractional part of u, so that (7.28) has the discrete translational ('Galilean') invariance under the transformation $u_i(n) \to u_i(n) + m$, for any integer m.

Figure 7.6 shows a snapshot of a chain of size $L = 1000$, with $f(x) = Rx(1-x)$, $R = 10.0$, and $\epsilon = 0.2$, after 250 000 iterates, starting from an almost flat state. Although the forcing is uniform (in the sense that the maps f are the same everywhere), structure has clearly developed on all scales from the smallest to the size of the system. Since $f(x) \neq f(-x)$, model (7.28) lacks $u \to -u$ symmetry. Thus, the maps drive the system predominantly in one direction and the mean displacement $\bar{u}$ increases steadily with time. In fact one could then argue that the system does not possess an attractor in phase space as it keeps expanding.

The state shown in Fig. 7.6 has positive Lyapunov exponents. To gain further insight into the dynamics and the size of the chaotic attractor, Bohr et al. [1992] computed the full spectrum of Lyapunov exponents for the parameter values given in Fig. 7.6, and for L's up to 100. For $L = 100$, all Lyapunov exponents are positive and the form of the spectrum is basically independent of the system size. Therefore, the number of positive Lyapunov exponents grows proportionally to the system size, as seen in other spatio-temporally chaotic systems, see chapters 1, 3 and 4. Since the sum of the Lyapunov exponents is positive, the system does not have an attractor of dimension less than the full phase space, but the width of the interface is still finite, for finite L, due to the diffusive term. However, the Lyapunov exponents are not all positive for arbitrary parameter values: when R decreases, so that the system is driven less strongly, the minimum value decreases too, eventually crossing zero to produce some negative exponents. For $R = 2.0$ and $L = 100$ there are several negative exponents. This is expected since by reducing R sufficiently, one can produce a non-chaotic state with no positive exponents. The local chaotic fluctuations of (7.28) mimic the effect of external noise in producing rough interfaces with power law correlations at long distances, as for the Kuramoto–Sivashinsky equation. The interface width

$$W(t) = \left(\langle |u(x,t) - \langle u(t) \rangle|^2 \rangle \right)^{\frac{1}{2}}$$

initially grows with time, but eventually reaches a steady state, saturating at a value proportional to $L^{\frac{1}{2}}$. Hence, $\chi = \frac{1}{2}$, as in the KPZ model. Thus, the dynamics of the interface model (7.28) produces a rough interface with power law correlations and positive Lyapunov exponents [Bohr et al. 1992].

The structure functions for the CML interface

$$W_q(r) = \langle |u(x+r) - u(x)|^q \rangle^{\frac{1}{q}}$$

scale in r. In analogy with the higher order structure functions for the velocity

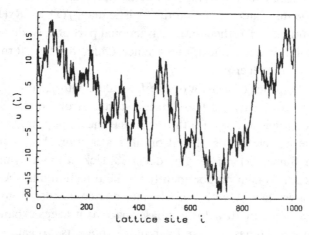

Figure 7.6 Snapshot of the turbulent interface resulting from (7.28) with $R = 10.0$, $\epsilon = 0.2$ and $L = 1000$ after $250\,000$ iterates starting from an almost flat state [Bohr et al. 1992].

field in turbulence, see sections 2.2 and 2.3, one can estimate the different moments for the interfaces and thereby check whether they exhibit spatial multiscaling, meaning that the various moments scale with different exponents. In these models one finds no multiscaling and $W_q(r)$ grows like $r^{1/2}$ for all q as in the KPZ and KS equations. It is not completely clear when and why multiscaling exists. Maybe some degree of non-locality is needed as in incompressible turbulence, where non-locality is present due to the pressure term. In a coupled map lattice introduced by Pikovsky [1992], non-local interactions, of a form which is difficult to justify physically, lead to large local derivatives of the field and this gives rise to spatial multiscaling. Krug [1994] introduced a version of a solid-on-solid model relating to epitaxial growth in which large gradients develop during the growth process. Also in this case, the presence of large derivatives leads to spatial multiscaling.

So, does the system (7.28) belong to the KPZ class of roughening exponents? Is β determined by the linear EW model or the nonlinear KPZ model? One would naturally expect the effective equation to be nonlinear because of the absence of $u \rightarrow -u$ symmetry of (7.28). Any such non-linearity will generate a $(\nabla u)^2$ term under the renormalization group, since no symmetry forbids this term. Fig. 7.7a shows $W(t)^2$ versus t on a log–log plot with $R = 10.0$, $L = 10\,000$, and $\epsilon = 0.2$. The data are consistent with $\beta \simeq 0.25$ rather than the anticipated value $\beta = 1/3$. Other values of the parameters lead to similar results. The simplest reason for this discrepancy in heuristic renormalization group arguments might be that while the chaotic fluctuations in (7.28) produce a large noise term in the effective equation, they generate an effective nonlinear $(\nabla u)^2$ term with a very small coefficient. Thus, one would have to study extremely large systems before the expected cross-over away from the EW exponent to the KPZ value would be observable. By adding a term $(\nabla u)^2$ to model (7.28) a nonlinear KPZ behaviour is immediately observed as is evident in Fig. 7.7b. In fact by using the method of tilting the interface, as described in section 7.4.1 for the KS equation, one can estimate the effective strength of the nonlinear term. It is found to be very small for model (7.28) leading to cross-over times beyond the range of the numerical calculations in Fig. 7.7 [Bohr et al. 1992].

Kapral et al. [1994] have introduced another kind of CML in order to study the dynamics of interfaces. They use local maps that are not chaotic but periodic (or oscillatory) with super-stable periodic cycles. The local maps are chosen to be piecewise linear. Using a diffusive coupling in 2D, Kapral et al. [1994] investigate the interfaces between two stable phases of the CML (i.e. one of the homogeneous and stable configurations). As the initial condition they use a randomized 'band' separating two stable phases and study how this interfacial band develops in time. Depending on the value of the diffusive coupling, many different behaviours are observed. For a relatively weak coupling, the interfacial region will, after developing small scale irregularities, freeze in space and will not propagate. For increasingly larger values of the coupling, the interfacial regime becomes more and more complex by creating, among other things, complex overhangs. Finally, at high values, the random interfacial region may spread and consume large portions of the regular phases. It is surprising that the largest Lyapunov exponent is negative and it is an open problem why these models appear turbulent even though all Lyapunov exponents are negative. By disregarding the overhangs, the authors compare the interface between the random band and the ordered region with the results obtained for the EW and KPZ equations as discussed in the previous sections. It is found that the behaviour is rather similar to the EW equation, although for large values of the

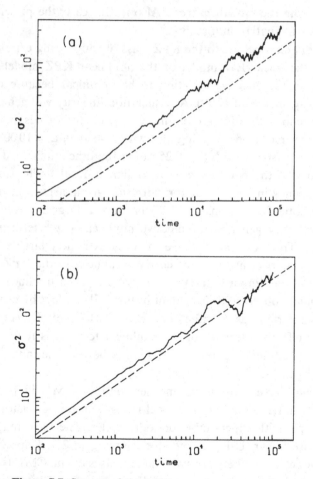

Figure 7.7 Square of the width as a function of time on a log–log plot. (a) Data for $L = 10\,000$ for the model in equation (7.28) showing an exponent $\beta \simeq 1/4$, which is the slope of the dotted line. (b) Data for $L = 10\,000$ for the modified model obtained from (7.28) adding a term $0.1(\nabla u)^2$. Here, the data are consistent with $\beta = 1/3$, which is the slope of the dotted line [Bohr et al. 1992].

coupling strength there is a clear deviation, which, however, does not approach KPZ behaviour. All the complex behaviour found in this model is a transient. In the end the random band will disappear and only the periodic states will remain. Nevertheless, the transient time grows exponentially with the system size and for large systems the transient behaviour can therefore be compared to the EW and KPZ equations.

7.5   Depinning models

7.5.1   Quenched randomness and directed percolation networks

In the KPZ equation, the noise is a fluctuating quantity in time and space. In many experimental situations, however, an interface will move in a medium which is static in space and time so the interface will be subjected to *quenched randomness*. The importance of this effect was first discussed by Koplik and Levine [1985], Kessler et al. [1991] and Parisi [1992]. Kessler et al. [1991] studied the EW equation with quenched noise and Parisi introduced a simple algorithm for interface motion in a quenched random field. Both found that there is a cross-over phenomenon where the interfaces are characterized by anomalous exponents for short time and small systems. If we try to write a PDE describing motion in a quenched random environment, the noise term will have the form $\eta(\mathbf{x}, h)$ meaning that the value of the noise will be fixed at a specific point in $(\mathbf{x}, h)$ space. If the noise η has a low value at a certain space point the interface will therefore be *pinned* for a long time at this point. To overcome the pinning, an external field is added. Denoting this field by F, the following PDE emerges for the motion of an interface in a quenched random environment [Parisi 1992, Leschhorn and Tang 1994]

$$\frac{\partial h(\mathbf{x}, t)}{\partial t} = \nu \nabla^2 h(\mathbf{x}, t) + \frac{\lambda}{2}(\nabla h(\mathbf{x}, t))^2 + \eta(\mathbf{x}, h) + F. \tag{7.29}$$

When the interface is pinned, the value of the field F must be increased adiabatically up to a critical value, F_c, where the interface starts to move. At that point, the interface exhibits a depinning transition, in analogy with charge-density waves [Coppersmith 1993] and motion of flux lines in high T_c superconductors [Jensen 1993]. In a porous medium as described in section 7.3, the interface will be pinned at sites with small pore sizes and an excess pressure is needed to overcome this pinning. The main message of this section is that the motion of the interface at this transition is very different from the standard dynamics characterized by the KPZ equation, and will in fact belong to a different universality class.

One way to describe pinning phenomena was proposed by Tang and Leschhorn [1992] and Buldyrev et al. [1992], who introduced interface models where the interface moves on a lattice prepared as a directed percolation network [Schögl 1972, Grassberger and Sundermeyer 1978, Stauffer and Aharony 1992]. Because this type of network, in a finite system, always has an absorbing state, the interface stops at its pinning point (although it could then be punctuated again [Barabasi et al. 1992]). Tang and Leschhorn [1992] and Buldyrev et al. [1992] found numerically the roughness exponent to be $\chi = 0.63...$ and it turns out that this value can be estimated in terms of exponents from theory of directed percolation. This value is in better agreement with the experimental results on porous media. In this type of motion, the possible overhangs were deleted to ensure a self-affine function [Tang and Leschhorn 1992, Buldyrev et al. 1992].

7.5.2   Self-organized-critical dynamics: the Sneppen model

Instead of using a lattice prepared directly at the directed percolation critical point, one could apply the invasion percolation rule [Wilkinson and Willemsen 1983] and punctuate the interface at its lowest pinning site [Sneppen 1992, Havlin et al. 1993]. To use such an invasion percolation rule for the dynamics of the self-affine interfaces, where overhangs are not allowed, it is of utmost importance to adjust the slopes in a simple way after the interface has been punctuated at its minimal pinning point. This was achieved by Sneppen [1992], who proposed a simple adjustment such that the slopes are never larger than 1 (on a lattice); i.e. every time the interface is punctuated, the slopes are adjusted to the left and right until the condition is fulfilled. An interesting point to mention here is that the invasion percolation rule is global; in fact, it is like a pressure term and this non-locality could be very important for the possibility of observing multiscaling in the growth (for a discussion of this point, see sections 7.4.3 and 7.5.6).

The model with slope adjustments is defined on a lattice where each lattice point (x, h) is assigned an uncorrelated random number $\eta(x, h)$ with a gaussian distribution. In the one dimensional version, a discrete interface $h(x)$ is defined on a discrete chain $x = 1, 2, 3, ..., L$ with periodic boundary conditions. The chain is updated by finding the site with the *smallest random number* $\eta(x, h(x))$ among all sites on the interface. On this site, one unit is added to h. Then neighbouring sites are adjusted upwards $(h \rightarrow h+1)$ precisely until all slopes $|h(x)-h(x-1)| \leq 1$, see Fig. 7.8a. When the site of the minimal random numbers is always chosen, the corresponding dynamics is called 'extremal dynamics', as introduced by Zaitsev [1992]. As discussed above, the dynamics mimics the motion of, say, the interface of oil penetrating a porous medium. Here, progress is often observed to occur in local avalanches at the point where the lowest pressure is needed to advance.

Sneppen [1992] found that the width $W = \langle |h - \langle h \rangle |^2 \rangle^{1/2}$ of a saturated interface scales with system size as L^χ with $\chi = 0.63 \pm 0.02$. That the exponent χ is larger than 0.5 is in agreement the experiments of the flow through porous media [Rubio et al. 1989, Kahanda et al. 1992, He et al. 1992, Kahanda and Wong 1993, Horvárth et al. 1991], although some of these experiments also reveal exponents larger than 0.63. Note that this scaling may be understood from a self-organization of the interface towards a 'critical' attractor which consists of an ensemble of directed percolating strings (of high values of η) at their critical point. Tang and Leschhorn [1993] demonstrated that the distribution of η along a saturated interface indeed agrees with this, see section 7.5.3 for details. It is interesting to note that the temporal drift of the interface towards a critical state

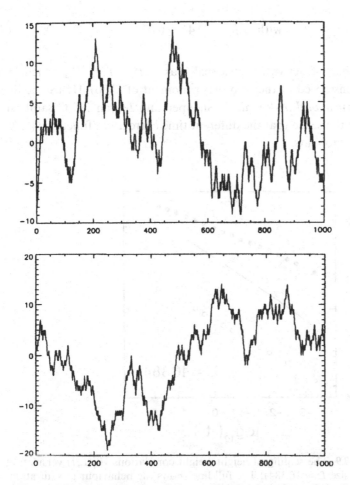

Figure 7.8 (a) A snapshot of the roughening interface of the Sneppen model. (b) A snapshot of the roughening string that does not break the up–down symmetry [Sneppen and Jensen 1994].

in this model can be viewed as an example of self-organized criticality [Bak et al. 1987].

The motion towards the critical state is very fast and might not even be described by a power law, see Fig. 7.9. In contrast, in the critical, saturated state the height–height time correlations

$$W_q(L,t) = \langle |h(x,t+\tau) - h(x,\tau)) - \langle h(x,t+\tau) - h(x,\tau)\rangle|^q \rangle^{1/q}, \qquad (7.30)$$

where the average is taken over $x \in [1, L]$ and the members of the ensemble, scale with non-trivial exponents

$$W_2(L,t) \sim t^{\beta_2} \quad \text{with} \quad \beta_2 = 0.69 \pm 0.02 \qquad (7.31a)$$

$$W_\infty(L,t) \sim t^{\beta_\infty} \quad \text{with} \quad \beta_\infty = 0.41 \pm 0.02 . \qquad (7.31b)$$

Also, the number of activated sites scales as $N(t) \sim t^{0.58 \pm 0.02}$, which in some sense can be considered as the zero order moment of (7.30). Hence the Sneppen model exhibits *temporal multiscaling* [Sneppen and Jensen 1993, Barabasi 1991, Zhang 1992] in the sense that the different time correlation functions (7.30) scale with different exponents.

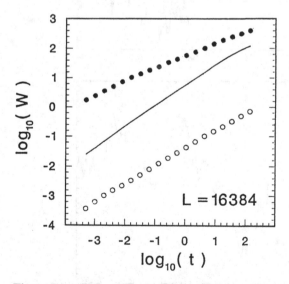

Figure 7.9 The scaling of height–height correlations $W_q(L,t)$ versus time (for system size $L = 16\,384$): The full line shows the behaviour at saturation of the second moment, W_2. The black circles reveal the infinite order moment W_∞ of the height–height correlations, whereas the open circles show the zero-order moment, counting the fraction of activated sites [Sneppen and Jensen 1993].

7.5.3 Coloured activity

The Sneppen model generates a 1D directed interface without overhangs, which makes the distribution of activities along this interface especially simple to study. One can describe the activity in terms of events on the interface: an event starts by finding the site with minimal η along the interface. This initiates a local avalanche which has a characteristic size (defined as the number of adjusted sites). Let $x(t)$ and $x(t + \Delta t)$ be the positions of the events at time t and $t + \Delta t$, respectively. With $\Delta t = 1/L$, we consider the proceeding event after the event occurring at t, if $\Delta t = 2/L$, the second event after t is considered, etc., and the probability distribution function $P(X)$ in the variable $X = |x(t) - x(t + \Delta t)|$ is calculated. In Fig. 7.9a, this distribution function is shown for different values of Δt, indicating a power law scaling

$$P(X) = A(X, \Delta t) X^{-\delta}, \tag{7.32}$$

where $\delta = 2.18 \pm 0.03$ independent of the value of Δt. The distribution (7.32) for various values of Δt can be cast into a scaling form

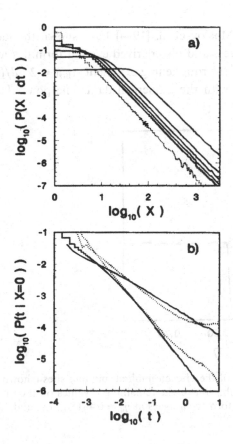

Figure 7.10 (a) Spatial distribution of activity centres, separated in time by $\Delta t = 1/L$, $2/L$, $4/L$, $10/L$ and $100/L$ where the system size $L = 8192$. (b) Probability for activity in a given site as a function of time: 'steep' full line: first return; 'flat' full line: all returns [Sneppen and Jensen 1993].

$$P(X) = X^{-\delta}g(X/\Delta t^{\alpha}),\tag{7.33}$$

where g is a scaling function and $\alpha = 0.58$ [Tang and Leschhorn 1993].

Strongly correlated spatial activity with a decaying power law of the form (7.32) will induce connected local activity. Therefore, the temporal behaviour of the activity is also correlated which can be tested by, in a given point, measuring the probability distribution function of time differences t between successive local events, i.e. the first returns to a point. This distribution, see Fig. 7.10b, behaves like

$$P_{first}(t) \propto t^{-\tau_{first}}\tag{7.34}$$

with the non-trivial exponent $\tau_{first} \approx 1.35 \pm 0.1$ (for times $1/L \ll t \ll 1$). If the activity along the interface exhibited a random walk with bounded step sizes, then $P_{first}(t) \propto t^{-1.5}$, corresponding to the distribution of first return times for random walkers. As $\tau_{first} < \frac{3}{2}$, the motion of activity is more correlated than a finite step random walker. Figure 7.10b displays the chance of activity in a point X, at a time $t + t'$, given there has been activity in X at time t', i.e. the distribution of all returns:

$$P_{all}(t) \propto t^{-\tau_{all}}\tag{7.35}$$

with exponent $\tau_{all} = 0.62 \pm 0.04$. Maslov et al. [1994] have shown the identity $\tau_{first} + \tau_{all} = 2$ for this kind of process and also derived explicit formulas for the values of the exponents in terms of the roughening exponent: $\tau_{first} = 2 - d/(d+\chi)$ and $\tau_{all} = d/(d + \chi)$ in agreement with the numerical data. The power laws in

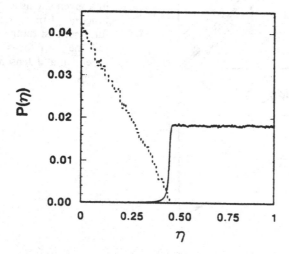

Figure 7.11 The distribution of the chosen minimal pinning sites (shown by the dotted line) and the distribution of the remaining η-values on the interface (shown by the full line). The former vanishes at $\eta_c = 0.4615$ [Tang and Leschhorn 1993].

(7.34–35) are in sharp contrast to the normally studied cases, KPZ and Kim–Kosterlitz (described in section 7.3), with uncorrelated noise implying that these distributions are flat. The Sneppen model breaks the Galilean invariance because $\chi + \chi/\beta_{crit} < 2$ which indicates yet one more difference to the KPZ equation.

The saturated state can be characterized by the distributions of the chosen minimal values $P_\eta(\eta_{min})$ and of η-value along the string, Fig. 7.11. This shows that the minimal η along the saturated strings does not exceed a certain threshold value, $\eta_c \approx 0.4615$, reflecting that all sites along the interface are assigned a value of the random number larger than the critical value, i.e. $\eta > \eta_c$.

Note, that we use the same letter η for the initially distributed random values on the lattice, $\eta(x, h(x))$, and the pinning numbers along the dynamical interface.

Because the η-values that are larger than the critical value are never chosen for growth, as they are never the global minimum, the sites where $\eta > \eta_c$ are effectively 'pinned' in the motion [Tang and Leschhorn 1993].

7.5.4 A scaling theory for the Sneppen model: mapping to directed percolation

A scaling theory for the behaviour of the Sneppen model has been put forward by various groups: Olami et al. [1994], Leschhorn and Tang [1994], Maslov and Paczuski [1994], Huber et al. [1995]. Even though the approaches varies, the basic idea concerns a mapping onto directed percolation (see Appendix H). As shown in Fig. 7.11, the distribution of the chosen punctuation values of η, $P_\eta(\eta_{min})$, vanishes at a critical point $\eta_c = 0.4615\ldots$ Tang and Leschhorn [1993] noted that $p_c = 1 - \eta_c = 0.5385\ldots$ is precisely the critical density of bonds in directed percolation for obtaining the percolating cluster. One can easily argue that the critical interface of the model corresponds to the 'backbone' of a directed percolation cluster (note here that the interface must be turned: the x-direction for the Sneppen interface corresponds to the time axis, or 'parallel' axis, for directed percolation, whereas the h-direction corresponds to the 'perpendicular' axis for directed percolation). Consider the points (x, h) on the interface for which $\eta \geq \eta_c$, which, as seen in Fig. 7.11, are precisely the 'pinned' sites that form the interface. Since the η-values are chosen uniformly in the interval [0,1], the density of these points is precisely equal to p_c and will therefore form a percolating cluster. In fact, we can say that the value η_c is the highest possible value of η for a connected interface to exist.

For directed percolation, see for instance Stauffer and Aharony [1992], it is well known that there is a correlation length parallel to the cluster

$$\xi_\| \sim |\, p - p_c\, |^{-\nu_\|} \tag{7.36}$$

and one perpendicular to the cluster

$$\xi_\perp \sim |\, p - p_c \,|^{-\nu_\perp} \, . \tag{7.37}$$

From this, the roughness scaling is estimated as

$$\xi_\perp \sim \xi_\parallel^{\nu_\perp/\nu_\parallel} \, , \tag{7.38}$$

leading to the value of the roughness exponent

$$\chi = \nu_\perp/\nu_\parallel = 1.097/1.733 = 0.633 \tag{7.39}$$

which agrees well with numerical simulations.

A central ingredient in the theory of Olami et al. [1994] is the notion of *associated processes* which are defined as follows. Suppose that at time $t = t_0$, the point $(x_0, h(x_0))$ was chosen for growth and that $\eta(x_0, h(x_0)) \le \eta_0$. The η_0-associated process, $AP_{\eta_0}(s)$, is the series of steps in which the points $x_1, x_2, \ldots, x_s$ were chosen for growth, if the following conditions are met:

$$\eta(x_1) < \eta_0, \quad \eta(x_2) < \eta_0, \quad \ldots, \quad \eta(x_s) < \eta_0$$

$$\eta(x_{s+1}) \ge \eta_0 \, . \tag{7.40}$$

Alternatively, we can call the process $AP_{\eta_0}(s)$ the η_0-punctuated avalanche. It is interesting to note that for a specific value of η, η_0, the union of associated processes, $AP_{\eta_0}(s)$ forms a 'fishnet' where each hole is described by an associated process, each one forms a compact object with a volume proportional to the time s and therefore given by $s \sim r_\perp r_\parallel^d$, where $r_\perp$ and $r_\parallel$ are perpendicular and parallel sizes in the fishnet.

The mechanism of self-organization to the critical state is elegantly described by the so-called 'gap-equation' [Paczuski et al. 1994, Maslov and Paczuski 1994]. In the initial state of the Sneppen model, all the random numbers along the interface are distributed uniformly. As the dynamics of the model proceeds, the minimal number η_{min} is always taken out leaving a small gap in the distribution, starting at $\eta = 0$. This gap, $G(s)$, increases gradually with time s and avalanches will separate instances when the gap $G(s)$ jumps to its next highest value. The average size of the jumps in the gap is $[1 - G(s)]/L^d$ and the growth of the gap is described by the following equation with $\epsilon = \eta_c - \eta$

$$\frac{\partial G(s)}{\partial s} = \frac{1 - G(s)}{L^d \langle s \rangle_\epsilon}, \tag{7.41}$$

where $\langle s \rangle_\epsilon$ is the average avalanche size, at a certain value of ϵ, which is the number of events occurring between one $G(s)$ blocking surface to another. The average avalanche size diverges as the critical state is approached

$$\langle s \rangle_\epsilon \sim \epsilon^{-\gamma}. \tag{7.42}$$

The distribution of avalanches scales as [Leschhorn and Tang 1994]

$$P(s) \sim s^{-\tau} F\left(\frac{s}{\epsilon^{-1/\sigma}}\right).$$ (7.43)

From (7.43) one has

$$\langle s \rangle_\epsilon \sim \epsilon^{\frac{\tau-2}{\sigma}}$$ (7.44)

leading to the identity

$$\gamma = \frac{2-\tau}{\sigma}.$$ (7.45)

Maslov and Paczuski [1994] obtain an identity for γ:

$$\gamma = 1 + \nu_\perp$$ (7.46)

near the critical point. Maslov et al. [1994] were also able to derive another equation, which directly relates the exponent γ to the average spatial extent of the avalanches $\langle r \rangle$ and the distance from criticality

$$\gamma = \frac{\langle r \rangle \epsilon}{1-\eta}, \quad \epsilon \to 0.$$ (7.47)

The most important consequence from this equation is that $\langle r \rangle \epsilon \to$ constant meaning that

$$\langle r \rangle \sim \epsilon^{-\gamma_\parallel} \quad \text{where} \quad \gamma_\parallel = 1.$$ (7.48)

For an avalanche, $r \sim s^{1/D}$, where D is the fractal dimension of the avalanche, leading to the following average

$$\langle r \rangle \sim \int^\epsilon s^{1/D} P(s) ds \sim \epsilon^{(\tau-1/D-1)/\sigma}.$$ (7.49)

Combining (7.48) and (7.49) one has

$$\sigma = 1 + 1/D - \tau.$$ (7.50)

The spatial correlation exponent is given by the avalanche correlation size exponent divided by its dimension: $\nu_\parallel = 1/(D\sigma) = 1/(D+1-d\tau)$. In this way Maslov et al. [1994] and Paczuski et al. [1995] found that all exponents for this type of SOC extremal dynamics can be obtained from two independent exponents, namely τ and D.

The equations can now be 'turned' around to obtain an equation for the avalanche exponent τ. The dimension of an avalanche is given by: $D = 1 + \chi = 1 + \nu_\perp/\nu_\parallel$. Furthermore $1/\sigma = \nu_\parallel + \nu_\perp$ which leads to [Maslov and Paczuski 1994]:

$$\tau = 1 + \frac{d - 1/\nu_\parallel}{d + \nu_\perp/\nu_\parallel}.$$ (7.51)

When one inserts the values for $\nu_\perp$ and $\nu_\parallel$ from directed percolation, see Stauffer and Aharony [1992], a good agreement with the numerically estimated values for τ is obtained.

7.5.5   A geometric description of the avalanche dynamics

In order to get a geometric description of the avalanche dynamics on the Sneppen interface and on directed percolation clusters, Huber et al. [1995] introduced a dimension formula giving a geometric and general relation between the size distribution and the fractal dimensions of a set of objects. They considered a set of self-similar objects, each with fractal dimension D whose union is a fractal of dimension D_{tot}. Define a subset of this total set, consisting of one point from each object. If this point set has dimension D_{num}, then the number of objects having sizes between s and $s + ds$, contained within a box of length L, is:

$$n_L(s)\, ds = L^{D_{num}}\, s^{-\tau}\, f(s/L^D)\, ds, \tag{7.52}$$

where the scaling function $f(x)$ approaches 1 for $x \ll 1$, and 0 for $x \gg 1$. Since $\int s n_L(s) ds \propto L^{D_{tot}}$, by matching powers of L, it follows that the number of objects of size s (for large, fixed L) is distributed as $n(s) \propto s^{-\tau}$ with:

$$\tau = 1 + \frac{D_{num} - D_{tot} + D}{D}. \tag{7.53}$$

In the case of $D_{num} = D_{tot}$, one obtains

$$\tau = 1 + \frac{D_{num}}{D}. \tag{7.54}$$

To apply these dimension formulas, Huber et al. [1995] considered voids on the backbone in $d = 1 + 1$ directed percolation (DP). As discussed in section 7.5.4, the strings on the backbone of a directed percolating cluster are equivalent to interface snapshots of the Sneppen model. A void on the backbone is a region completely enclosed by the DP network and the geometric counterpart of an avalanche. The size of a backbone void is the number of non-cluster sites enclosed by two merging branches.

Consider the distribution of voids touching an interface identified with the one-dimensional outer boundary of the backbone of the DP network. The distribution of the 1D border voids will differ from the overall distribution of voids because larger voids will have a larger probability of touching the 1D interface. Along a 1D path on the backbone, small voids sit densely, implying $D_{num} = 1$; as before, $D = D_{tot} = 1 + \chi$. Thus the distribution of backbone voids along the interface scales with exponent

$$\tau_{1d} = 1 + \frac{1}{1 + \chi}. \tag{7.55}$$

To relate to the Sneppen model, note that a void of length ℓ that borders the backbone hull, has a probability to be punctuated proportional to the number of singly-connected sites $n_{\mathrm{red}}(\ell)$ on the length ℓ of the hull (i.e. the *red* bonds introduced by Stanley [1977] and Pike and Stanley [1981]). Coniglio [1981] found $n_{\mathrm{red}}(\ell) = \ell^{1/\nu_{\parallel}}$. Weighting the void-size distribution by the chance that a given void is punctuated ($\propto n_{\mathrm{red}}(\ell)$) and noting that the void area is $s = \ell^{1+\chi}$, one obtains the distribution of the areas of the interface avalanches as a power law with exponent

$$\tilde{\tau}_{1d} = 1 + \frac{1}{1+\chi} - \frac{1}{\nu_{\parallel}}\frac{1}{1+\chi} = 1 + \frac{\nu_{\parallel}-1}{\nu_{\parallel}+\nu_{\perp}} \tag{7.56}$$

which is identical to the earlier formula of Maslov and Paczuski [1994] for $d = 1$.

7.5.6 Multiscaling

To obtain a theoretical estimate of the exponents for the temporal height–height correlations as described in section 7.5.2, the following quite elegant calculation was proposed both by Olami et al. [1994] and by Tang and Leschhorn [1994]. As discussed in section 7.5.4, the area of the associated process is covered in some time interval t after the process is initiated at time τ, $r_{\parallel}r_{\perp} \sim t$. As $r_{\perp} \sim r_{\parallel}^{\chi}$, one obtains

$$r_{\parallel} \sim t^{1/(1+\chi)} , \tag{7.57a}$$

$$r_{\perp} \sim t^{\chi/(1+\chi)} . \tag{7.57b}$$

To proceed, the height–height correlations $W_q(L,t)$, defined in (7.30), are approximated by the activity functions

$$A_q(L,t) = \langle | h(x,t+\tau) - h(x,\tau) |^q \rangle^{1/q} . \tag{7.58}$$

As long as the avalanches do not cover the whole system, this approximation is well justified. In fact, the scaling of the activity, (7.58), is much more extended than when the average is subtracted as in (7.30), see Sneppen and Jensen [1994]. Inserting $| h(x,t+\tau) - h(x,\tau) | \simeq r_{\perp}$ and averaging over the interface leads to

$$A_q(L,t) \simeq \left(\frac{1}{L} r_{\parallel} r_{\perp}^q\right)^{1/q} \sim r_{\parallel}^{1/q} r_{\perp} \sim t^{\beta_q} . \tag{7.59}$$

Inserting equations (7.57), one obtains the interesting relationship

$$\beta_q = \frac{q\chi+1}{q(\chi+1)} \quad \text{for } q > 0 \quad \text{and} \quad \beta_0 = \frac{1}{\chi+1} , \tag{7.60}$$

in agreement with the numerical results (7.31). The temporal multiscaling is thus determined by the roughening exponent χ. As argued in Olami et al. [1994], it is maybe a little misleading to call this multiscaling because it is determined by only one exponent. It is like a two-scale Cantor set [Halsey et al. 1986] which is an example of a multifractal, even though it is determined by just one scaling number.

Falk et al. [1994] have studied the Sneppen model in 2+1 dimensions and have generalized the arguments of Olami et al. [1994] and Leschhorn and Tang [1994] for the temporal multiscaling. Figure 7.12 shows the height–height temporal correlations, (7.30), calculated for 2+1 dimensions. The scaling exponents are $\beta_0 = 0.80 \pm 0.01$, $\beta_2 = 0.60 \pm 0.01$, $\beta_4 = 0.40 \pm 0.02$, $\beta_\infty = 0.20 \pm 0.02$. In 2+1 dimensions, an associated process forms a volume of base-area $r_\parallel^2$ and height $r_\perp$; in general, the base will fill $r_\parallel^d$. By inserting $\mid h(x, t + \tau) - h(x, \tau) \mid \simeq r_\perp$ into the activity and averaging over the string, one obtains $A_q(L, t) \simeq (\frac{1}{L} r_\parallel^d r_\perp^q)^{1/q} \sim r_\parallel^{d/q} r_\perp \sim t^{\beta_q}$, and arrives at the generalized expression of β_q for all dimensions

$$\beta_q = \frac{q\chi + d}{q(\chi + d)} \text{ for } q > 0 \text{ and } \beta_0 = \frac{d}{\chi + d}. \tag{7.61}$$

The numerically estimated value of χ is 0.50 ± 0.03 in 2+1 dimensions which, inserted into (7.61), describes the obtained results for β_q well. Equation (7.61) could be a general result for intermittent interface motion. Indeed, even in cases where the interface has an infinite slope or when χ is larger than 1, one can simply identify the value of χ as the ratio of the infinite moment exponent to the zero-order moment exponent, β_∞/β_0 (note from (7.61) that this ratio is equal to χ).

An expression for the exponent for the jump correlations (7.32), has been derived by Leschhorn and Tang [1994] and Maslov and Paczuski [1994]. Using a conditional distribution for fixed $\eta_{\min}$ of the first event, $P(X)$ can be expressed

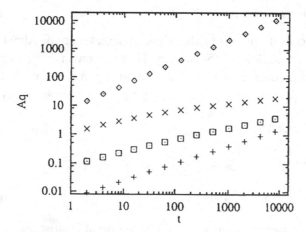

Figure 7.12 Scaling of the height–height correlations of the Sneppen model in $d = 2 + 1$ as a function of time, for a system of size 512×512. The diamonds show zeroth moment, plusses the second moment, squares the fourth moment and crosses the infinite moment [Falk et al. 1994].

as an integral over the conditional distribution folded with the distribution of the η_{min}'s which gives the following expression for the exponent

$$\delta = 1 + (\gamma/\nu_\parallel) \simeq 2.21 ,\tag{7.62}$$

in good agreement with the numerical estimate.

7.6   Dynamics of a membrane

The dynamical motion of roughening interfaces usually breaks the symmetry because the interface penetrates a medium from one end to the other, as e.g. in the experiment in a porous medium, or the experiment on burning paper. The models discussed so far also break the symmetry in the h-direction: Kuramoto–Sivashinsky and KPZ include a term $(\nabla h)^2$, and in deposition models particles move in one direction only. In contrast, one can imagine a dynamical roughening of a string in 1D, or membrane-surface in 2D, that just wobbles up and down without breaking the symmetry. Such a model was proposed in Sneppen and Jensen [1994] using the non-local rule discussed in section 7.5.2, the only difference being that one chooses with equal probability to *either add or subtract* 1 from h.

Initially, the string roughens in a transient towards the saturated state with a width $w = \langle (h - \langle h \rangle)^2 \rangle^{1/2} \propto L^\chi$ where $\chi = 0.46 \pm 0.04$ similar to the KPZ and EW models. At times larger than the transient time τ, moments of the height–height time correlations are measured by ensemble averaging over saturated states leading to $W_2(L,t) \propto t^{\beta_2}$ with $\beta_2 = 0.35 \pm 0.02$ and for the infinite moment $W_\infty(t) = \langle | \max_x \{ h(x, t+\tau) - h(x,\tau) \} - \min_x \{ h(x, t+\tau) - h(x,\tau) \} | \rangle_\tau \sim t^{\beta_\infty}$ with $\beta_\infty = 0.20 \pm 0.03$. The zero order moment, defined as the number of sites where changes have occurred, behaves as $N(t) \propto t^{\beta_0}$ with $\beta_0 = 0.43 \pm 0.03$. Therefore, this symmetric string model does also show temporal multiscaling due to the avalanche dynamics.

This dynamics of a symmetric 'membrane' generates also non-trivial correlations in space and time. Both the distribution of the jumps between activity centres, see (7.33), and the time distributions for first and all returns in a given point are power law distributed with a set of exponents. There is no clear connection between the exponents of this model and a well-known class of critical exponents. It is not clear yet whether the exponents eventually fall in the class of an 'evolution model' [Bak and Sneppen 1993].

Chapter 8

Lagrangian chaos

The purpose of this chapter is a brief review of the problem of Lagrangian chaos (i.e. of the chaotic motion of fluid particles) using methods and tools borrowed from the theory of dynamical systems. Particular attention is given to the relation of the Lagrangian chaos with mixing and transport diffusion phenomena in fluids. We discuss a possible general mechanism for the onset of Lagrangian chaos in two-dimensional fluids, and we show that, for many properties of the particle motion, the presence of a chaotic velocity field (Eulerian chaos) is not relevant.

In addition, we study the small scale structure of passive scalars in incompressible fluids. In particular we show that the classical Batchelor law (which gives a k^{-1} scaling for the power spectrum of the passive scalar in a suitable range of wave numbers k) is strictly related to the Lagrangian chaos and, hence, it is also valid for fluids which are not turbulent in the Eulerian sense. We also discuss the very irregular small-scale structure of magnetic fields due to the intermittency of the dynamical system which rules the evolution of a fluid particle.

8.1 General remarks

A problem of great interest in the physics of fluids concerns the study of the spatial and temporal structure of the so-called passive fields, indicating by this term passive quantities driven by the flow, such as the temperature under certain conditions. The paradigmatic equation for the evolution of a passive scalar field $\Theta(\mathbf{x}, t)$ is given by

$$\partial_t \Theta + \nabla \cdot (\mathbf{v}\,\Theta) = \chi \nabla^2 \Theta \tag{8.1}$$

where $\mathbf{v}(\mathbf{x}, t)$ is a given velocity field and χ is the molecular diffusion coefficient.

The hydrodynamic equation (8.1) can be studied through two different approaches. Either one deals at any time with the field Θ in the space domain covered by the fluid, or one deals with the trajectory of each fluid particle. The two approaches are usually designated as 'Eulerian' and 'Lagrangian', although both of them are due to Euler [Lamb 1945]. The two points of view are in principle equivalent. Indeed, if we denote by $\mathbf{v}(\mathbf{x}, t)$ the Eulerian velocity field, then the motion of a fluid particle initially located at $\mathbf{x}(0)$ is determined by the differential equation

$$\frac{d\mathbf{x}}{dt} = \mathbf{v}(\mathbf{x}, t), \tag{8.2}$$

with the initial condition $\mathbf{x}(0)$.

Equation (8.2) also describes the motion of test particles, for example a powder embedded in the fluid, provided that the particles are small enough not to perturb the velocity field, although large enough not to perform a Brownian motion. Particles of this type are the tracers used for flow visualization in fluid mechanics experiments [Tritton 1988]. We note that the true equation for the motion of a material particle in a fluid is rather complicated, see Maxey and Riley [1983].

In spite of the formal relation between Eulerian and Lagrangian approaches, it is difficult to extract information on one description starting from the other. For instance, there are situations where the velocity field is regular – i.e. absence of Eulerian chaos – but the corresponding motion of fluid particles is chaotic [Hénon 1966]. If the solution of (8.2) has a sensitive dependence on initial conditions, i.e. initially nearby trajectories diverge exponentially, one speaks of *Lagrangian chaos*. As mentioned, Lagrangian chaos can be present even in regular velocity fields. This may appear as a paradox, but is not too surprising. Since (8.2) is a nonlinear dynamical system, in general chaotic behaviours arise for time-dependent velocity fields in two dimensions, or even for stationary velocity fields in three dimensions.

If $\chi = 0$, it is easy to realize that (8.1) is equivalent to (8.2). In fact, we can write

$$\Theta(\mathbf{x}, t) = \Theta_o(\mathbf{T}^{-t}\mathbf{x}), \tag{8.3}$$

where $\Theta_o(\mathbf{x}) = \Theta(\mathbf{x}, t = 0)$ and $\mathbf{T}$ is the formal evolution operator of (8.2),

$$\mathbf{x}(t) = \mathbf{T}^t \mathbf{x}(0). \tag{8.4}$$

Taking into account the molecular diffusion χ, (8.1) is the Fokker–Planck equation of the Langevin equation [Chandrasekhar 1943]

$$\frac{d\mathbf{x}}{dt} = \mathbf{v}(\mathbf{x}, t) + \eta(t), \tag{8.5}$$

where η is a gaussian process with zero mean and variance

$$\langle \eta_i(t)\,\eta_j(t') \rangle = 2\chi \delta_{ij}\,\delta(t - t').$$

The dynamical system (8.2) is conservative for an incompressible fluid:

$$\nabla \cdot \mathbf{v} = 0. \tag{8.6}$$

In two dimensions, the constraint (8.6) is automatically satisfied assuming

$$v_1 = \frac{\partial \psi}{\partial x_2}, \qquad v_2 = -\frac{\partial \psi}{\partial x_1}, \tag{8.7}$$

where $\psi(\mathbf{x}, t)$ is called the *stream function*, and $\mathbf{x} = (x_1, x_2)$. Inserting (8.1) into (8.2) the evolution equations become

$$\frac{dx_1}{dt} = \frac{\partial \psi}{\partial x_2}, \qquad \frac{dx_2}{dt} = -\frac{\partial \psi}{\partial x_1}. \tag{8.8}$$

Formally (8.8) is a Hamiltonian system with the Hamiltonian given by the stream function ψ.

The presence of Lagrangian chaos indicates that some gross properties of mixing and diffusion are not strongly related to the presence of the diffusive term $\chi\nabla^2\Theta$. Therefore, in many cases the Lagrangian description permits a better understanding of the physics [Ottino 1989, 1990], than the Eulerian one.

It is also interesting to consider the advection of vector fields. For instance, the behaviour of a magnetic field $\mathbf{B}(\mathbf{x}, t)$ in electric conducting fluids is of great physical relevance in astrophysics and geophysics [Weiss 1985, Zel'dovich and Ruzmaikin 1986]. With appropriate approximations, one has the so-called kinematic dynamo problem described by the following equations

$$\left. \begin{aligned} \partial_t \mathbf{B} + (\mathbf{v} \cdot \nabla)\,\mathbf{B} &= (\mathbf{B} \cdot \nabla)\,\mathbf{v} + \chi\nabla^2\mathbf{B} \\ \nabla \cdot \mathbf{B} &= 0 \end{aligned} \right\} \tag{8.9}$$

where χ is now the magnetic diffusion coefficient of the fluid. The above equation is obtained from the magnetohydrodynamics for incompressible fluids, Maxwell equations and Ohm's law, assuming a prescribed velocity field as known input, unaffected by the electromagnetic field.

Again, in the kinematic dynamo, the role of Lagrangian chaos is prominent [Bayly 1986, Bayly and Childress 1987, Falcioni et al. 1989]. We shall see in section 8.3.5 that in the limit $\chi \to 0$ one has a close connection between the kinematic dynamo equation and the dynamical system (8.2).

8.1.1 Examples of Lagrangian chaos

As a first case, we consider the following 3D solenoidal stationary velocity field:

$$\mathbf{v} = (A \sin z + C \cos y, \; B \sin x + A \cos z, \; C \sin y + B \cos x) \qquad (8.10)$$

where A, B and C are non-zero real parameters. According to (8.10) the dynamical system associated with the motion of fluid particles is:

$$\left. \begin{aligned} \frac{dx}{dt} &= A \sin z + C \cos y \\[2mm] \frac{dy}{dt} &= B \sin x + A \cos z \\[2mm] \frac{dz}{dt} &= C \sin y + B \cos x. \end{aligned} \right\} \qquad (8.11)$$

Because of the incompressibility condition, the evolution $\mathbf{x}(0) \to \mathbf{x}(t)$, given by (8.11), defines a conservative, i.e. volume preserving, dynamical system.

Arnold [1965] argued that (8.11) is a good candidate for chaotic motion. Let us briefly repeat his argument. In a steady-state solution of the 3D Euler equation one has:

$$\left. \begin{aligned} \nabla \cdot \mathbf{v} &= 0 \\ \mathbf{v} \times (\nabla \times \mathbf{v}) &= \nabla \alpha \\ \alpha &= \frac{P}{\rho} + \frac{v^2}{2} \end{aligned} \right\} \qquad (8.12)$$

where P is the pressure and ρ the density. As a consequence of the Bernoulli theorem [Landau and Lifshitz 1987], $\alpha(\mathbf{x})$ is constant along a streamline – that is a trajectory of the system $d\mathbf{x}/dt = \mathbf{v}$. One can easily verify that chaotic motion can appear only if $\alpha(\mathbf{x})$ is constant (i.e. $\nabla \alpha(\mathbf{x}) = 0$) in a part of the space. Otherwise the trajectory would be confined on a 2D surface $\alpha(\mathbf{x}) = $ constant, where the motion must be regular as a consequence of general arguments (Bendixson–Poincaré theorem [Aggarwal 1972]). In order to satisfy such a constraint, one has to impose the Beltrami properties:

$$\nabla \times \mathbf{v} = \gamma(\mathbf{x}) \mathbf{v}, \qquad \mathbf{v} \cdot \nabla \gamma(\mathbf{x}) = 0. \qquad (8.13)$$

The field $\mathbf{v}$ given by (8.10) is a simple case of velocity possessing the Beltrami properties (in this case $\gamma(\mathbf{x}) = $ constant). It is possible to show [Galloway and Frisch 1987] that (8.10) is the unique stable solution of the Navier–Stokes equations for large v and with the forcing

$$\mathbf{f} \equiv v \, (A \sin z + C \cos y, \; B \sin x + A \cos z, \; C \sin y + B \cos x). \qquad (8.14)$$

Equation (8.11) was independently introduced by Childress [1970] as a model for kinematic dynamos, and now these flows are usually called ABC flows (for

Arnold, Beltrami, Childress). Numerical experiments by Hénon [1966] provided evidence that they are chaotic for special values of the parameters A, B and C. An extensive numerical and analytical study of the ABC model can be found in [Dombre et al. 1986].

Without entering in a detailed analysis, we show the Poincaré section at $z = 0$ of several trajectories in Fig. 8.1. The regions of ordered and chaotic motions are evident: the picture recalls the typical features of non-integrable Hamiltonian systems with two degrees of freedom. For two-dimensional incompressible stationary flows, the stream function ψ does not depend on time. As a consequence, the motion of fluid particles is given by a time-independent Hamiltonian system with one degree of freedom, where it is impossible to sustain chaos. However, for explicit time-dependent ψ the system (8.8) can exhibit chaotic motion. For example, many authors have studied the 2D chaotic advection in Stokes flows (i.e. in a fluid between two eccentric cylinders which rotate with a given time-dependent angular velocity) [Aref and Balachandar 1986, Chaiken et al. 1987] or in a simple model which provides an idealization of a stirred tank [Aref 1984].

In order to emphasize one of the basic mechanisms for the chaotic advection in 2D incompressible flow we discuss the following example [Kuznetsova et al.

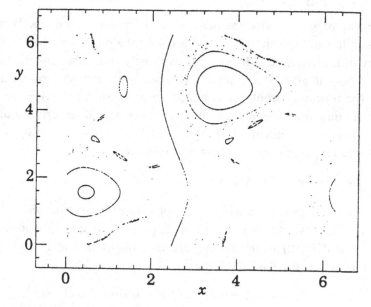

Figure 8.1 Poincaré section (plane $z = 0$) for eight trajectories of the ABC flow, equation (8.11) with parameters $A = 2.0$, $B = 1.70$, $C = 1.50$ [Crisanti et al. 1991].

1983]. An exact solution of the 2D Euler equation which mimics large scale vortex structure is given by:

$$\psi_0(x, y) = \frac{\gamma}{2\pi} \ln \left(\cosh \frac{2\pi y}{l} - \alpha \cos \frac{2\pi x}{l} \right). \tag{8.15}$$

Here $l/2\pi$ is the period of the vortices row in the x direction, and α describes the vorticity distribution: $\alpha = 0$ corresponds to a uniform distribution and $\alpha = 1$ gives a series of point vortices. In realistic situations there is a rather concentrated vorticity distribution, so it is natural to assume $\alpha \leq 1$. We discuss the case $\alpha = 1$, although the qualitative results will be valid also for $\alpha < 1$. A physically realistic perturbation to the steady solution (8.15) is an external wave propagating perpendicularly to the vortices row, i.e.

$$\delta \mathbf{v} = (0, v_0 \sin(2ky - 2\omega t)). \tag{8.16}$$

Note that this system has a periodic structure in the x direction. For $v_0 = 0$ the fixed points are on the line $y = 0$ at $x = nl$, $n = 0, \pm 1, \pm 2, \ldots$ (stable points), and $x = (n + 1/2)l$, $n = 0, \pm 1, \pm 2, \ldots$ (unstable points). The trajectories are qualitatively similar to those of the nonlinear pendulum, i.e. closed orbits (rotations) and open ones (librations) separated by orbits of infinite period (separatrices) linking two unstable fixed points. It is well known that in one-dimensional time-dependent Hamiltonian systems, the onset of chaos typically takes place around the separatrices by unfolding and crossing of stable and unstable manifolds. In some cases, a method, due to Melnikov [1963], see appendix B.2, permits the proof that the motion is chaotic in a stochastic layer around a separatrix. Kuznetsova et al. [1983] were able to compute Melnikov's integral explicitly, and thus to prove the existence of Lagrangian chaos. Figure 8.2 illustrates, by means of a Poincaré section, the typical behaviour of the system at a small value of the perturbative parameter $\mu_0 = v_0 k/\omega$:

(a) chaotic behaviour for trajectories with initial conditions close to a separatrix;

(b) regular motion if the initial conditions are far from the separatrices. The area of the chaotic layer (the region of chaotic behaviour around the separatrix) increases with μ_0, and for large μ_0-values, it is practically impossible to distinguish between regular and chaotic regions.

We shall see that such a scenario is not particular for a stream function whose time dependence is explicitly known, but it is also valid when the Eulerian equation passes from a steady solution to a time periodic solution, via a Hopf bifurcation.

For the case of time periodic velocity fields, i.e. $\mathbf{v}(\mathbf{x}, t + T) = \mathbf{v}(\mathbf{x}, t)$, where T is the period, the differential equation (8.2) can be studied in terms of discrete dynamical systems. From standard theorems of the theory of differential equations, one can determine the position $\mathbf{x}(t + T)$ in terms of $\mathbf{x}(t)$. Moreover, for a periodic velocity field, the map $\mathbf{x}(t) \to \mathbf{x}(t+T)$ does not depend on t. The above arguments allows us to write (8.2) in the form:

$$\mathbf{x}(n + 1) = \mathbf{F}[\mathbf{x}(n)], \tag{8.17}$$

where now the time is measured in units of the period T. If the incompressibility condition holds, the map (8.17) is conservative, i.e.

$$\left|\det \mathbf{A}[\mathbf{x}]\right| = 1, \qquad \text{where} \qquad A_{ij}[\mathbf{x}] = \frac{\partial F_i[\mathbf{x}]}{\partial x_j}.$$

An explicit deduction of the form of $\mathbf{F}$ for a general 2D or 3D flow is usually a very difficult task. However, in some simple cases, reasonable models, containing the main physical features, can be deduced from the physics of the system. The reader can find this type of derivation in Chaiken et al. [1987] for the Stokes flow, and in Lupini and Siboni [1989] for the free oscillation in a basin using the shallow water approximation.

If one is not interested in a specific problem, but just in some 'generic' properties, the study of 3D or 2D conservative maps allows one to determine many of the basic mechanisms of the chaotic advection. A quite general case, obtained as a continuous perturbation of the identity map, is given by maps of the form

$$\left.\begin{aligned} x(n + 1) &= x(n) + f\left[y(n), z(n)\right] \\ y(n + 1) &= y(n) + g\left[x(n + 1), z(n)\right] \\ z(n + 1) &= z(n) + h\left[x(n + 1), y(n + 1)\right]. \end{aligned}\right\} \tag{8.18}$$

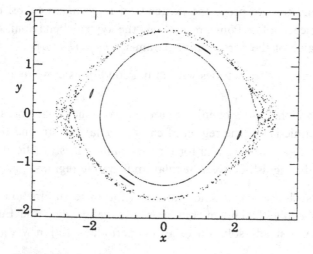

Figure 8.2 Iterations of the Poincaré map $\mathbf{x} = \mathbf{x}(n\pi)$, for three trajectories of system (8.15), (8.16) with $\mu_o = 0.01$ and $l = 2\pi$. The trajectories started from $x_0 = 1.0$ and $y_0 = 1.5$ (the chaotic one), 1.4 (the four components one), 1.2 (the inner one) [Crisanti et al. 1991].

A straightforward calculation shows that regardless of the form of the three functions f, g and h, (8.18) is volume-preserving. A simple non-trivial expression which captures the main features of 3D maps of the form (8.18) can be obtained by confining the three-dimensional dynamics on a 3D torus and retaining only the lowest terms of the Fourier expansion of f, g and h. This procedure leads to

$$
\left.
\begin{aligned}
x(n+1) &= x(n) + A_1 \sin z(n) + C_2 \cos y(n) \\
y(n+1) &= y(n) + B_1 \sin x(n+1) + A_2 \cos z(n) \\
z(n+1) &= z(n) + C_1 \sin y(n+1) + B_2 \cos x(n+1).
\end{aligned}
\right\}
\tag{8.19}
$$

Unfortunately, the knowledge of the properties of three-dimensional volume-preserving maps is not so rich as for symplectic maps [Sun 1983, 1984]. However, one may try to use some ideas and methods developed for Hamiltonian systems [Lichtenberg and Liebermann 1983], see, for example, Feingold et al. [1988].

8.1.2 Stretching of material lines and surfaces

The problem of the dynamical stretching of material lines and surfaces is directly related to features of Lagrangian chaos. We now discuss how some results obtained with phenomenological arguments in the theory of turbulence, or with the help of stochastic models [Drummond and Münch 1990, 1991], can be derived using methods of the theory of dynamical systems. This approach stresses the role of the Lagrangian chaos, and permits the extension of these results to include situations without Eulerian turbulence.

The linearization of (8.2) around a solution $\mathbf{x}(t)$ leads to

$$
\frac{dz_i}{dt} = \sum_{j=1}^{d} \left. \frac{\partial v_i}{\partial x_j} \right|_{\mathbf{x}(t)} z_i.
\tag{8.20}
$$

If we consider the vector $\delta \mathbf{x}$ joining two close fluid particles, (8.20) describes the evolution of a material line element $\delta \mathbf{x}$, if $|\delta \mathbf{x}|$ is small enough. Batchelor [1952] argued that for large times and small $|\delta \mathbf{x}(0)|$

$$
|\delta \mathbf{x}(t)| \sim |\delta \mathbf{x}(0)| \, e^{\alpha_1 t} \quad \text{with} \quad \alpha_1 > 0.
\tag{8.21}
$$

This result has been criticized by Cocke [1969, 1971]. The origin of this dispute stems from an exchange of limits. If $\delta \mathbf{x}(0)$ is small enough α_1 is nothing but the effective Lyapunov exponent, whose most probable value is the first (maximum) Lyapunov exponent of (8.2) (see appendix C). Cocke's remark is based on his result that, under general assumptions, for large times in a turbulent fluid

$$
l(t) = |\delta \mathbf{x}(t)| \sim t^{1/2}
\tag{8.22}
$$

so that $t^{-1} \ln |\delta \mathbf{x}(t)| \to 0$ as $t \to \infty$, and therefore (8.21) cannot be valid. However (8.22) is derived for large times with fixed $|\delta \mathbf{x}(0)|$, i.e. for $t \gg (1/\alpha_1) \ln(L/|\delta \mathbf{x}(0)|)$ where L is a typical length. On the other hand, in (8.21) one takes the limit of small $|\delta x(0)|$ first and then large t, i.e. $|\delta x(t)|$ is infinitesimal at any time, and hence there is no contradiction between (8.21) and (8.22).

The problem of the growth of the area of a material surface element can be treated in a similar way. A small material surface is spanned by two non-parallel vectors $\delta x_1(t)$ and $\delta x_2(t)$ starting from the same point, and hence by two lengths $l_{1,2}(t) = |\delta x_{1,2}(t)|$ and the angle between the two vectors. If l_i is small enough $\delta x_i(t)$ obeys (8.20) and the area of the material surface is given by

$$\delta \mathscr{S}(t) = |\delta x_1(t) \times \delta x_2(t)| \sim \delta \mathscr{S}(0)\, e^{\alpha_2 t} \tag{8.23}$$

where '$\times$' denotes the external product. By denoting with $\lambda_1 \geq \lambda_2 \geq \lambda_3$ the characteristic Lyapunov exponents of (8.2) we have that the most probable value of α_2 is $\lambda_1 + \lambda_2$, see appendix B.

A general theorem states that for steady velocity fields at least one of the characteristic Lyapunov exponents should be zero. In 3D the incompressibility condition requires that $\lambda_1 + \lambda_2 + \lambda_3 = 0$ so that in a chaotic system $\lambda_1 > 0$, $\lambda_2 = 0$ and $\lambda_3 = -\lambda_1 < 0$. In this case, the typical growth of a surface element coincides with that of a line element.

Feingold et al. [1988] found $\lambda_2 = 0$ also for a generic three-dimensional time periodic velocity field described by a conservative three-dimensional map. The validity of this result has been checked introducing a probabilistic version of the map (8.18) considering the evolution law:

$$x(t+1) = F^{(i_t)}[x(t)] \tag{8.24}$$

where $F^{(i_t)}$ is a sequence of maps generated according to a stochastic rule. For instance a Bernoullian trial, in which $F^{(i_t)} = F^{(j)}$ with probability P_j independent of $F^{(i_{t-1})}$, mimics a situation with strong Eulerian chaos. In order to mimic weak Eulerian chaos it is suitable to use a Markovian rule i.e. if $i_{t-1} = i$ then $F^{(i_t)} = F^{(j)}$ with probability P_{ij} [Galluccio and Vulpiani 1994].

The problem of the probability distribution of $l(t)$ and $\delta \mathscr{S}(t)$ can be analysed in terms of dynamical systems. The probability distribution $\mathscr{P}[l(t)]$ of $l(t)$ is not universal and its details strongly depend on the features of the dynamical system (8.2). Nevertheless for large times, $\mathscr{P}[l(t)]$ is close to a log–normal distribution (see appendix C)

$$\mathscr{P}[l(t)] \simeq \frac{1}{l(t)\,\sqrt{2\pi\mu t}} \exp\left[-\left(\ln\left[\frac{l(t)}{l(0)}\right] - \lambda_1 t \right)^2 \middle/ 2\mu t \right] \tag{8.25}$$

where

$$\mu = \lim_{\tau \to \infty} \frac{1}{\tau} \overline{\left(\ln\left[\frac{l(\tau)}{l(0)}\right] - \lambda_1 \tau \right)^2}.$$

A similar result can be obtained for $\mathscr{P}[\delta\mathscr{S}(t)]$:

$$\mathscr{P}[\delta\mathscr{S}(t)] \simeq \frac{1}{\delta\mathscr{S}(t)\sqrt{2\pi\mu_2 t}} \exp\left[-\left(\ln\left[\frac{\delta\mathscr{S}(t)}{\delta\mathscr{S}(0)}\right] - \alpha_2 t\right)^2 \Big/ 2\mu_2 t\right]$$

where $\alpha_2 = \lambda_1 + \lambda_2$ and

$$\mu_2 = \lim_{\tau\to\infty} \frac{1}{\tau}\overline{\left(\ln\left[\frac{\delta\mathscr{S}(\tau)}{\delta\mathscr{S}(0)}\right] - \alpha_2\tau\right)^2}.$$

Drummond and Münch [1990, 1991] showed that for 3D isotropic turbulence under realistic assumptions one has

$$\alpha_2 = \lambda_1 \quad \text{and} \quad \mu_2 = \mu. \tag{8.26}$$

These results, derived in the context of turbulence, seem to be strongly related to the presence of Lagrangian chaos and not to Eulerian turbulence. In fact, they can be obtained as a simple application of rather general methods of the theory of dynamical systems. In particular the identities (8.26) have been checked for the system (8.24) using different rules for the generation of the sequence i_t (i.e. periodical, Markovian, Bernoullian) corresponding to different Eulerian behaviour (periodic, weakly chaotic, strongly chaotic), see Galluccio and Vulpiani [1994] for details.

8.2 Eulerian versus Lagrangian chaos

The relation between Lagrangian and Eulerian chaos is a very complicated issue, also in extreme cases, such as fully developed turbulence, see Corrsin [1962], Lumley [1962]. In this section we discuss two different problems: the onset of Lagrangian chaos in relation to the features of the velocity field, and the effects of the presence of Eulerian chaos on the motion of fluid particles. In principle, the evolution of the velocity field $\mathbf{v}$ is described by partial differential equations. However, a good approximation can be obtained by using a Galerkin approach, and reducing the Eulerian problem to a system of F ordinary differential equations. The motion of a fluid particle is then described by the $(d + F)$-dimensional dynamical system

$$\frac{d\mathbf{Q}}{dt} = \mathbf{f}(\mathbf{Q}, t) \quad \text{with } \mathbf{Q}, \mathbf{f} \in \mathbb{R}^F \tag{8.27a}$$

$$\frac{d\mathbf{x}}{dt} = \mathbf{v}(\mathbf{x}, \mathbf{Q}, t) \quad \text{with } \mathbf{x}, \mathbf{v} \in \mathbb{R}^d \tag{8.27b}$$

where d is the space dimensionality and $\mathbf{Q} = (Q_1, \ldots, Q_F)$ are the F variables, usually normal modes, which describe the evolution of the velocity field $\mathbf{v}$. Note that the Eulerian equations (8.27a) do not depend on the Lagrangian part (8.27b) and can be solved independently.

In order to characterize the degree of chaos, three different Lyapunov exponents can be considered:

(a) λ_E for the Eulerian part (8.27a);
(b) λ_L for the Lagrangian part (8.27b), where the evolution of the velocity field is assumed to be known;
(c) λ_T for the total system of the $d + F$ equations.

These Lyapunov exponents are defined as:

$$\lambda_{E,L,T} = \lim_{t\to\infty} \frac{1}{t} \ln \frac{|\mathbf{z}(t)^{(E,L,T)}|}{|\mathbf{z}(0)^{(E,L,T)}|} \tag{8.28}$$

where the evolution of the three tangent vectors $\mathbf{z}$ is given by the linearized stability equations for the Eulerian part, for the Lagrangian part and for the total system, respectively:

$$\frac{dz_i^{(E)}}{dt} = \sum_{j=1}^{F} \frac{\partial f_i}{\partial Q_j}\bigg|_{\mathbf{Q}(t)} z_j^{(E)}, \qquad \mathbf{z}^{(E)} \in \mathbb{R}^F \tag{8.29a}$$

$$\frac{dz_i^{(L)}}{dt} = \sum_{j=1}^{d} \frac{\partial v_i}{\partial x_j}\bigg|_{\mathbf{x}(t)} z_j^{(L)}, \qquad \mathbf{z}^{(L)} \in \mathbb{R}^d \tag{8.29b}$$

$$\frac{dz_i^{(T)}}{dt} = \sum_{j=1}^{d+F} \frac{\partial G_i}{\partial y_j}\bigg|_{\mathbf{y}(t)} z_j^{(T)}, \qquad \mathbf{z}^{(T)} \in \mathbb{R}^{F+d} \tag{8.29c}$$

where $\mathbf{y} = (Q_1,\ldots,Q_F,x_1,\ldots,x_d)$ and $\mathbf{G} = (f_1,\ldots,f_F,v_1,\ldots,v_d)$. The meaning of these Lyapunov exponents is quite evident:

(a) λ_E is the mean exponential rate of the increase in the uncertainty in the knowledge of the velocity field;
(b) λ_L estimates the rate at which the distance $\delta x(t)$ between two fluid particles initially close increases with time, when the velocity field is given, i.e. a particle pair in the same Eulerian realization: $\delta x(t) \sim \exp(\lambda_L t)$;
(c) λ_T is the rate of growth of the distance between initially close particle pairs, when the velocity field is not known with infinite precision.

There is no general relation between λ_E and λ_L. One could expect that in the presence of a chaotic velocity field the particle motion has to be chaotic. However, the inequality $\lambda_L \geq \lambda_E$ – even if generic – sometimes does not hold, e.g. in some systems like the Lorenz model [Falcioni et al. 1988] and in generic 2D flows when the Lagrangian motion happens around well-defined vortex structures [Babiano et al. 1994] as discussed in section 8.2.2. On the other hand, it is possible to prove [Crisanti et al. 1991] that

$$\lambda_T = \max(\lambda_E, \lambda_L). \tag{8.30}$$

In section 9.2 we give an example where $\lambda_E > 0$ and $\lambda_L < 0$, which can occur in a compressible fluid.

8.2.1 Onset of Lagrangian chaos in two-dimensional flows

In 2D incompressible fluids, the continuity equation $\nabla \cdot \mathbf{v} = 0$ is automatically satisfied by writing the velocity components as derivatives of a stream function $\psi(x, y)$. The stream function is formally the Hamiltonian of a one-dimensional system whose generalized coordinate and momentum are x and y. This allows us to connect the Lagrangian chaos with the behaviour of the velocity field. In fact, it is well known that one-dimensional systems are integrable if the energy is conserved. Chaotic behaviour is therefore possible only for time-dependent Hamiltonians when there are no integrals of motion. There exists a wide literature on quasi-integrable systems. In particular, the effects of time-dependent perturbations on autonomous one-dimensional Hamiltonian systems are well understood.

For our purposes, we shall consider the Navier–Stokes equations at low Reynolds numbers, in two dimensions and with periodic boundary conditions. A convenient way to study the Lagrangian behaviour is by means of truncated models of these equations. They are obtained by expanding the stream function ψ in a Fourier series and taking into account only the first F terms [Boldrighini and Franceschini 1979, Lee 1987]

$$\psi = -i \sum_{j=1}^{F} k_j^{-1} Q_{k_j} e^{i\mathbf{k}_j \mathbf{x}} + \text{c.c.}, \qquad (8.31)$$

where c.c. indicates the complex conjugate term and $\mathbf{Q} = (Q_1, \ldots, Q_F)$ are the F variables (normal modes) which describe the Eulerian field evolution. Inserting (8.31) into the Navier–Stokes equations and by an appropriate time rescaling, we obtain the system of F ordinary differential equations

$$\frac{dQ_j}{dt} = -k_j^2 Q_j + \sum_{l,m} A_{jlm} Q_l Q_m + f_j, \qquad (8.32)$$

with $j = 1, \ldots, F$ and f_j represents the effect of the external constant forcing on the j mode. For an explicit form of the coefficients A_{jlm} see appendix F.

Franceschini and co-workers have studied this truncated model with $F = 5$ and $F = 7$ [Boldrighini and Franceschini 1979, Lee 1987]. Their choice of the parameters corresponds to force on the 3rd mode and $f_j = Re \, \delta_{j,3}$ [Lee 1987]. Let us now discuss the Lagrangian behaviour of a fluid particle at varying Reynolds number in the case of $F = 5$ modes. For $Re < Re_1 = 22.8537\ldots$, there are four stable stationary solutions, say $\hat{\mathbf{Q}}$, so that $\lambda_E < 0$. At $Re = Re_1$, these solutions become unstable, via a Hopf bifurcation [Marsden and

McCracken 1975], and four stable periodic orbits appear, implying $\lambda_E = 0$. For $Re_1 < Re < Re_2 = 28.41\ldots$, one thus finds the stable limit cycles:

$$\mathbf{Q}(t) = \hat{\mathbf{Q}} + (Re - Re_1)^{1/2}\delta\mathbf{Q}(t) + O(Re - Re_1) \tag{8.33}$$

where $\delta\mathbf{Q}(t)$ is periodic with period

$$T(Re) = T_0 + O(Re - Re_1) \qquad T_0 = 0.732\,77\ldots \tag{8.34}$$

At $Re = Re_2$, these limit cycles lose stability and there is a period doubling cascade toward Eulerian chaos.

For $Re < Re_1$, the stream function is asymptotically stationary, $\psi(\mathbf{x}, t) \to \hat{\psi}(\mathbf{x})$, and the corresponding one-dimensional Hamiltonian is time independent. The Bendixson–Poincaré theorem [Aggarwal 1972] and the continuity equation $\nabla \cdot \mathbf{v} = 0$ imply that the solutions of the Lagrangian equation (8.27b) are regular: fixed points and periodic or unbounded orbits. Figure 8.3 reports the structure of the separatrices, i.e. orbits of infinite period, at $Re = Re_1 - 0.05$. One can observe the presence of hyperbolic fixed points and of two kinds of separatrices: the isolated 'eights', labelled by A, and the connected periodic separatrices, labelled by B. For $Re = Re_1 + \epsilon$ the stream function becomes time-dependent

$$\psi(\mathbf{x}, t) = \hat{\psi}(\mathbf{x}) + \sqrt{\epsilon}\,\delta\psi(\mathbf{x}, t) + O(\epsilon), \tag{8.35}$$

where $\hat{\psi}(\mathbf{x})$ is given by $\hat{\mathbf{Q}}$ and $\delta\psi$ is periodic in $\mathbf{x}$ and in t with period T. The region of phase space, here the real two-dimensional space, adjacent to a separatrix is very sensitive to perturbations, even of very weak intensity. Indeed, generically in one-dimensional Hamiltonian systems, a periodic perturbation

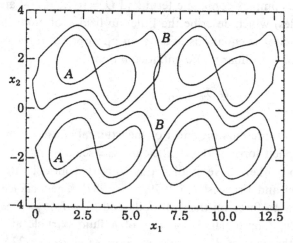

Figure 8.3 Structure of the separatrices of equation (8.8), with ψ given by equations (8.32) and (8.33), in the 5-modes truncation case (see appendix G) with $Re = Re_1 - 0.05$ [Falcioni et al. 1988].

gives rise to stochastic layers around the separatrices where the motion is chaotic, as consequence of unfolding and crossing of the stable and unstable manifolds in domains centred at the hyperbolic fixed points (see appendix B). The Melnikov method [Melnikov 1963] allows one to prove the existence and to estimate the size of a stochastic layer for Hamiltonian systems close to integrability, if the unperturbed solution on the separatrix is known. However, for the truncated models of Navier–Stokes equations, we are not able to use this method, since the structure of the Hamiltonian, i.e. of the stream function ψ, is rather complicated. One can only provide numerical evidence for the existence of the chaotic regions, by computing the maximum Lyapunov exponent. We show in Figure 8.4 a picture of the chaotic and regular motion for small ϵ by means of the Poincaré map

$$\mathbf{x}(nT) \to \mathbf{x}(nT + T). \tag{8.36}$$

The period $T(\epsilon)$ is computed numerically. The size of the stochastic layers rapidly increase with ϵ. At $\epsilon \approx 0.7$ they overlap and it is practically impossible to distinguish between regular and chaotic zones. Four different behaviours are observed at increasing ϵ:

(a) chaotic motion bounded in the stochastic layers originated by the separatrices of type A;

(b) unbounded chaotic motion in the stochastic layers originated by the separatrices of type B;

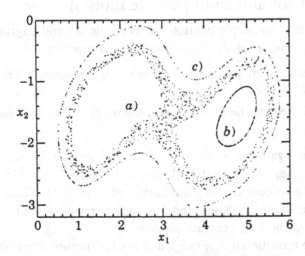

Figure 8.4 Poincaré map for three trajectories of system (8.8) in the 5-modes model with $Re = Re_1 + 0.05$. The initial conditions are selected close to a separatrix, case (a) $(x_1(0) = 3.2, x_2(0) = -1.6)$, or far from the separatrices, cases (b) $(x_1(0) = 4.3, x_2(0) = -2.0)$ and (c) $(x_1(0) = 4.267, x_2(0) = -3.009)$ [Falcioni et al. 1988].

(c) unbounded regular motion in the regions separated by stochastic layers of type B;

(d) bounded regular motion in the regions inside stochastic layers of type B but far enough from the stochastic layers of type A.

At $\epsilon > \epsilon_c \simeq 0.7$ the chaotic regions overlap and the motion is always diffusive motion.

This scenario for the onset of Lagrangian chaos in two-dimensional fluids is generic and does not depend on the particular truncated model. In fact, it is only related to the appearance of stochastic layers under the effects of small time-dependent perturbations in one-dimensional integrable Hamiltonian systems, and it can thus occur even in potential flow as e.g. produced by two or more (linear) surface waves [Bohr and Lundbek Hansen 1996]. As a consequence of a general feature of one-dimensional Hamiltonian systems we expect that a stationary stream function becomes time periodic through a Hopf bifurcation. This is the case for all known truncated models of the Navier–Stokes equations. By contrast the diffusive behaviour is closely related to the detailed structure of the velocity fields (for more details see Crisanti et al. [1991]).

These considerations do not apply to three-dimensional fluids, where stationary velocity fields may support Lagrangian chaos. We do not expect that the presence of a Hopf bifurcation in the solution of the Eulerian equation plays any role in three dimensions, except perhaps in quite particular situations [Feingold et al. 1988].

8.2.2   Eulerian chaos and fluid particle motion

We have seen that there is no simple relation between Eulerian and Lagrangian behaviour. In the following, we shall discuss two important points:

(i) what are the effects on the Lagrangian chaos of the transition to Eulerian chaos, i.e. from $\lambda_E = 0$ to $\lambda_E > 0$.

(ii) whether a chaotic velocity field ($\lambda_E > 0$) should imply a stochastic motion of fluid particles.

The first point can again be studied by truncated models of the two-dimensional Navier–Stokes equations. For the model (8.32) with $F = 5$ modes, the limit cycles bifurcate to new double period orbits at $Re = 28.41$. Then, there is a period doubling transition to chaos, and a strange attractor in the $\mathbf{Q}$ space appears at $Re_c \approx 28.73$, where λ_E becomes positive.

Note that, unlike the transition to Lagrangian chaos, the transition to Eulerian chaos is strongly model dependent. For example, in the $F = 7$ model there is a transition to chaos via collapsing of periodic orbits [Lee 1987].

In order to investigate the relation between the Lagrangian behaviour and that of the Eulerian field, one can study λ_E and λ_L as functions of Re near the onset of Eulerian chaos in the two models. One sees that λ_L is not affected by the sharp increase of λ_E at Re_c, see Falcioni et al. [1988]. The same qualitative behaviour has been observed in the 7-mode model around the corresponding critical value $Re_c \approx 555$.

These results provide numerical evidence for the conjecture that the onset of Eulerian chaos has no influence on the Lagrangian properties. This feature should be valid in most situations, since it is natural to expect that in generic cases there is a strong separation of the characteristic times for Eulerian and Lagrangian dynamics.

The second point – the conjecture that a chaotic velocity field implies chaotic motion of particles – could also look very natural. Indeed, it appears to hold in many systems, and in particular in the truncated models. Nevertheless, one can find a class of systems where it is false, e.g. the system (8.27) may exhibit a chaotic behaviour with $\lambda_E > 0$, even if $\lambda_L = 0$.

This is also the case of one of the most famous chaotic system, the Lorenz model [Lorenz 1963]. Indeed, it exhibits the apparently surprising feature of Eulerian chaos without Lagrangian chaos. The Lorenz model is obtained by strong simplifications of the equations for the convection of the temperature in a slice of fluid, say x-y plane, with a gravitational field and a temperature gradient directed along y. Here there are $F = 3$ degrees of freedom and, hence, three variables Q_1, Q_2 and Q_3 related to the stream function and to the displacement of temperature δT from the linear behaviour, in y, in the absence of convection,

$$
\begin{aligned}
\psi &= \sqrt{2}\, Q_1 \sin(x) \sin(y), \\
\delta T &= Q_2 \cos(x) \sin(y) - Q_3 \sin(2y).
\end{aligned}
\tag{8.37}
$$

The Eulerian equation for the normal modes reads

$$
\left.
\begin{aligned}
\frac{dQ_1}{dt} &= Pr\,(Q_2 - Q_1) \\
\frac{dQ_2}{dt} &= Q_1(Ra - Q_3) - Q_2 \\
\frac{dQ_3}{dt} &= Q_1 Q_2 - (8/3)Q_3
\end{aligned}
\right\}
\tag{8.38}
$$

where Ra is the Rayleigh number and Pr is the Prandtl number. The Lagrangian equations are:

$$
\left.
\begin{aligned}
\frac{dx}{dt} &= \frac{\partial \psi}{\partial y} = \sqrt{2}\, Q_1(t) \sin(x) \cos(y) \\
\frac{dy}{dt} &= -\frac{\partial \psi}{\partial x} = -\sqrt{2}\, Q_1(t) \cos(x) \sin(y)
\end{aligned}
\right\}
$$

where $x, y \in [0, \pi]$ and t are rescaled dimensionless variables.

It is possible to show that $\lambda_L = 0$ for all Ra and that the orbits are always closed and depend only on the initial conditions but not on Ra [Falcioni et al.

1988]. This can be explained by some straightforward considerations by noting that the equation for the orbit is time-independent:

$$\frac{dx}{dy} = g(x, y)$$

and that

$$\sin(x) \sin(y) = \text{const}$$

is an integral of motion. For the sake of simplicity we limit the discussion to an initial condition $(\pi/2 + q_1, \pi/2 + q_2)$ close to the Lagrangian fixed point $(\pi/2, \pi/2)$. One thus has

$$\left.\begin{aligned}
\frac{dq_1}{dt} &= -\sqrt{2}\, Q_1(t)\, q_2 \\
\frac{dq_2}{dt} &= \sqrt{2}\, Q_1(t)\, q_1.
\end{aligned}\right\} \tag{8.39}$$

By integrating (8.39) in polar coordinates, one obtains:

$$r(t) = \text{const} \qquad \text{and} \qquad \phi(t) = \phi(0) + \sqrt{2} \int_0^t d\tau\, Q_1(\tau). \tag{8.40}$$

This result implies that the particles move on a circle of radius r with angular velocity $\sqrt{2}\, Q_1(t)$. The motion can be chaotic for appropriate values of the control parameter Ra since $Q_1(t)$ itself can be chaotic. However the Lagrangian Lyapunov exponent is $\lambda_L = 0$ for all Ra since both δr and $\delta\phi$ are constant in time, and hence two particles initially close remain close in the same realization of the velocity field. In spite of this, the Lagrangian motion of the particles

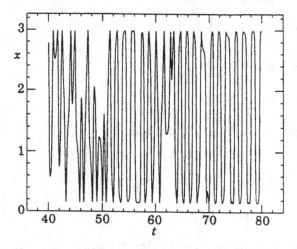

Figure 8.5 The Lagrangian coordinate $x(t)$ for the Lorenz model with $Ra = 26.24$ and $Pr = 10$, the Eulerian part (8.38), is chaotic [Falcioni et al. 1988].

looks chaotic when $\lambda_E > 0$ (see Fig. 8.5), if the velocity field is not known with infinite precision, since in this case the total Lyapunov exponent $\lambda_T = \lambda_E$, see (8.30).

Indeed, the phase difference $\delta\phi(t)$ between a particle in the velocity field described by $\mathbf{Q}$ and one in the velocity field described by $\mathbf{Q} + \delta\mathbf{Q}$ increases as

$$\delta\phi(t) = \sqrt{2} \int_0^t d\tau \, \delta Q_1(\tau) \sim e^{\lambda_E t}.$$

The fact that even in the presence of Eulerian chaos one can have $\lambda_L = 0$, is not a peculiar property of the Lorenz model but it is present in 2D systems in the smooth regions around well-defined vortex structures.

In order to show this let us consider the motion in the open plane of N point vortices [Aref 1983] with vorticities $\Gamma_1, \Gamma_2, \ldots, \Gamma_N$ and positions $(x_i(t), y_i(t))$, $i = 1, \ldots, N$, which is described by the Hamiltonian system

$$\Gamma_i \frac{dx_i}{dt} = \frac{\partial H}{\partial y_i} \tag{8.41a}$$

$$\Gamma_i \frac{dy_i}{dt} = -\frac{\partial H}{\partial x_i} \tag{8.41b}$$

where

$$H = -\frac{1}{4\pi} \sum_{i \neq j} \Gamma_i \Gamma_j \ln r_{ij} \tag{8.42}$$

and $r_{ij}^2 = (x_i - x_j)^2 + (y_i - y_j)^2$. In the case of point vortices on the torus or on the sphere, it is sufficient to substitute the $\ln r_{ij}$ in equation (8.42) with the appropriate Green's function $G(r_{ij})$, see Weiss and McWilliams [1991].

The motion of N point vortices is described in an Eulerian phase space with $2N$ dimensions. A system of three vortices with arbitrary values of Γ_i is integrable. The vortex motion is in this case regular (i.e. there is no exponential divergence of nearby trajectories in phase space). For $N \geq 4$, apart from nongeneric initial conditions and/or values of the parameters Γ_i, the system appears in general to be chaotic [Aref 1983, Novikov and Sedov 1978]. The motion of a passively advected particle located in $(x(t), y(t))$ in the velocity field defined by equations (8.41), is given by

$$\frac{dx}{dt} = -\sum_i \frac{\Gamma_i}{2\pi} \frac{y - y_i}{R_i^2} \tag{8.43a}$$

$$\frac{dy}{dt} = \sum_i \frac{\Gamma_i}{2\pi} \frac{x - x_i}{R_i^2} \tag{8.43b}$$

where $R_i^2 = (x - x_i)^2 + (y - y_i)^2$.

Let us first consider the motion of advected particles in a three-vortices (integrable) system in which $\lambda_E = 0$. In this case, the stream function for the advected particle is periodic in time and the expectation is that the advected particles may display chaotic behaviour. The typical trajectories of passive particles which have initially been placed respectively in close proximity to a vortex centre or in the background field between the vortices display a very different behaviour. The particle seeded close to the vortex centre displays a regular oscillatory motion around the moving vortex and $\lambda_L = 0$; by contrast, the particle in the background field undergoes an irregular and aperiodic trajectory, and λ_L is positive.

We now discuss a case where the Eulerian flow is chaotic. To this end it is sufficient to consider system (8.42) with $N \geq 4$. Figures 8.6a,b show the trajectories of two passive particles deployed respectively in proximity to a vortex centre and in the background field between the vortices, for $N = 4$. In the first case, the particle rotates around the moving vortex. The vortex motion is chaotic; consequently, the particle position is unpredictable at large times as is the vortex position. Nevertheless, the Lagrangian Lyapunov exponent for this trajectory is zero. Note that in this case the Eulerian Lyapunov exponent is positive. The regular motion of the advected particle around an irregularly moving centre is not in contradiction with its zero Lagrangian Lyapunov exponent, as the positive Eulerian Lyapunov exponent refers to the vortex motion and the null Lagrangian Lyapunov exponent implies that two nearby particles remain close during their motion, performing a regular orbit around the chaotically moving vortex.

Analogously to what has been observed for the three-vortex system, for $N = 4$ there are also chaotic particle trajectories. These results indicate once more that there is no strict link between Eulerian and Lagrangian chaoticity. In the present situation, the discriminator between regular and chaotic behaviour is the distance of the advected particles from the vortices. In general, we note that there is no particle ejection from the regular islands for (numerically) arbitrarily large integration times. This behaviour suggests a true asymptotic nature of the regular islands. A posteriori, the above results seem to be qualitatively analogous to the existence of invariant curves separating regular and chaotic regions in non-integrable Hamiltonian systems. The difference with the present case comes from the fact that the Hamiltonian is not time periodic, and the size, position and shape of the regular islands may aperiodically vary with time.

In order to further characterize the coexistence of regular and chaotic Lagrangian behaviour in vortex systems, we have evaluated the Lagrangian Lyapunov exponents for an ensemble of initial conditions chosen at random in the vortex domain. The trajectories appear to be roughly divided in two classes: the regular trajectories where $\lambda_L = 0$ and the chaotic ones where $\lambda_L > 0$.

One may wonder whether a much more complex Eulerian flow, such as 2D turbulence, may give the same scenario for particle advection: i.e. regular trajectories close to the vortices and chaotic behaviour between the vortices.

We show now that this is indeed the case and that the chaotic nature of the trajectories of advected particles is not strictly determined by the complex time evolution of the turbulent flow.

To this end, one has to study the Lagrangian dynamics by integrating the equations of motion (8.2) of an ensemble of passively advected particles in the stream function obtained by numerically integrating the forced-dissipative

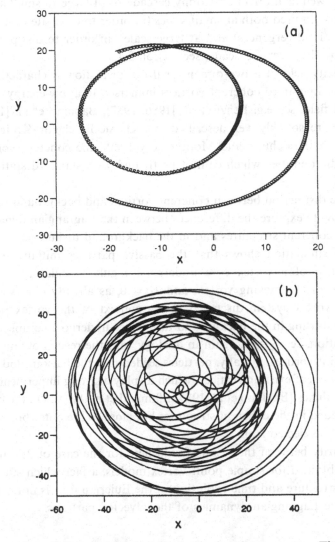

Figure 8.6 Particle trajectories in the four-vortex system. The Eulerian dynamics is in this case chaotic. Panel (a) shows a regular Lagrangian trajectory and panel (b) shows a chaotic Lagrangian trajectory. The different behaviour of the two particles is due to different initial conditions [Babiano et al. 1994].

Navier–Stokes equations in two dimensions [Babiano et al. 1994]. The Eulerian equations of motion are given by

$$\frac{\partial \omega}{\partial t} + J(\omega, \psi) = F + D \tag{8.44}$$

where $\omega = \Delta \psi$ is the vorticity and J is the two-dimensional Jacobian. The terms F and D indicate forcing and dissipation respectively. In the case of two-dimensional turbulent flows, there is an *inverse* energy cascade from small to large scales, as well as a direct enstrophy cascade from large to small scales. Dissipation is thus required both at small scales (in order to dissipate enstrophy and avoid ultraviolet divergences) and at large scales, in order to dissipate the energy piled up by the inverse cascade, see section 2.4.

A typical vorticity field of a two-dimensional turbulent flow is characterized by the presence of long-lived coherent vortices immersed in a low-energy background turbulent field, see e.g. Benzi et al. [1986, 1987], Babiano et al. [1987]. While the latter is reasonably well described by a classic Batchelor–Kraichnan statistical approach [Kraichnan and Montgomery 1980], the coherent vortices behave as individual entities which cannot be treated with standard statistical approaches.

Given the basic distinction between coherent vortices and background turbulence, we want now to explore the differences between the Lagrangian dynamics in the interior of coherent structures and in the background turbulence.

The numerical simulation shows that the passive particles initially seeded in the deep interior of a vortex core undergo a regular oscillatory motion around the centre of the moving vortex, and $\lambda_L = 0$, as already observed for the simple point vortex system discussed in the context of the vortex points systems. Particles starting in the background turbulence undergo a complex and unpredictable motion, i.e. $\lambda_L > 0$, again in analogy with the results obtained in the case of the point vortices. Usually, particles seeded into the background field do not enter the cores of existing vortices, which display a strong impermeability to inward particle fluxes [Babiano et al. 1987, Elhmaidi et al. 1993]. On the other hand, particles seeded inside coherent structures but close to the outer boundary of the vortex may be ejected from it.

Note the similarity between the results found here for the case of 2D turbulence and those obtained for simple point vortex models, a fact which suggests that the detailed structure and time evolution of the Eulerian flow is not crucial for determining the Lagrangian dynamics of the advected particles.

8.2.3   A comment on Lagrangian chaos

In conclusion we can say that there is no general relation between Lagrangian and Eulerian chaos. In the typical situation Lagrangian chaos may appear also for velocity fields with a regular behaviour, like in the case of the 2D incompressible fluid (8.37). However, it is also possible to have the opposite situation, with Eulerian chaos but $\lambda_L = 0$, as in the Lorenz model and in 2D systems well inside vortex structures. We finally stress that, in any case, one is not able to separate the Lagrangian from the Eulerian properties by the knowledge of the motion of only one trajectory, e.g. a buoy in the oceanic currents [Osborne et al. 1986]. Indeed, using the standard data treatment methods [Grassberger and Procaccia 1983] on $\mathbf{x}(t)$, one extracts the total Lyapunov exponent λ_T and not λ_L or λ_E.

Nevertheless, one can detect λ_L by looking at the time behaviour of a scalar field in two close points $\mathbf{x}$ and $\mathbf{x}'$. From (8.3) one has

$$\Theta(\mathbf{x}, t) - \Theta(\mathbf{x}', t) \simeq \sum_{i=1}^{d} \frac{\partial \Theta_o}{\partial (\mathbf{T}^{-t}\mathbf{x})_i} \left(\mathbf{T}^{-t}\mathbf{x} - \mathbf{T}^{-t}\mathbf{x}' \right)_i \sim e^{\lambda_L t}.$$

Note once more that the knowledge of the scalar field in one point $\mathbf{x}$ in principle is not sufficient to decide whether the Lagrangian chaos is present or not. One needs at least a two-point statistics. For example in the Lorenz model one observes a very irregular behaviour of $\Theta(\mathbf{x}, t)$ while $\Theta(\mathbf{x}, t) - \Theta(\mathbf{x}', t)$ increases only polynomially with t.

8.3   Statistics of passive fields

8.3.1   The growth of scalar gradients

Let us discuss the evolution of the gradients of passive scalars, $\nabla\Theta$, by using concepts borrowed from the theory of dynamical systems.

In Lagrangian terms, the solution of (8.1) with $\chi = 0$ is

$$\Theta(\mathbf{x}, t) = \Theta_o \left(\mathbf{T}^{-t}\mathbf{x} \right) \tag{8.45}$$

where $\Theta_o(\mathbf{x}) \equiv \Theta(\mathbf{x}, 0)$ is the initial distribution, and $\mathbf{T}^t$ is the operator which gives the evolution $\mathbf{x}(t) = \mathbf{T}^t\mathbf{x}(0)$, for equation (8.2).

One can write the spatial derivative of a passive scalar as

$$\frac{\partial \Theta(\mathbf{x}, t)}{\partial x_k} = \sum_{j=1}^{d} \frac{\partial \Theta_o(\mathbf{y})}{\partial y_j} \frac{\partial y_j}{\partial x_k}, \tag{8.46}$$

where the variable $\mathbf{y} = \mathbf{T}^{-t}\mathbf{x}$ is given by the time-reversed equation:

$$\frac{d\mathbf{y}}{dt} = -\mathbf{v}(\mathbf{y}, -t) \qquad \mathbf{y}(0) = \mathbf{x}. \tag{8.47}$$

The terms $\partial y_j/\partial x_k$ are strictly connected to the stability properties of (8.47). Indeed, after a time t, an uncertainty $\delta\mathbf{y}(0)$ in the initial condition becomes

$$\delta\mathbf{y}(t) = \mathbf{A}(t)\,\delta\mathbf{y}(0),$$

where the matrix elements of the linear operator **A** are $A_{jk} = \partial y_j / \partial x_k$. Then, the behaviour of $|\nabla\Theta|$ is essentially the same as that of $|\delta y(t)/\delta y(0)|$: if the system (8.2) is chaotic, this is the case also for the solutions of (8.47), due to the volume preserving nature of the dynamics. The global chaotic behaviour is characterized by the maximum Lyapunov exponent of the time-reversed flow

$$\lambda_1 = \lim_{t\to\infty} \frac{1}{t} \ln \left| \frac{\delta y(t)}{\delta y(0)} \right| = \lim_{t\to\infty} \frac{1}{t} \ln |\nabla\Theta|.$$

Therefore, λ_1 gives the typical exponential growth of the gradients.

It is worth stressing that the maximum Lyapunov exponent λ_1 of the reverse motion is equal to the opposite of the minimum Lyapunov exponent of the direct motion. The sum of the Lyapunov exponents is zero for conservative systems (in our context defined by the flows of incompressible fluids). Therefore, the absolute values of the maximum and minimum Lyapunov exponents are always equal in two dimensions. In three dimensions this could be false. It is thus conceivable that values for the maximum Lyapunov exponents of the direct and of the reversed flows may be different, although they seem to be equal in most situations [Feingold et al. 1988]. However, even if in three dimensions the statistics of contractions and expansions were not identical, these arguments should be still valid, since the only important feature is the presence of Lagrangian chaos, i.e. $\lambda_1 > 0$ and hence $\lambda_d < 0$ in the direct flow.

8.3.2   The multifractal structure for the distribution of scalar gradients

The maximum Lyapunov exponent is the typical rate of increase of the distance of nearby trajectories, although there exist fluctuations around λ_1 of the effective Lyapunov exponent γ computed for a finite-time trajectory. In fact, the local growth rate $\gamma(\mathbf{x}, t)$ is defined by

$$|\nabla\Theta| \propto e^{\gamma t} \tag{8.48}$$

and $\lim_{t\to\infty} \gamma(\mathbf{x}, t) = \lambda_1$, for almost all initial conditions $\mathbf{x}$. The generalized Lyapunov exponents, $L(q)$, are a natural way of characterizing these finite time fluctuations. They are defined as follows

$$\overline{|A_{jk}|^q} = \overline{\left| \frac{\partial (\mathbf{T}^{-t}\mathbf{x})_j}{\partial x_k} \right|^q} \sim e^{L(q)t} \qquad \text{for large } t \tag{8.49}$$

where $\overline{(\cdots)}$ means time average. The $L(q)$ will be discussed in more detail in appendix B. The relation between the maximum Lyapunov exponent and $L(q)$ is given by

$$\lambda_1 = \lim_{q\to 0} \frac{L(q)}{q}.$$

From general theorems of probability theory [Feller 1971], $L(q)$ is shown to be a convex function of q, i.e.

$$\frac{L(q')}{q'} \geq \frac{L(q)}{q} \qquad \text{for } q' > q$$

implying

$$L(q) \geq q\,\lambda_1.$$

The equals sign holds only when the effective Lyapunov exponent has negligible finite-time fluctuations $O(1/t)$. In most cases, it is sensible to assume that the probability of measuring a given effective Lyapunov exponent decays as $\exp(-S(\gamma)\,t)$ for $\gamma \neq \lambda_1$. It can be shown that $L(q) = \max_\gamma [\gamma q - S(\gamma)]$, the Legendre transform of $S(\gamma)$. Gaussian γ-fluctuations (i.e. log–normal fluctuations of the gradients) correspond to $S(\gamma) = (\gamma - \lambda_1)^2/(2\sigma^2)$ and $L(q) = \lambda_1 q + (\sigma^2/2)q^2$. Absence of γ-fluctuations corresponds to the linear behaviour $L(q) = \lambda_1 q$.

Using (8.46) and (8.49), for the moments of order q we have

$$\langle |\nabla\Theta(\mathbf{x},t)|^q \rangle \sim e^{L(q)\,t} \tag{8.50}$$

where $\langle \cdots \rangle$ means spatial average and, if the system is chaotic, we made the ergodic assumption $\overline{(\cdots)} = \langle (\cdots) \rangle$.

Let us consider now the probability measure $d\mu(\mathbf{x}) \propto |\nabla\Theta(\mathbf{x})|d\mathbf{x}$, and its coarse-graining over boxes Λ_i of size ℓ: $p_i(\ell) = \int_{\Lambda_i} d\mu(\mathbf{x})$. If one looks at the scaling properties of the moments of the discretized measure with respect to the size of the boxes, one expects $\langle (p_i(\ell))^q \rangle \propto \ell^{(q-1)d_q+d}$, where d is the spatial dimension and d_q are the Renyi dimensions [Halsey et al. 1986, Paladin and Vulpiani 1987a]. In the literature the interested reader can find the relations between the generalized dimensions d_q and $L(q)$ [Bessis et al. 1988]. Here it is sufficient to note that the intermittency in the chaotic behaviour, i.e. a nonlinear dependence of L on q, leads to a multifractal structure of the probability measure μ. That is to say, the coarse-grained probability scales with a position-dependent index $p_i(\ell) \propto \ell^{\alpha(i)}$, where $\alpha(i)$ is the singularity of the point $\mathbf{x}$, the centre of the box Λ_i, with respect to the Lebesgue measure, which would give $p_i(\ell) \propto \ell^d$. The dominant contribution to the moment of order q is given by an appropriate value $\bar{\alpha}(q)$, that is by the boxes with $\alpha(i) \in [\bar{\alpha}, \bar{\alpha} + d\alpha]$. The more relevant the multifractality, the larger the deviations of d_q from a constant value, see appendix E.

Neglecting the diffusion coefficient, the conservative nature of (8.1) with $\chi = 0$, implies that the support of the measure μ is the full space, i.e. the fractal dimension $d_0 = d$. Moreover, it has been shown for rather generic cases that a non-uniform growth of $\nabla\Theta$ leads to an information dimension (fractal dimension of the set of full probability measure) $d_1 < d_0$, and to an anomalous scaling with a non-constant d_q [Paladin and Vulpiani 1987a].

Such a multifractal object is observable only on length scales

$$\ell > \ell_2 \sim \ell_0 e^{-\gamma_{\max} t} \tag{8.51}$$

where ℓ_0 is the initial scale of $\nabla\Theta$ and $\gamma_{\max}$ is the maximum local growth rate. This is due to the fact that for $\ell < \ell_2$, the volume stretching which causes the increase of the Θ gradients, has not had time enough to act. One can also obtain an upper limit for the scale where multifractality can be observed [see section 8.3.4, remark IV].

Molecular diffusion is expected to be effective in smoothing the scalar field distribution at scale

$$\ell_d \sim \sqrt{\chi/\lambda_1}. \tag{8.52}$$

Therefore, the approximate solution (8.45) breaks down and has to be replaced by the solution of the full equation (8.1) after the typical time needed by the stretching mechanism to generate a scale of order ℓ_d, i.e. $t_d \sim (1/\lambda_1)\ln\left(\ell_0\sqrt{\lambda_1/\chi}\right)$. For $t > t_d$, the gradients typically stop their exponential growth. Actually, as the highest γ-value in (8.48) begins to satisfy the relation $\gamma t \sim \ln\left(\ell_0\sqrt{\lambda_1/\chi}\right)$, a gradual smoothing of the highest gradients should appear with corresponding modifications of the dimension d_q for the q-order moments to which they give the main contribution.

8.3.3   The power spectrum of scalar fields

We discuss the fluctuations of passive scalars on length scales that are small enough, such that the convected quantity is able to forget its initial state at large scales, but large enough, in order to safely neglect the molecular diffusion effects. Useful information on the small-scale properties of $\Theta(\mathbf{x})$ is provided by the structure function [Monin and Yaglom 1975]

$$S_\theta(r) = \left\langle |\Theta(\mathbf{x}+\mathbf{r}) - \Theta(\mathbf{x})|^2 \right\rangle = 2\int_0^\infty \Gamma(k)\left(1 - \frac{\sin kr}{kr}\right) dk \tag{8.53}$$

where $\Gamma(k)$ is the power spectrum of $\Theta(\mathbf{x})$, which gives the Θ^2 amount contained in the wave vectors whose moduli are around k.

If $\Gamma(k)$ and $S_\theta(r)$ follow power laws i.e.

$$\Gamma(k) \propto k^{-\sigma} \tag{8.54}$$

and

$$S_\theta(r) \propto r^\delta \tag{8.55}$$

one has $\delta = \sigma - 1$ if $1 < \sigma < 3$, see, for example, Rose and Sulem [1978], Babiano et al. [1984] for a detailed derivation of this scaling relation.

In three-dimensional turbulent fluids, for large Prandtl number v/χ, the power spectrum obeys the Obukhov–Corrsin scaling law $\Gamma(k) \propto k^{-5/3}$ in the inertial range (where the nonlinear transfer of the fluid energy overwhelms its molecular dissipation) [Obukhov 1949, Corrsin 1951], and the Batchelor law $\Gamma(k) \propto k^{-1}$ in the viscous convective range (where the velocity field behaviour is dominated by viscous effects but the molecular diffusion of the passive scalar is still negligible) [Batchelor 1959].

It is interesting to observe that Batchelor's original result was obtained by depicting the turbulent fluid, in the viscous convective range, as composed of blobs, with linear size of the order of the Kolmogorov microscale, that are subjected to the straining action of a velocity field considered regular on these length scales. Then, it is not surprising that one can derive Batchelor's result by means of ideas borrowed from dynamical systems theory, without using phenomenological assumptions. In fact, the k^{-1} regime is a spectral property of an advected scalar which follows directly from Lagrangian chaos, and does not depend on the dimensionality d, for $d \geq 2$. Turbulent fluids and very simple flows, for instance periodic in time, exhibit this same property.

Let us consider a smooth initial condition $\Theta_o(\mathbf{x})$ and $N \gg 1$ particles, convected according to (8.2) by a smooth velocity field. The ith particle is initially in $\mathbf{x}^{(i)}(0)$ and transports its own $\Theta^{(i)} = \Theta_o(\mathbf{x}^{(i)}(0))$. At time t one has

$$S_\theta(r) \approx N^{-2} \sum_{i,j} (\Theta^{(i)} - \Theta^{(j)})^2 P(\mathbf{x}^{(i)}, \mathbf{x}^{(j)}, t) \tag{8.56}$$

where $P(\mathbf{x}^{(i)}, \mathbf{x}^{(j)}, t)$ is the probability density to have $|\mathbf{x}^{(i)} - \mathbf{x}^{(j)}| = r$ at time t. The dominant contribution to the sum (8.56) comes from the pairs of particles with large $|\Theta^{(i)} - \Theta^{(j)}|$, that is from initially distant particles, since $\Theta_o(\mathbf{x})$ is smooth. In the presence of Lagrangian chaos, two particles which are initially at distance ℓ_0 can reach a distance $O(r)$ after a time $t_r \sim \lambda_1^{-1} \ln(\ell_0/r)$. Here λ_1 is again the maximum Lyapunov exponent of the time reversed flow (8.47). In other words, t_r is the typical time necessary to get a 'good mixing' on scale r: the $\Theta^{(i)}$ of the particles in a box of edge r, after this time, are distributed in the whole range of accessible values. Therefore $P(\mathbf{x}^{(i)}, \mathbf{x}^{(j)}, t)$ for $t > t_r$ becomes independent of r, (apart from logarithmic corrections) so that:

$$S_\theta(r) \sim r^\delta \quad \text{with } \delta = 0. \tag{8.57}$$

One cannot directly use the dimensional relation $\delta = \sigma - 1$ because the value $\delta = 0$ is out of the validity range for locality of the power spectrum. One can only assert that

$$\Gamma(k) \propto k^{-\sigma} \quad \text{with } \sigma \leq 1.$$

However, the conservation law $\int \Gamma(k)dk = $ constant implies $\sigma \geq 1$, so that the Batchelor scaling $\sigma = 1$ follows.

With a more detailed argument Ott and Antonsen [1990] find for $\Gamma(k,t)$ the following expression:

$$\Gamma(k,t) \sim \frac{1}{kt} P\left(\frac{1}{t}\ln(k/k_0),\ t\right);$$ (8.58)

where $P(\gamma,t)$ is the probability density to get an effective Lyapunov exponent γ in a time interval t. $P(\gamma,t)$ can be written in terms of the function $S(\gamma)$, related via a Legendre transformation to the $L(q)$ of the dynamical system (8.47), as:

$$P(\gamma,t) \sim \exp(-S(\gamma)t).$$ (8.59)

The reader can find in appendix B that for $\gamma \simeq \lambda_1$:

$$S(\gamma) \simeq \frac{(\gamma - \lambda_1)^2}{2\sigma^2}$$

in a typical dynamical system. Then, from a simple computation, one has that in the range

$$k \in [x_1 k_0 \exp(\lambda_1 t),\ x_2 k_0 \exp(\lambda_1 t)], \qquad \text{with } x_1 < 1 \text{ and } x_2 > 1,$$

the leading term of (8.58) is of order k^{-1} if $(1/t)\ln(\max[x_2, 1/x_1])$ is small. This shows that, for large t, (8.58) gives a small modification of the k^{-1} law in a very large range of k values.

Moreover, Horita et al. [1990] recently found numerical evidence that, in typical two-dimensional maps showing strong intermittency, one has:

$$S(\gamma) = 0, \qquad \text{for } 0 < \gamma < \lambda_1.$$

In this case (8.58) gives exactly the k^{-1} behaviour for $k < k_0 \exp(\lambda_1 t)$.

8.3.4 Some remarks on the validity of the Batchelor law

(I) To obtain the result $\Gamma(k) \sim k^{-1}$ one has to linearize the stability equation which rules the separation between two fluid particles; and this separation has to increase at an exponential rate so as to produce a true Lagrangian chaos. These are sufficient conditions for the existence of the k^{-1} power-law regime.

This implies that for turbulence in three dimensions the k^{-1} law holds only in the viscous convective subrange, where the velocity field is smoothed by the viscosity, i.e. $|v(x+r) - v(x)| \sim r$. In the inertial range, by contrast, the velocity field is highly irregular and the typical velocity difference is not linear in r, i.e. $|v(x+r) - v(x)| \sim r^h$ with $h < 1$ ($h = 1/3$ in the Kolmogorov theory).

In two dimensions we are faced with a quite different scenario. It is believed that in this case there exists an inertial range where enstrophy (the integral of the square of the vorticity) cascades from larger to smaller scales. Furthermore, the inviscid Euler equations in $d = 2$ have some regularity properties, absent in $d = 3$, which also leads to $|v(x+r) - v(x)| \propto r$ in the inertial range [Rose and Sulem 1978]. Therefore the power spectrum of a scalar field, passively convected by a two-dimensional turbulent flow, should behave as k^{-1} at all length scales, and there are not the two different regimes exhibited by three-dimensional turbulence. In two-dimensional fluids the k^{-1} law is not restricted to high Prandtl numbers: due to the smoothness of the velocity field, the Batchelor law holds in the range $[k_0, k_B]$, where $k_0 \approx \ell_0^{-1}$. The Prandtl number only modifies the value of k_B. Some closure approximations [Lesieur and Herring 1985] and numerical simulations [Babiano et al. 1987] of the two-dimensional Navier–Stokes equations, provide evidence for a k^{-1} power law in the inertial range.

(II) The value of the scaling exponent -1 of the Batchelor law is not modified by intermittency. The k^{-1} law is an exact result [Kraichnan 1974], but the constant in front of this scaling law may be sensitive to the details of the Lagrangian motion.

(III) The scaling law (8.55) admits an intuitive interpretation in terms of the fractal structure of the iso-Θ surfaces. The fractal dimension D_Θ of these surfaces and the exponent δ are related by [Mandelbrot 1975, Procaccia 1984]:

$$D_\Theta = d - \frac{\delta}{2}.$$

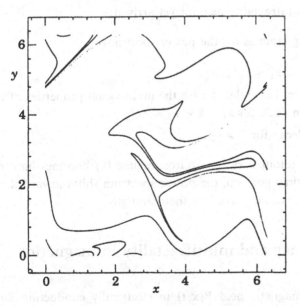

Figure 8.7 Particles advected by the velocity field obtained from a 5-mode truncation of the 2D Navier–Stokes equations (8.32), with $Re = 26.06$ (periodic regime), after ≈ 8 characteristic times defined as $\langle (\nabla \mathbf{v})^2 \rangle^{-1/2}$. At $t = 0$ the particles are uniformly distributed on the circle $(x - 3.20)^2 + (y - 3.10)^2 = (3.05)^2$ [Crisanti et al. 1990a].

The k^{-1} law corresponds to surfaces (lines if $d = 2$) invading the whole space, as $D_\Theta = d$, that is the most chaotic situation. We have described a scenario where Lagrangian chaos is the fundamental ingredient. This idea is illustrated in Fig. 8.7 which shows one iso-Θ line in a two-dimensional fluid, with a time-periodic velocity field obtained by a five-mode truncation of the Navier–Stokes equations, whose Lagrangian behaviours have been studied in section 8.2, see also appendix E. There is an impressive similarity with the iso-Θ line computed by numerical calculations of two-dimensional fully developed turbulent flows with $\approx 10^4$ modes (see Fig. 4b of Babiano et al. [1987]).

(IV) The multifractal nature of $\nabla\Theta$ is revealed on scales of length $\ell > \ell_2 \sim \ell_0 e^{-\gamma_{max}t}$. For wave numbers corresponding to these scales, the scaling index of the structure function $S_\theta(r)$, and thus the exponent of the power spectrum $\Gamma(k)$, is dictated by the (multifractal) behaviour of the measure $d\tilde{\mu} = |\nabla\Theta|^2 dx$. However, the onset of the k^{-1} regime for $\Gamma(k)$ occurs at length scales $\ell > \ell_1 \sim \ell_0 e^{-\lambda_1 t}$, on which a good mixing has taken place and $d\tilde{\mu}$, when coarse grained with such resolutions, most likely reached the uniform state. Therefore one expects the following different regions of scales:

(a) $\ell > \ell_1$, where multifractality has been lost, because 'good mixing properties' allow the system to forget the details of the dynamics;

(b) $\ell_2 < \ell < \ell_1$, where multifractality is at work;

(c) $\ell < \ell_2$, where multifractality has not yet arrived;

with the corresponding regimes for the power spectrum:

(a) $\Gamma(k) \propto k^{-1}$ for $k < k_1 \sim \ell_1^{-1}$;

(b) a non-trivial power law, related with the multifractal properties of the dynamical system (8.2), for $k_1 < k < k_2 \sim \ell_2^{-1}$;

(c) an exponential decay for $k > k_2$.

Let us stress that without a steady source, regime (a) becomes limited also from below, since, as time goes on, the power spectrum shifts toward the large k region and the low wave numbers lose their strength.

8.3.5   Intermittency and multifractality in magnetic dynamos

The behaviour of a magnetic field $\mathbf{B}(\mathbf{x}, t)$ in electrically conducting fluids is of great physical relevance both in astrophysics and geophysics [Weiss 1985, Zel'dovich and Ruzmaikin 1986]. In principle, one should consider the whole set of magnetohydrodynamic equations for the velocity field $\mathbf{v}(\mathbf{x}, t)$ and the magnetic field $\mathbf{B}(\mathbf{x}, t)$. However, a simpler manner to attack the problem is to assume a prescribed velocity field as known input, unaffected by the electromagnetic field. This is the so-called kinematic dynamo problem, described by the equations:

$$\left.\begin{aligned} \partial_t \mathbf{B} + (\mathbf{v} \cdot \nabla) \mathbf{B} &= (\mathbf{B} \cdot \nabla) \mathbf{v} + \chi \nabla^2 \mathbf{B} \\ \nabla \cdot \mathbf{B} &= 0 \end{aligned}\right\} \tag{8.60}$$

obtained from the Maxwell equations and the magnetohydrodynamic equations, where χ is the magnetic diffusivity coefficient of the fluid, inversely proportional to the conductivity. In this problem, the basic question is whether the motion can enhance a very small initial magnetic field, in the absence of external electromotive forces.

In the following, the statistical properties of the magnetic field on small scales, for a fluid with high conductivity, will be characterized by exploiting the connection between the kinematic dynamo equations and the dynamical system (8.2), which describes the Lagrangian motion of marker particles in that fluid.

Consider (8.60) for $\chi = 0$

$$\partial_t \mathbf{B} + (\mathbf{v} \cdot \nabla)\mathbf{B} = (\mathbf{B} \cdot \nabla)\mathbf{v}, \tag{8.61}$$

and the equation which rules the evolution of the distance between two fluid particles, say $\mathbf{R} = \mathbf{x}^{(2)} - \mathbf{x}^{(1)}$, where $\mathbf{x}^{(1)}$ and $\mathbf{x}^{(2)}$ are the positions of the particles whose time behaviour is given by (8.2). In the limit $\mathbf{R} \to 0$, for the interparticle distance one obtains the linear equation for the tangent vector $\mathbf{z}$

$$\frac{dz_k}{dt} = \sum_{j=1}^{d} \frac{\partial v_k}{\partial x_j} z_j. \tag{8.62}$$

Equation (8.62) is formally identical to (8.61), since $d/dt = \partial_t + (\mathbf{v} \cdot \nabla)$. By considering the flow of a fluid particle, $\mathbf{x}(t) = \mathbf{T}^t \mathbf{x}(0)$, one can write for the evolution of the magnetic field:

$$B_i(\mathbf{x}, t) = \sum_{j=1}^{d} A_{ij}(\mathbf{x}, t) \cdot B_j(\mathbf{T}^{-t}\mathbf{x}, 0),$$

with

$$A_{i,j}(\mathbf{x}, t) = \frac{\partial x_i}{\partial (\mathbf{T}^{-t}\mathbf{x})_j}.$$

The behaviour of the magnetic field is thus determined by two factors: (1) the evolution of the fluid particles and (2) the rate of growth of the tangent vector, i.e. the separation of particle pairs.

The equivalence of (8.62) and (8.61) allows us to use standard methods of dynamical systems. A magnetic field, convected by a fluid which exhibits Lagrangian chaos, will be exponentially amplified and the spatial variations in the magnetic field amplification can be described by the generalized Lyapunov exponents $L(q)$. The generalized Lyapunov exponents characterize the fluctuations around λ_1 of the effective Lyapunov exponent $\gamma(\mathbf{x}, t) = (1/t) \ln |z(t)|$, obtained by a measuring on long but finite times. In the present context this set of exponents may be defined by the spatial averages for the **B**-moments:

$$\langle |\mathbf{B}(\mathbf{x}, t)|^q \rangle \sim e^{L(q)t} \qquad \text{for} \qquad t \to \infty. \tag{8.63}$$

This definition is equivalent to the standard definition in terms of time averages for ergodic systems. Besides (8.63) is more appropriate for practical purposes.

We remark that an idea of generalized Lyapunov exponents has been introduced in the context of the magnetic dynamo, although limited to random velocity field, in Molchanov et al. [1984] and Zel'dovich et al. [1984]. For a study of the magnetic dynamo in terms of a periodic orbits expansion, see Aurell and Gilbert [1993], Balmforth et al. [1993].

Let us stress that since, in general, chaotic dynamical systems have a nonlinear shape of $L(q)$, for an intermittent behaviour in Lagrangian chaos, it is not necessary to consider highly irregular velocity fields, as in Zel'dovich et al. [1984], or strongly chaotic maps, as in Ott and Antonsen [1988, 1989]. As an example in Fig. 8.8 we show the nonlinear shape of $L(q)$ obtained for the ABC map (8.19) corresponding to a three-dimensional time-periodic velocity field. As it was pointed out in the case discussed in section 8.2, a nonlinear $L(q)$ involves the multifractality of the probability measure given by

$$d\mu(x) = \frac{|\mathbf{B}(\mathbf{x}, t)| d^d x}{\int |\mathbf{B}(\mathbf{x}, t)| d^d x}, \tag{8.64}$$

where the normalization integral is over the space domain covered by the fluid.

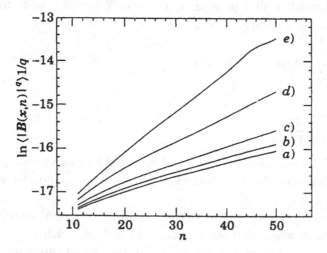

Figure 8.8 $(1/q) \ln \langle |\mathbf{B}(\mathbf{x}, n)|^q \rangle$ versus the number of steps performed, n, for the ABC map (8.19) with parameters $A = 0.5, B = 0.08, C = 0.16$. $q = 0.10$ (a), 0.25 (b), 0.50 (c), 1.00 (d), 1.50 (e) [Falcioni et al. 1989].

The multifractal structure of the magnetic field is quite similar to the one of the passive scalar gradients. Multifractality and fluctuations in time of the degree of chaos are two aspects of the same problem. The relation between the spatial multifractality (i.e. non-constant d_q) and temporal intermittency (i.e. nonlinear $L(q)$) is not simple: a detailed discussion for some chaotic systems may be found in Ott and Antonsen [1989]. However, a nonlinear $L(q)$ corresponds to a highly irregular spatial distribution for $\mathbf{B}(\mathbf{x}, t)$. Indeed, the volume of the regions where $\mathbf{B}(\mathbf{x}, t) \sim \exp(\gamma t)$, if $\gamma \neq \lambda_1$, vanishes as $\exp(-S(\gamma) t)$, although they give the leading contribution to $\langle |\mathbf{B}(\mathbf{x}, t)|^q \rangle$, for $q \neq 0$. Therefore, the multifractal structure of the probability measure (8.64) can be observed only on very small scale, which decreases exponentially in time [Falcioni et al. 1989].

This scenario is quite particular of fluids with zero magnetic diffusivity coefficient. We expect that for times larger than: $t^* \sim -(1/\lambda_1) \ln(\chi)$ the magnetic diffusivity cannot be neglected, since its effect becomes relevant on scales where $\exp(-\lambda_1 t) \sim \chi^{1/2}$. For instance, it is well known that in two dimensions the magnetic field cannot increase for $\chi \neq 0$. Let us note that the magnetic field is a vector quantity. Even in 2D it has a sign and there are cancellations between opposite sign. Therefore, for many aspects it is much more relevant to look at the growth of

$$| \langle \mathbf{B}(\mathbf{x}, t) \rangle | \sim e^{\alpha t}.$$

The exponent α can be much smaller than any generalized Lyapunov exponent $L(q)/q$. For instance in a time-dependent 2D flow, α is identically zero [Zel'dovich 1957] although the Lyapunov exponent might be positive. In order to take into account the cancellations, one should consider the signed measure obtained by $\mathbf{B}$ instead of the multifractal probability measure (8.64) [Du et al. 1994].

Let us conclude with a brief discussion on the case $\chi \neq 0$. In this case, (8.64) becomes [Molchanov et al. 1984, Childress and Gilbert 1995]

$$\mathbf{B}(\mathbf{x}, t) = M_{\mathbf{x}} \big[\mathbf{A}(\mathbf{x}, \mathbf{y}, t) \, \mathbf{B}(\mathbf{y}, 0) \big],$$

where $M_{\mathbf{x}}$ denotes the mean value over the Wiener paths starting from $\mathbf{y}$ at time 0 and ending in $\mathbf{x}$ at time t, i.e. the solutions of

$$\frac{d\mathbf{x}}{dt} = \mathbf{v}(\mathbf{x}, t) + \eta, \qquad \mathbf{x}(0) = \mathbf{y}, \qquad \mathbf{x}(t) = \mathbf{x};$$

η is a gaussian process of zero mean and correlation

$$\langle \eta_i(t) \, \eta_j(t') \rangle = 2 \chi \delta_{ij} \, \delta(t - t');$$

and

$$\mathbf{A}(\mathbf{x}, \mathbf{y}, t) = \exp \left[\int_0^t d\tau \, \mathbf{a}[\mathbf{z}(\tau)] \right]$$

with

$$a_{ij} \equiv \frac{\partial v_i}{\partial x_j} \quad \text{and} \quad \mathbf{z}(0) = \mathbf{y}, \quad \mathbf{z}(\tau) = \mathbf{x}.$$

In the case of a three-dimensional velocity field which is renewed after a finite time, and for which there exists a time τ such that the velocity field at time $t + \tau$ is completely uncorrelated from that at time t, it is possible to show [Molchanov et al. 1984] that $L(q, \chi) \to L(q, 0)$ for $\chi \to 0$. Moreover $L(q, 0)$ is a nonlinear function of q, indicating that, in this extreme case, the intermittency does not disappear for finite magnetic diffusivity coefficient.

Note that in two dimensions the antidynamo theorem [Zel'dovich 1957] implies $L(q, \chi) = 0$ for all $\chi \neq 0$. In terms of probability theory this is due to the fact that in two dimensions the Wiener trajectories are reflexive whereas in three dimensions they are not [Molchanov et al. 1984]. On the other hand, some results [Finn and Ott 1990] for three-dimensional regular velocity fields indicate that for small χ and $t > t^*$, $L(q, \chi) = \omega q$, where ω is the exponential growth rate of the magnetic flux, $\omega = 0$ in two dimensions. Let us stress that in general the limit of ω for $\chi \to 0$ does not coincide with the Lyapunov exponent λ_1 of equation (8.2), where $\chi = 0$. For a detailed treatment of Anosov systems and maps on a 2D torus, which are in general nonlinear, see Collet [1989] and Oseledec [1993].

Chapter 9

Chaotic diffusion

The dispersion of a contaminant in a fluid is the result of two different effects: advection and molecular diffusion. At the fundamental level it is of interest to understand the mechanisms that lead to transport enhancement as a fluid is driven farther from the motionless state. The main reason for the enhancement is that the trajectories of individual fluid elements, or tracers, can be quite complex even in simple laminar flows [Moffatt 1983]. This issue is strongly related to the Lagrangian chaos discussed in chapter 8. Indeed the diffusive properties are very often determined by the presence of chaos in the motion of test particles.

Since the diffusion process takes place over long times and large spatial scales, the physics of passive diffusive transport can be characterized in terms of effective diffusion coefficients D_{ij} which contain the cumulative effects of advection and molecular diffusion. The effective equation at long times, and large scale, can be derived by the usual multiple scale or 'hydrodynamic' analysis [Maxwell 1890] and is:

$$\frac{\partial}{\partial_t} \langle \Theta \rangle = D_{ij} \frac{\partial^2}{\partial x_i \, \partial x_j} \langle \Theta \rangle + O\left(|\nabla^2 \langle \Theta \rangle|^2\right), \qquad i, j = 1, \ldots, d, \qquad (9.1)$$

where $\langle \Theta \rangle$ is the concentration averaged locally over a volume of linear dimension much larger than the typical length l of the velocity field. Equation (9.1) is a gradient expansion valid when $|\nabla \langle \Theta \rangle| / \langle \Theta \rangle \ll l^{-1}$. If we neglect the higher order terms, (9.1) is the diffusion equation with an effective diffusion tensor D_{ij}.

The diffusion coefficients D_{ij} have direct practical importance since they measure the spreading over very long times of a spot of tracer particles evolving according to (8.5). Therefore, it is possible to compute D_{ij} directly from the covariance tensor

$$D_{ij} = \lim_{t \to \infty} \frac{1}{2t} \langle\, (x_i(t) - \langle x_i \rangle)\, (x_j(t) - \langle x_j \rangle)\, \rangle, \qquad i, j = 1, \ldots, d. \qquad (9.2)$$

Here the average is taken over the initial positions or, equivalently, over an ensemble of test particles.

Equations (9.1) and (9.2) imply that the diffusion process is gaussian, at least on large time and space scales. This is the typical situation. There exist, nevertheless, cases where anomalous diffusion is observed [Pasmanter 1988, Zaslavsky et al. 1993], i.e. $\langle |x_i(t) - \langle x_i(t) \rangle|^2 \rangle \sim t^{2v}$ with $v \neq 1/2$. In addition, it is possible to have apparent superdiffusion, i.e. $v > 1/2$, up to a certain typical time after which there is a standard diffusion, such as the relative diffusion in 3D turbulence where the average square distance between two test particles increases with an exponent $v = 3/2$ in the inertial range [Richardson 1926].

It is easy to realize that there exists a relation between the diffusion coefficients and the correlation function of the Lagrangian velocity. A simple calculation gives:

$$\langle |x_i(t) - x_i(0)|^2 \rangle = \int_0^t \int_0^t \langle v_i(\mathbf{x}(t_1)) v_i(\mathbf{x}(t_2)) \rangle \mathrm{d}\, t_1 \mathrm{d}\, t_2 \simeq 2t \int_0^t C_{i,i}(\tau) \mathrm{d}\, \tau$$

where

$$C_{i,i}(\tau) = \langle v_i(\mathbf{x}(t)) v_i(\mathbf{x}(t + \tau)) \rangle.$$

If $\int_0^\infty C_{i,i}(\tau) \mathrm{d}\tau < \infty$, one has thus the Taylor relation [Monin and Yaglom 1975]

$$D_{i,i} = \int_0^\infty C_{i,i}(\tau) \mathrm{d}\, \tau \qquad (9.3)$$

This is a particular case of the well-known Kubo formula [Kubo 1957], derived by Taylor [1921] in the fluid dynamics context.

Avellaneda and Majda [1989, 1991] obtained a very general result; if the molecular diffusivity χ is non-zero and the infrared contribution to the velocity field is weak enough, namely

$$\int \frac{\langle |\hat{\mathbf{v}}(\mathbf{k})|^2 \rangle}{k^2} \mathrm{d}\mathbf{k} < \infty \qquad (9.4)$$

where $\langle (\cdot) \rangle$ is a time average, then one has standard diffusion, i.e. the coefficients in (9.2) are finite.

From the above results one has just two possible origins for the superdiffusion:

(a) strong correlation between $\mathbf{v}(\mathbf{x}(t))$ and $\mathbf{v}(\mathbf{x}(t + \tau))$ for large τ (from (9.3))

(b) velocity field with very long spatial correlation violating (9.4)

We shall discuss superdiffusion in sections 9.1.3, 9.1.4 and 9.2.

Real fluids always have a certain degree of Lagrangian chaos as we have seen in the previous chapter. The understanding of the diffusion process is therefore a hard task since it may depend in a complicated way on the detailed structure of the Eulerian velocity field. In the presence of Lagrangian chaos the diffusion properties are rather peculiar, even in the absence of molecular diffusion.

Due to its interplay with advection the effect of molecular diffusion can be non-trivial. To illustrate this point, we shall consider simple examples of diffusion of tracer particles in a specified flow, without considering the origin of this field.

9.1   Diffusion in incompressible flows

9.1.1   Standard diffusion in the presence of Lagrangian chaos

If the stream function is time-dependent, the system (8.2) is not autonomous so that, in general it is non-integrable and chaotic particle trajectories may exist. Because of the chaos, test particles may undergo diffusion processes even in very simple velocity fields, e.g. time-periodic laminar velocity fields, without the help of molecular diffusion. In order to illustrate this point we consider the following stream function [Solomon and Gollub 1988a,b]

$$\psi = \frac{A}{k} \sin[k(x + B \sin(\omega t))] W(y) \tag{9.5}$$

where A is the maximum y-velocity, $k = 2\pi/\lambda$ the wave number and $W(y)$ is a function that satisfies rigid boundary condition at $y = 0$ and $y = a$. The stream function (9.5) describes single-mode, time-dependent, two-dimensional Rayleigh–Bénard convection between two rigid boundaries. The direction y is

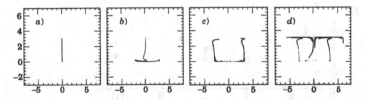

Figure 9.1 A line of 6000 particles, at the initial time (panel (a)) uniformly distributed on the line $x = 0$ between $y = 0$ and $y = \pi$, is transported according to equations (8.2) and (9.5) with $A = 0.2$, $B = 0.4$, $\omega = 0.5$, $\lambda = 2\pi$ and $W(y) = \sin(y)$. Panels (b), (c) and (d) show the spreading after, respectively, 2, 4 and 6 periods of the convecting field.

the vertical one and the two surfaces $y = 0$ and $y = a$ are the bottom and top of the convective cell respectively.

Because of the roll oscillation, the trajectory of a particle near the separatrices is chaotic. If one follows the evolution of a line of tracers initially located along the separatrices, see Fig. 9.1, one may observe that it is stretched and folded, while spreading among the rolls, and the particles diffuse in the x-direction. Numerical computations show that the diffusion coefficient D_{11} in the x-direction is proportional to B/λ, i.e. it is proportional to the width of the chaotic layer near the separatrices. It is relevant to note that, in spite of the simplicity of this model, the numerical results are in nice agreement with the experimental data on Rayleigh–Bénard convection [Solomon and Gollub 1988a,b].

Let us now briefly discuss the diffusion properties of test particles passively advected by the velocity field obtained with the 5-modes system discussed in section 8.2.1. We are interested in the range $\epsilon > \epsilon_c \simeq 0.7$. The motion is chaotic but one has long intervals of regular motion (the almost ballistic motion in the x-direction among the separatrices). So there are long-time correlations that induce a large value of D_{11}. The scenario is the following: the particles undergo regular ballistic motion in the x-direction for a certain time (diverging as $\epsilon \to \epsilon_c^+$), after which they are trapped inside a small highly chaotic region, then again ballistic motion and so on. This gives a strong time correlation in v_1 which decays on times much longer than $1/\lambda$ and induces a large value of D_{11}, as shown in Fig. 9.2 [Falcioni and Vulpiani 1989]. We shall see in section 9.1.4 that this mechanism of trapping for long times in a ballistic motion can induce superdiffusion.

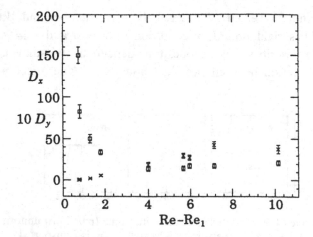

Figure 9.2 Diffusion coefficients $D_x = D_{11}$ (squares) and $10D_y = 10D_{22}$ (crosses) as functions of $Re - Re_1$, see section 8.2.1, for particles convected by the 5-modes truncation of the Navier–Stokes equations.

9.1.2 Standard diffusion in steady velocity fields

Real fluids always exhibit a (small) degree of molecular diffusion. The relevant dimensionless parameter which measures the relative importance of advection over molecular diffusion is the Peclet number

$$Pe = \frac{V L}{\chi}$$

where V is the typical velocity and L the typical length of the convective flow, e.g. the size of the vortex cell. The Peclet number can also be regarded as the ratio of the convection time L/V over the diffusion time L^2/χ across a distance L.

The physically interesting situation happens for large Peclet numbers. In this context we discuss the widely studied two-dimensional convective velocity field given by

$$\mathbf{v} = \left(\frac{\partial}{\partial y} \psi, -\frac{\partial}{\partial x} \psi \right) = (B \cos y, A \sin x),$$

where ψ is the stream function $\psi = A \cos x + B \sin y$.

For $|A| = |B|$ one has a model for Rayleigh–Bénard convection, since the phase space consists of square cells separated by lines (separatrices) where the rotation time diverges. The dispersion of a passive impurity on large scale is impossible without molecular diffusion that allows the jumping among different rolls. For a large Peclet number the effective diffusion coefficient has been found, theoretically [Shraiman 1987] and experimentally [Solomon and Gollub 1988b] to scale as

$$D_{ij} \sim \chi Pe^{1/2} \sim \chi^{1/2}. \tag{9.6}$$

D_{ij} is thus much larger than χ for $\chi \to 0$. A simple way to obtain (9.6) is the following [Pomeau 1985]. In the vicinity of the separatrix between two rolls, the component of the flow perpendicular to the separatrix vanishes, and the only mechanism of transport from one roll to another is molecular diffusion. Therefore, the only particles which can leave the roll, and contribute to transport, are those 'not too far' from the boundary. As a consequence, the transport is entirely due to particles in a layer of width δ near the separatrix. D can be estimated by noting that the particles close to the separatrices perform a random walk with diffusion coefficient $\sim L^2/\tau$ where $\tau = L/V$. The fraction of particles in the 'active' layer is $\sim \delta/L$, so that the effective diffusion coefficient is $D \sim (\delta/L)(L^2/\tau)$. The width δ depends on the molecular diffusion, and is roughly given by $\delta^2 \sim \chi\tau$. This immediately leads to the conclusion that the effective diffusion coefficient scales as $D \sim \chi Pe^{1/2}$. This result is obtained in a more rigorous way in Shraiman [1987].

When $|A| \neq |B|$, narrow channels arise among the convective cells. The motion of test particles inside a channel appears to be ballistic for long times and this enormously enhances the transport along the channel direction. The process is strongly anisotropic and can be regarded as due to long runs in the channel interrupted by trapping periods inside the rolls. We thus introduce two effective diffusion coefficients, D_l along the channel direction and D_t along the direction transversal to the channel. In the limit of small molecular diffusion one has

$$D_l \propto \frac{L^2}{V}\Big||A| - |B|\Big|^3 \chi^{-1} \quad \text{and} \quad D_t \propto V\Big||A| - |B|\Big|^{-1}\chi. \tag{9.7}$$

As in the case $|A| = |B|$, (9.7) can be derived by simple arguments. Without loss of generality, by a suitable choice of length and time units, we have $L = O(1)$ and $V = O(1)$ so that we can set $B = 1$ and $A = -(1 + \delta)$. The stream function becomes $\psi = \sin y - (1 + \delta)\cos x$, which for $\delta = 0$ describes convection cells of width 2π, where in the absence of a noise term the motion of a test particle is always periodic. The separatrices cross at the unstable hyperbolic fixed points of the flow. When $\delta > 0$ the border lines between cells do not coincide and there appear channels along the y-direction. For small δ the width of a channel is $\sim \delta$, although the maximum distance between the separatrices increases up to $\sim \delta^{1/2}$ near the unstable fixed points. The motion of a particle inside the channels, neglecting molecular diffusion, is ballistic and the velocity field changes sign between neighbouring channels. For small χ, a test particle can jump into a channel, because of molecular diffusion. Then, one has ballistic motion inside the channel with velocity $V_c \sim O(1)$ either in the up or in the down y-direction stopped by a capture from a cell after a time $T_c \sim \delta^2/\chi$, and so on. Let us consider the case for which

$$T_c/T_r \gg 1 \quad \text{i.e.} \quad \delta^2/\chi \gg 1,$$

since the circulation time $T_r \sim V/L \sim O(1)$. The typical length of a run along a channel is

$$L_c \sim T_c V_c \sim \delta^2/\chi.$$

The probability p to find a particle in a channel is proportional to its width $\sim \delta$, and thus

$$D_l \sim p\frac{L_c^2}{T_c} \sim \frac{\delta^3}{\chi}. \tag{9.8}$$

With similar arguments one can show:

$$D_t \sim \frac{\chi}{\delta}. \tag{9.9}$$

The results (9.8) and (9.9) are well confirmed by numerical simulations [Crisanti et al. 1990a,b].

We stress that in spite of the apparently 'anomalous' diffusion process – long runs interrupted by trappings – for large t the kurtosis of $(x(t) - \langle x \rangle)$ tends to the gaussian value 3 so that the diffusion is standard and gaussian.

The situation here is similar to the one obtained for the truncated Navier–Stokes equations discussed in section 9.1.1. There the 'jumping' was due to the Lagrangian chaos, here to the molecular diffusion. However, the physical mechanism which rules the diffusion is the same. The tracer is trapped for long times in a small limited region of space and then escapes along ballistic channels until a subsequent trap.

It is easy to realize that similar behaviour may be also found in three-dimensional flows. In the case of longitudinal dispersion of a contaminant for laminar flows in long straight tubes or channels Taylor [1953, 1954] obtained:

$$D_1 \sim \frac{1}{\chi}$$

similar to (9.8). The same result holds for ABC flow, where there are regular trajectories with a ballistic motion [Biferale et al. 1995b].

9.1.3 Anomalous diffusion in random velocity fields

Consider a 2D velocity field of the form

$$\mathbf{v} = (u(y), 0) \tag{9.10}$$

where $u(y)$ is a quenched random function with a given spectrum $S(k)$, i.e.

$$u(y) = \int_{-\infty}^{+\infty} dk\, e^{iky}\, U(k), \qquad \langle U(k)\, U(k') \rangle = S(k)\, \delta(k - k'). \tag{9.11}$$

The average is over realizations of the field.

Metheron and De Marsily [1980] showed that anomalous diffusion in the x-direction occurs if

$$\int_0^\infty dk\, \frac{S(k)}{k^2} = \infty. \tag{9.12}$$

On the contrary, if $\int dk\, k^{-2} S(k) < \infty$ the diffusion is standard and

$$D_{11} \sim \frac{1}{\chi} \int_0^\infty dk\, \frac{S(k)}{k^2}. \tag{9.13}$$

If for simplicity we assume that

$$S(k) \sim k^\gamma \quad \text{for } k \to 0 \tag{9.14}$$

then standard diffusion occurs if $\gamma > 1$. On the other hand if $-1 \le \gamma \le 1$ the particles will perform anomalous diffusion with [Young and Jones 1991]

$$\langle |x(t) - x(0)|^2 \rangle \sim t^{2\nu}, \qquad \nu = \frac{3 - \gamma}{4} \ge \frac{1}{2}. \tag{9.15}$$

Note that $\int dk\, k^{-2} S(k) \sim \langle u^2 \rangle\, \delta^2$ where δ is the typical length of the velocity field $u(y)$, e.g. the typical distance between two zeros of $u(y)$. Therefore (9.13) is essentially the result (9.8). The dependence on δ^2 instead of δ^3 is due to the fact that in (9.13) all the particles contribute to the diffusion, while in the (9.8) only a fraction $O(\delta)$.

If (9.14) only holds for $k \geq k_0$ and $S(k) = 0$ for $k < k_0$, then the anomalous diffusion law (9.15) is valid only for $t \leq 1/(k_0^2 \chi)$ [Young and Jones 1991]. For larger times the diffusion is standard with a diffusion coefficient given by (9.13).

The condition $\int dk\, k^{-2}S(k) = \infty$ for anomalous diffusion can be understood with a physical argument. In fact $\int dk\, k^{-2}S(k) < \infty$ means that the typical distance δ between two zeros of $u(y)$ is finite. In this case the process is similar to that of a velocity field given by a sequence of strips of size δ and velocity $\pm V_o$ alternatively, where V_o is the typical velocity $\sqrt{\int dk\, S(k)}$. One can then repeat the arguments of section 9.1.2 and obtain (9.13).

Obviously if $\int dk\, k^{-2}S(k) = \infty$, i.e. $\delta = \infty$, the approach of section 9.1.2 cannot be used and the problem must be treated in a more careful way.

The generalization to higher dimensions is straightforward for shear flow. For a general discussion of anomalous diffusion in random media see Redner [1989], Zumofen et al. [1990] and Isichenko [1992]

9.1.4   Anomalous diffusion in smooth velocity fields

In the previous section the origin of superdiffusion was related to the spatial structure of the velocity field. We turn now to the interesting case of deterministic velocity fields with very long Lagrangian correlation times.

It is easy to realize that, in order to have anomalous diffusion in a regular velocity field one needs $\chi = 0$ and

$$\int_0^\infty \langle \mathbf{v}(\mathbf{x}(t + \tau)) \cdot \mathbf{v}(\mathbf{x}(t)) \rangle d\tau = \infty.$$

This can happen in periodic flows whose Lagrangian phase space has a complicated self-similar structure of islands and cantori [Zaslavsky et al. 1993]. For the velocity field

$$\mathbf{v} = (\partial_y \psi + \epsilon \sin z, -\partial_x \psi + \epsilon \cos z, -\psi) \tag{9.16}$$

where

$$\psi(x, y) = 2 \left[\cos x + \cos\left(\frac{x + \sqrt{3}\,y}{2}\right) + \cos\left(\frac{x - \sqrt{3}\,y}{2}\right) \right] \tag{9.17}$$

numerical simulations [Zaslavsky et al. 1993] have given strong evidence, that the flow exhibits anomalous diffusion in the x–y plane for some intervals of ϵ-values in the range $[0, 5]$. The anomalous diffusion is essentially due to the almost trapping of the trajectories, for arbitrarily long time, close to the cantori which are organized in a complicated self-similar structure. Of course if $\nu > 1/2$, equation (9.2) does not hold, and Zasvalsky [1994] proposed a diffusion equation involving fractional derivatives.

In a beautiful experiment, Cardoso and Tabeling [1988] studied diffusion in a linear array of counter rotating vortices. In order to keep the roll structure regular, the experiments were performed in an electrolyte in which a longitudinal electric current was maintained together with a vertical, stationary, spatially periodic magnetic field. In this way it was possible to observe a regime of anomalous diffusion with $v = 2/3$ in agreement with the theory of Pomeau et al. [1989].

Let us now introduce a small molecular diffusivity and calculate the diffusion coefficents of the flow (9.16–9.17). From the result of Avellaneda and Majda [1991] we know that at any finite χ the asymptotic diffusion has to be standard. Nevertheless, one can expect that in a generic deterministic flow, anomalies in the zero-diffusivity dynamics could be captured by introducing a small χ and looking for a singular behaviour of D_{ij}. One still has anomalous diffusion with $v > 1/2$ up to a typical diffusive time $T_D = O(1/\chi)$, while at very long time the diffusion is standard, with

$$D \sim \chi^{-\beta}.$$

Matching the standard diffusion with the anomalous behaviour at the typical diffusive time T_D one obtains

$$\beta = 2v - 1.$$

The above argument has been tested with accurate computation with direct numerical simulations and multiscale techniques [Biferale et al. 1995b].

9.2 Anomalous diffusion in fields generated by extended systems

The study of diffusion in time-dependent, non-periodic flows is complicated by the need to simulate the full Navier–Stokes equations. In the spirit of this book, it would seem natural to consider diffusion in field theories simpler than the Navier–Stokes equations, but complicated enough to generate turbulence, e.g. with power laws and/or positive Lyapunov exponents.

In this section we briefly discuss the behaviour of passively advected test particles in velocity fields generated by turbulent dynamical systems.

9.2.1 Anomalous diffusion in the Kuramoto–Sivashinsky equation

As discussed earlier (in chapters 5 and 7) the Kuramoto–Sivashinsky equation is perhaps the simplest example of a partial differential equation with turbulent behaviour. Restricting our attention to one spatial dimension, one finds, by differentiating (5.99) and letting $u = -2\nabla\phi$

$$\partial_t u + u\nabla u = -\nabla^2 u - \nabla^4 u, \tag{9.18}$$

and now one notes the close resemblance to the Navier–Stokes equations, inasmuch as the left-hand side is the time derivative of u in a frame moving with the velocity u. This makes the 'Galilean invariance' used in chapter 5 obvious and stresses the connection to the Burgers equation (7.1). It is thus natural to treat $u(x, t)$ as a velocity field responsible for advecting particles, and thus take the equation for a test particle as

$$\frac{\mathrm{d}x}{\mathrm{d}t} = u(x(t), t). \tag{9.19}$$

In the following we shall study the behaviour of diffusing particles in one dimension described by equations (9.18) and (9.19) [Bohr and Pikovsky 1993]. The velocity field is, as we know from chapters 5 and 7, turbulent with an irregular cellular structure. The test particle is attracted to the stagnation points where the (strongly compressible) fluid converges, i.e. the zeros in the velocity field, where u is positive to the left and negative to the right. The other zeros are repelling.

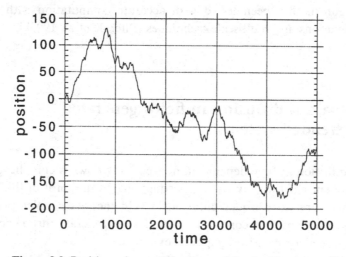

Figure 9.3 Position of a particle advected by the Kuramoto–Sivashinsky equation.

In Bohr and Pikovsky [1993] periodic boundary conditions on a system of length L are used for (9.18), and the dynamics is followed until a statistically steady turbulent state is obtained. Then a particle, obeying (9.19) is started at a random position and the subsequent evolution of $x(t)$ and u is followed. A typical trajectory $x(t)$ is shown in Fig. 9.3. Although the motion appears random, the particle moves almost ballistically for long stretches – of the order of several thousands of time units in a system of size $L = 1000$. This anomalous appearance can be quantified by looking at the mean square of the distance $\langle |\Delta x(t)|^2 \rangle$, which is defined by averaging the square of the displacement $\Delta x(t) = x(t' + t) - x(t')$, over the starting time t'. In Fig. 9.4 a typical plot of $\langle |\Delta x(t)|^2 \rangle$ is shown for a numerical solution of (9.18–19) using a spectral method [Zaleski 1989] and averaging over t' in several histories $x(t)$. One sees that, for times t up to around ten thousand, a good fit is obtained by assuming an anomalous diffusive behaviour with exponent $v \simeq 0.7$. For the largest times normal diffusion is regained.

To understand this result we shall make use of the Taylor formula derived in section 9.1

$$\langle |x(t) - x(0)|^2 \rangle \simeq 2t \int_0^t \langle u(x(\tau))u(x(0)) \rangle \mathrm{d}\tau. \tag{9.20}$$

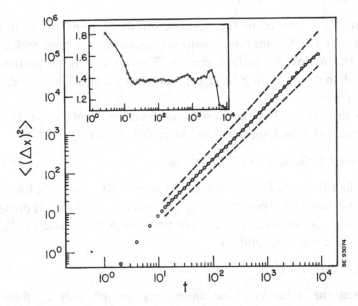

Figure 9.4 $\langle |\Delta x(t)|^2 \rangle$ versus t obtained by averaging over t' in several histories $x(t)$ of 50 000 time units in a system of effective size $L = 4277.5$ The two dashed lines have slopes 3/2 and 4/3 respectively. The best fit gives $\langle \Delta x^2 \rangle \sim t^{2v}$ with $2v = 1.38$. The inset shows the logarithmic derivative [Bohr and Pikovsky 1993].

The intergral on the right-hand side cannot, however, be evaluated directly even if the Eulerian field $u(x,t)$ is a known function of x and t. We must also know the trajectories $x(t)$ over which to average.

In order to solve (9.20) one needs to write the Lagrangian correlation function $\langle u(x(\tau),\tau)\,u(x(0),0)\rangle$ in terms of $\langle|\Delta x(\tau)|^2\rangle$. This difficulty is similar to the closure problem for turbulence and cannot be solved in an exact way. Even in the case of very simple velocity fields one has to introduce rather crude assumptions in order to obtain a closed equation for $\langle|\Delta x(\tau)|^2\rangle$. For a general discussion on closure techniques for diffusion problems see [Misguich et al. 1987, McComb 1991, Lynov et al. 1991].

A standard 'mean-field' approach is to replace the Lagrangian correlation function $|\langle u(x(\tau),\tau)\,u(x(0),0)\rangle|$ by the Eulerian correlation function by $\langle u(x+r,t+\tau)u(x,t)\rangle$ where $r^2 = \langle|x(t+\tau)-x(t)|^2\rangle$. Although this approximation will certainly give wrong numerical factors, one can hope that the scaling exponent η will be correctly reproduced.

We now invoke the well-known statistical equivalence, already discussed in chapter 7, between the Kuramoto–Sivashinsky equation and the Burgers equation with noise (the so-called KPZ equation). By differentiating the KPZ equation (7.11) and introducing the field $u(x,t) = -2\lambda\nabla h(x,t)$ one gets the equivalent form

$$\partial_t u + u\nabla u = \nu\nabla^2 u + \eta(x,t)$$

where the instability supplied by the 'negative surface tension' (i.e. the negative Laplacian term of the Kuramoto–Sivashinsky equation) has been replaced by the noise η and a positive surface tension. The noise η is proportional to the gradient of the η appearing in (7.11), and thus satisfies $\langle\eta(x,t)\eta(x',t')\rangle = -\Gamma\nabla^2\delta(x-x')\delta(t-t')$.

The precise form of the Eulerian correlation function is not known. However the scaling form, valid for long times and large distances, is:

$$\langle u(x+r,t+\tau)u(x,t)\rangle = \tau^{-1/z}H(r^z/\tau), \tag{9.21}$$

where the scaling function $H(\xi) \to 0$ for $\xi \to \infty$ and $H(\xi) \to$ const for $\xi \to 0$, and where the dynamical exponent $z = 3/2$ [Forster et al. 1977]. Applying the mean field approach introduced above we note, from (9.20) and (9.21), that there exists a self-consistent solution

$$\langle|x(t)-x(0)|^2\rangle \sim t^{4/3}.$$

We thus expect the diffusion to be anomalous, cut off only by finite size effects. It is interesting that, assuming the scaling (9.21), the asymptotic power law behaviour is only possible when $z = 3/2$. Recently Wang and Wang [1994] have shown that the above result can be obtained also by assuming a gaussian distribution for the trajectory of the particle at long times and explicitly averaging over this distribution before solving for $r(t)$. In the linear case, however, obtained by neglecting the nonlinear term in the KPZ equation, Bohr and Pikovsky predicted $\langle |x(t) - x(0)|^2 \rangle \sim t^{3/2}$ up to some cut-off time (independent of L) where it crosses over to $t \ln t$ behaviour, but there the method by Wang and Wang again gives $t^{4/3}$. At present it is not known under which conditions (e.g. time-scales for the field and the particle) each of these two approaches should be valid. It is interesting to note that the 'particle model' mentioned in chapter 7, for the Kuramoto–Sivashinsky equation [Rost and Krug 1995a] also gives a number very close to 4/3 for the exponent.

If the flow is seeded with several test particles, again evolving according to (9.18–19), we find that they coalesce in time. Figure 9.5 shows the fate of 46 particles: after 2000 iterations only 5 remain. From Fig. 9.5 we do indeed see some indication that the long stretches of almost ballistic motion are caused by coherent motion of patches in the cellular structure, since nearby walkers tend to follow each other. The figure is reminiscent of paths formed by directed polymers in a random environment [Kardar and Zhang 1987]. In fact, the scaling exponent $v = 2/3$ is precisely the one found there. The fact that particles coalesce seems to be a very general feature of *compressible flows*. The particles sit close to the stable zeros of the field and thus nearby particles move exponentially closer to each other. The chance that a subsequent fluctuation

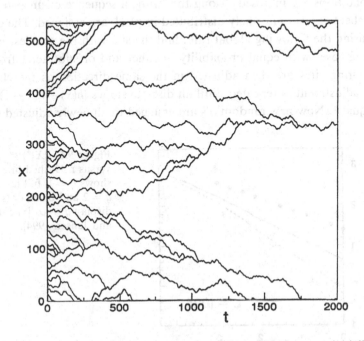

Figure 9.5 Space-time plot of many walkers. The flow was initially seeded with 46 walkers [Bohr and Pikovsky 1993].

will separate the particles therefore rapidly becomes very small. One can say that we have a case, analogous to the systems discussed in section 8.2.2, in which the largest Eulerian Lyapunov exponent (i.e. that of the velocity field) is positive, whereas the Lagrangian one, characterizing the relative position of two particles, is negative, although the particle moves in a chaotic field, the fact that the Lagrangian Lyapunov exponent can be negative is related to the compressibility of the field. In two dimensions it is possible to have one positive and one negative Lagrangian Lyapunov exponent and thus the particles will aggregate on a strange attractor [Lundbeck Hansen and Bohr 1998].

9.2.2 Multidiffusion along an intermittent membrane

We conclude this chapter with a discussion of the behaviour of passively advected test particles moving along the dynamical membrane, or string in 1D, introduced in section 7.6. The membrane fluctuates around zero and, in a way similar to that of the previous section, we consider the motion of passive particles which follow a velocity field given by the gradient of the membrane field.

For completeness let us briefly recall the model. In the 1D version, the discrete membrane field $h(x, t)$ is defined on $x = 1, 2, 3, \ldots, L$ and periodic boundary conditions are imposed. Along the string, a sequence of uncorrelated random numbers $\eta(x, t)$ uniformly distributed in $[0, 1]$, are assigned. The chain is updated, using the following global rule: find the site with the smallest η. On this site one chooses with equal probability to either add or subtract 1 from h. The neighbouring sites are then adjusted in the same direction as the chosen site, and the adjustment is repeated until all discrete slopes $|h(x) - h(x - 1)|$ are smaller or equal 1. Now new random η's are assigned to all newly adjusted sites.

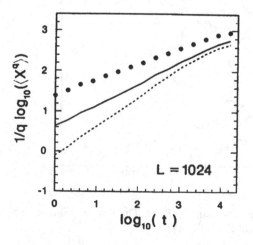

Figure 9.6 $\langle |X|^q \rangle^{1/q}$ versus t. Dashed line shows $q = 1$, full line $q = 2$ and black dots display $q \to \infty$ [Sneppen and Jensen 1994].

This dynamics will create localized bursts of activity, i.e. intermittent motion, as discussed in section 7.6.

To investigate the diffusive properties, we consider passive particles which follow the string $h(x,t)$ according to a discrete version of:

$$\frac{dx(t)}{dt} = \partial_x h(x(t),t).$$

In practice at each discrete time t, a particle moves $\delta x(t) = h(x+1,t) - h(x-1,t)$. Figure 9.6 displays the moments of the absolute displacement $X(t) = x(t) - x(0)$ as a function of time t. The average is over an ensemble of many particles. One observes that $\langle |X| \rangle \propto t^{0.74 \pm 0.03}$, $\langle X^2 \rangle^{1/2} \propto t^{0.52 \pm 0.03}$, and $\langle |X_{max}| \rangle \propto t^{0.38 \pm 0.03}$. These scaling laws imply that dynamical exponents Z_q, defined by $\langle |X|^q \rangle^{1/q} \propto t^{Z_q}$, depend on the moment q. This behaviour is called *multidiffusion* [Sneppen and Jensen 1994], and it is intimately related to the intermittent dynamics of the string.

In strongly turbulent flows, Richardson [1926] predicted that the distance between two passively advected particles would behave superdiffusively and grow as t^ν with $\nu = 3/2$. Richardson did not consider the possibility of multidiffusion but in the case of a strongly intermittent velocity field one could expect to observe multidiffusion.

Appendix A

Hopf bifurcation

Consider a general d-dimensional system

$$\frac{d\mathbf{x}}{dt} = \mathbf{f}^{(\mu)}(\mathbf{x}), \tag{A.1}$$

where μ is a control parameter. We assume that there is a fixed point $\mathbf{x} = \mathbf{x}_0$, for simplicity let $\mathbf{x}_0 = \mathbf{0}$. The Hopf bifurcation at the critical value μ_0, for simplicity let $\mu_0 = 0$, refers to the case where the Jacobian $\mathbf{A}^{(\mu)}$ of the system (A.1) in the fixed point

$$A_{i,j}^{(\mu)} = \frac{\partial f_i^{(\mu)}}{\partial x_j}\Big|_0$$

has the following structure for small values of $|\mu|$. The eigenvalues $\lambda_1, \ldots, \lambda_d$ of $\mathbf{A}^{(\mu)}$ vary continuously with μ. For $\mu < 0$ the real parts of all the eigenvalues are negative, for $\mu > 0$ two eigenvalues have positive real part and imaginary part $\pm\omega$, while the other eigenvalues have negative real parts. Of course the fixed point is stable for $\mu < 0$ and unstable for $\mu > 0$. For small $|\mu|$ one has a limit cycle around the fixed point. The size of the limit cycle is $O(\sqrt{|\mu|})$ and its period is close to $2\pi/\omega$.

For the stability of the limit cycle one has to discuss the nonlinear terms. Denoting by x and y the variables corresponding to the eigenvalues with positive real part at small positive μ the general two-dimensional expansion is [Guckenheimer and Holmes 1983, Iooss and Joseph 1990]

$$\frac{dx}{dt} = (d\mu + a(x^2 + y^2))x - (\omega + c\mu + b(x^2 + y^2))y,$$

$$\frac{dy}{dt} = (\omega + c\mu + b(x^2 + y^2)x + (d\mu + a(x^2 + y^2)))y. \tag{A.2}$$

The equation is more transparent in complex coordinates $z = x + iy$ where it becomes

$$\frac{dz}{dt} = (d + ic)\mu z + i\omega z + (a + ib)|z|^2 z.$$

If $a < 0$ then the periodic solution is stable, while if $a > 0$ the periodic solution is repelling.

To discuss the scenario of a Hopf bifurcation, we consider the following simplification of (A.2):

$$\frac{dx}{dt} = (\mu + a(x^2 + y^2))x - \omega y,$$
$$\frac{dy}{dt} = (\mu + a(x^2 + y^2))y + \omega x. \tag{A.3}$$

The Jacobian matrix in the fixed point $(0,0)$ has eigenvalues $\mu \pm i\omega$.

Equation (A.3) in polar coordinates can be expressed as

$$\frac{dr}{dt} = (\mu + ar^2)r,$$
$$\frac{d\theta}{dt} = \omega. \tag{A.4}$$

Lets us consider two cases for the above equation. The first, is the 'mexican hat' – with $a < 0$, and $\omega \neq 0$. As the value of the parameter μ passes through zero, a Hopf bifurcation occurs and a stable limit cycle appears with a radius $r = \sqrt{-\mu/a}$ and a frequency ω. For $a > 0$ one has an unstable periodic orbit for negative values of μ with $r = \sqrt{-\mu/a}$. The limit cycle appears with a constant frequency, which does not depend on the other variable r. For the general expansion (A.2) the frequency depends on r, which is the case discussed in chapter 5 for the stability analysis of the uniform solution to the Ginzburg–Landau equation.

Appendix B

Hamiltonian systems

B.1 Basic elements

The evolution of a Hamiltonian system is ruled by the equations:

$$\frac{\mathrm{d}q_i}{\mathrm{d}t} = \frac{\partial H}{\partial p_i}, \qquad \frac{\mathrm{d}p_i}{\mathrm{d}t} = -\frac{\partial H}{\partial q_i}, \qquad i = 1, \ldots, N, \tag{B.1}$$

where $H(q_1, \ldots, q_N, p_1, \ldots, p_N, t)$ is the Hamiltonian and q_i and p_i are conjugate variables. The number N is traditionally called the number of degrees of freedom, while the dimension of the manifold where the motion develops is $2N$ in the non-autonomous case, and $2N - 1$ in the autonomous case. If H does not depend explicitly on time, then H is a constant of motion, in addition one has the Liouville theorem which ensures the conservation of the volume in phase space.

A Hamiltonian system is a dynamical system of particular type, since a small generic perturbation of (B.1) destroys the Hamiltonian structure. Therefore, it is not structurally stable and has properties very different from those of a generic, i.e. non-Hamiltonian, system. In spite of this fact, Hamiltonian systems have a relevant role in many applications, e.g. celestial mechanics, particle accelerators, fluid mechanics and plasma physics [Gallavotti 1983, Lichtenberg and Liebermann 1983].

A completely integrable system is a limit case of an autonomous Hamiltonian system, where N independent constants of motion, F_i, exist and are in involution, i.e.

$$\{H, F_i\} = \{F_i, F_j\} = 0,$$

where $\{\cdot\}$ indicates the Poisson brackets. In such a system there exists [Arnold 1976] a canonical transformation into action-angle variables,

$$(q_i, p_i) \to (I_i, \theta_i)$$

which reduces the Hamilton equations to the form

$$\frac{dI_i}{dt} = 0, \qquad \frac{d\theta_i}{dt} = \omega_i(\mathbf{I}) = \frac{\partial H(\mathbf{I})}{\partial I_i}. \tag{B.2}$$

In the general integrable case, it can however be very difficult to find such a transformation. The solution of (B.2) is:

$$I_i(t) = I_i(0), \qquad \theta_i(t) = \theta_i(0) + \omega_i(\mathbf{I}(0))t;$$

and the motion is confined on N-dimensional tori. A small generic perturbation destroys the complete integrability, i.e. if H has the form

$$H = H_0(\mathbf{I}) + \epsilon H_1(\mathbf{I}, \theta), \tag{B.3}$$

then, apart from very particular cases, for $\epsilon \ll 1$, the only surviving integral of the motion is H. This result, due to Poincaré, can be briefly sketched as follows, see Lichtenberg and Liebermann [1983]. Let us try to find the constants of motion, for the system (B.3), in the form, for each j

$$F = F_0 + \epsilon F_1 + \epsilon^2 F_2 + \dots, \qquad \text{where} \qquad F_0 = I_j.$$

We have to impose $\{H, F\} = 0$, in order to determine $F_1, F_2, \dots$ The equation for F_1 has the form

$$\{H_0, F_1\} = -\{H_1, F_0\}. \tag{B.4}$$

It is easy to see that (B.4) has no global solution. Let us consider the Fourier expansion of H_1 and F_1:

$$H_1 = \sum_{\mathbf{k}} c_{\mathbf{k}}(\mathbf{I}) e^{i\mathbf{k}\cdot\theta}, \qquad F_1 = \sum_{\mathbf{k}} f_{\mathbf{k}}(\mathbf{I}) e^{i\mathbf{k}\cdot\theta}, \tag{B.5}$$

where the vector $\mathbf{k} = (k_1, \dots, k_N)$ has integer components. Inserting (B.5) into (B.4) and noting that

$$\{H_0, (\cdot)\} = \sum_{l=1}^{N} \omega_l(\mathbf{I}) \frac{\partial(\cdot)}{\partial \theta_l} \qquad \text{and} \qquad \{H_1, F_0\} = -i \sum_{\mathbf{k}} k_j c_{\mathbf{k}}(\mathbf{I}) e^{i\mathbf{k}\cdot\theta},$$

one has

$$f_{\mathbf{k}}(\mathbf{I}) = \frac{k_j c_{\mathbf{k}}(\mathbf{I})}{\sum_{l=1}^{N} \omega_l(\mathbf{I}) k_l}. \tag{B.6}$$

From the above equation it is easy to show that the problem admits no solution because the scalar product $\sum_l \omega_l \cdot k_l$, that appears in the denominator of (B.6), can assume arbitrarily small values. This is the famous small divisor problem.

One of the most relevant results, in the field of Hamiltonian systems, is given by the KAM theorem (from Kolmogorov [1954], Arnold [1963] and Moser [1962]): for the Hamiltonian (B.3), under the hypothesis

(a) ϵ sufficiently small,

(b) $\det \Omega \neq 0$, where $\Omega_{ij}(\mathbf{I}) = \partial^2 H_0/\partial I_i \partial I_j$,

the measure of the invariant tori, on a fixed constant energy surface, is positive and goes to 1 for $\epsilon \to 0$. Roughly speaking, one has that the motion on the invariant tori is regular, i.e. similar to that of the integrable systems.

Let us note that the invariant tori have dimension N while the available phase space has dimension $2N - 1$. Thus in the case $N = 2$, the tori separate regions that can exhibit different chaotic behaviours. On the contrary, for $N \geq 3$ the complement of the set of invariant tori is connected. This fact allows for Arnold diffusion: the system can move on the whole surface of constant energy by diffusing among the unperturbed tori [Arnold and Avez 1968]. The presence of the Arnold diffusion has been proved for particular systems, but it is believed to hold in a generic system. Unfortunately it is not easy to give theoretical estimates of Arnold diffusion time scales in the general case [Malagoli et al. 1986].

The most relevant result in this direction is due to Nekhoroshev [1977] who showed that in a generic Hamiltonian system, under a technical hypothesis similar to those of the KAM theorem, a trajectory starting close enough to an unperturbed torus remains in its vicinity for extremely long times $O(\exp(c/\epsilon^\alpha))$, where the parameters c and α depend on the details of H.

The continuous time dynamics given by (B.1) can be reduced to a discrete map. For the sake of simplicity, we discuss an autonomous Hamiltonian system with two degrees of freedom. Instead of studying the continuous flow

$$\Gamma(0) \to \Gamma(t)$$

where $\Gamma(t) = [\mathbf{q}(t), \mathbf{p}(t)] \in R^4$, one can investigate the system (B.1) in terms of a discrete map in R^2.

The energy is a constant of motion, so only three variables x_1, x_2 and x_3 are sufficient for a complete specification of the state of the system. Consider a plane S defined by $x_3 = h = \text{constant}$ and denote by $\mathbf{P}(0), \mathbf{P}(1), \mathbf{P}(2), \ldots$ the intersections of the flow $\Gamma(t)$ with the plane S and $dx_3/dt < 0$, at the successive times $t_0, t_1, t_2, \ldots$ The plane S is called the *Poincaré section* of the flow (B.1). Due to the deterministic nature of the flow $\Gamma(0) \to \Gamma(t)$, the intersection $\mathbf{P}(k+1) \in R^2$ can be obtained by the previous one via a map, called a Poincaré map:

$$\mathbf{P}(k + 1) = \mathbf{g}\Big[\mathbf{P}(k)\Big]. \tag{B.7}$$

The knowledge of $\mathbf{P}(k)$ is equivalent to the knowledge of $\Gamma(t_k)$.

The above argument can be repeated for a generic Hamiltonian system with N degrees of freedom. The energy conservation reduces the dimension of the phase space to $2N-1$. Introducing a section S, the intersection $\mathbf{P}(k+1) \in R^{2N-2}$ of the trajectory $\Gamma(t)$ with S is related to $\mathbf{P}(k)$ via a map of the form (B.7). Moreover the Poincaré map associated with the Hamiltonian system is symplectic [Lichtenberg and Liebermann 1983], i.e. the matrix $\mathbf{A}$, defined by $A_{ij} = \partial g_i/\partial P_j$, has the following property

$$\mathbf{A}^T \mathbf{J} \mathbf{A} = \mathbf{J} \quad \text{where} \quad \mathbf{J} = \begin{pmatrix} 0 & 1 \\ -1 & 0 \end{pmatrix}.$$

Another relevant case in which one can reduce the continuous problem to a symplectic map is given by time-periodic Hamiltonians: $H(\mathbf{q},\mathbf{p},t+\tau) = H(\mathbf{q},\mathbf{p},t)$. In this case the analogue of $\mathbf{P}(k)$ is just $(\mathbf{q}(k\tau),\mathbf{p}(k\tau))$.

The Poincaré map simplifies the study of a continuous dynamical system because it makes it easier to visualize whether the motion is regular or chaotic. For instance, for $N=2$, the track on the Poincaré section of a regular trajectory is given either by a set of isolated points (periodic motion) or by a regular curve (quasi-periodic motion). On the contrary, a chaotic trajectory fills out a finite region. An example of these different features can be found in Hénon and Heiles [1964].

It is worth stressing that chaos in Hamiltonian systems has very different features from dissipative systems.

B.2 Motion near a separatrix: Melnikov method

Here we briefly discuss a method, due to Melnikov [1963], for studying the motion close to a separatrix of a quasi-integrable Hamiltonian system. This method yields a criterion for the existence of chaotic motion. We concentrate on the one-dimensional time-dependent Hamiltonian case, but this approach can also be applied to multidimensional systems [Wiggins 1988].

Consider a Hamiltonian function of the form:

$$H(q,p,t) = H_0(q,p) + \epsilon H_1(q,p,t),$$

where H_1 is periodic in time, with period τ. The evolution equation for $\mathbf{x} = (q,p)$ is

$$\frac{d\mathbf{x}}{dt} = \mathbf{f}^{(0)}(\mathbf{x}) + \epsilon \mathbf{f}^{(1)}(\mathbf{x},t), \quad \text{with} \quad \mathbf{f}^{(1)}(\mathbf{x},t+\tau) = \mathbf{f}^{(1)}(\mathbf{x},t). \tag{B.8}$$

The unperturbed system is integrable and is assumed to possess a hyperbolic fixed point $\tilde{\mathbf{x}}_0$ and a separatrix orbit $\mathbf{x}_0(t)$ such that $\lim_{t\to\pm\infty}\mathbf{x}_0(t) = \tilde{\mathbf{x}}_0$. See Fig. B.1 for an illustration of the phase space. The stable and the unstable orbits, x_0^s and x_0^u respectively, are smoothly joined. Let us consider the Poincaré map of the perturbed dynamics (B.8), i.e.

$$\mathbf{x}(t_0) \rightarrow \mathbf{x}(t_0 + \tau) = \mathbf{T}\mathbf{x}(t_0),$$

where $t_0 \in [0, \tau]$ denotes a shift in the initial condition. The hyperbolic fixed point, $\tilde{\mathbf{x}}_\epsilon$, of the map $\mathbf{T}$ stays close to $\tilde{\mathbf{x}}_0$: $\tilde{\mathbf{x}}_\epsilon = \tilde{\mathbf{x}}_0 + O(\epsilon)$. Moreover the stable (unstable) orbit $\mathbf{x}_\epsilon^s$ ($\mathbf{x}_\epsilon^u$), lying on the stable (unstable) manifold W_{ϵ,t_0}^- (W_{ϵ,t_0}^+), is close to $\mathbf{x}_0(t)$ for $t \rightarrow \infty$ ($-\infty$), that is

$$\left.\begin{array}{ll} \mathbf{x}_\epsilon^s(t, t_0) = \mathbf{x}_0(t - t_0) + \epsilon\, \mathbf{x}_1^s(t, t_0) + O(\epsilon^2) & t \in [t_0, \infty], \\ \mathbf{x}_\epsilon^u(t, t_0) = \mathbf{x}_0(t - t_0) + \epsilon\, \mathbf{x}_1^u(t, t_0) + O(\epsilon^2) & t \in [-\infty, t_0]. \end{array}\right\} \quad (B.9)$$

In the generic case, Fig. B.2, W_{ϵ,t_0}^+ intersects W_{ϵ,t_0}^- in a homoclinic point P, and it is not difficult to see that this implies a chaotic motion around the unperturbed separatrix. For simplicity we do not consider here the non-generic

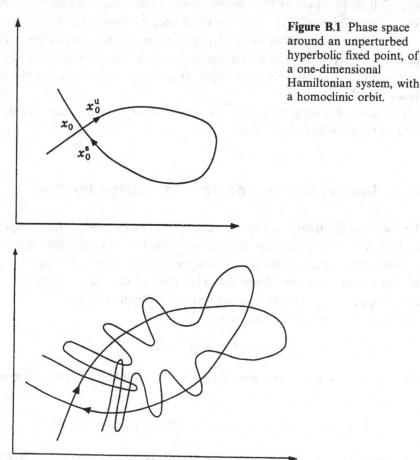

Figure B.1 Phase space around an unperturbed hyperbolic fixed point, of a one-dimensional Hamiltonian system, with a homoclinic orbit.

Figure B.2 Some homoclinic intersections due to a perturbation of the hyperbolic fixed point of Fig. B.1.

case of a tangent contact. First, let us observe that one intersection between W_{ϵ,t_0}^+ and W_{ϵ,t_0}^- implies an infinite number of intersections. Indeed, the forward or backward iterations $\mathbf{T}^{\pm n}(P)$, of the Poincaré map starting from P, belong both to W_{ϵ,t_0}^+ and to W_{ϵ,t_0}^-. Thus W_{ϵ,t_0}^+ intersects W_{ϵ,t_0}^- in an infinite number of homoclinic points, although W_{ϵ,t_0}^+ and W_{ϵ,t_0}^- cannot have self-intersections. Poincaré [1899] writes:

> The intersections form a kind of trellis, a tissue, an infinite tight lattice; each of curves must never self-intersect, but it must fold itself in a very complex way, so as to return and cut the lattice an infinite number of times.

The existence of one, and therefore infinite, homoclinic intersection implies chaos. This has been discussed in a rigorous way by Birkhoff [1927] and Holmes [1990]. Without entering into technical details, we observe that, because of the area conservation, the successive loops formed between homoclinic points must have the same area. Moreover, the distance between successive homoclinic points decreases exponentially, as the fixed point is approached. Therefore, one has an exponential growth of the length of the loops and a strong bending of W_{ϵ,t_0}^- near the saddle point. One has that a small part of the plane around $\tilde{\mathbf{x}}_\epsilon$ will be stretched and folded, and two points, initially close, will be found far apart after few iterations. Guckenheimer and Holmes [1983] proved rigorously the above scenario, by showing the equivalence between the iterated horseshoe map and the dynamics near $\tilde{\mathbf{x}}_\epsilon$.

The existence (or not) of the homoclinic intersection can be determined by a method, due to Melnikov, that is a perturbative computation using the properties (B.9). Basically, in order to prove the existence of a homoclinic point, one has to estimate the 'signed' distance

$$\mathbf{d}(t, t_0) = \mathbf{x}_\epsilon^s(t, t_0) - \mathbf{x}_\epsilon^u(t, t_0) = \mathbf{x}_1^s(t, t_0) - \mathbf{x}_1^u(t, t_0).$$

A direct computation gives

$$\frac{d}{dt}\mathbf{x}_1^{s,u}(t, t_0) = \mathbf{M}(\mathbf{x}_0)\mathbf{x}_1^{s,u}(t, t_0) + \mathbf{f}^{(1)}\left(\mathbf{x}_0(t - t_0), t\right),$$

where

$$M_{ij}(\mathbf{x}_0) = \left.\frac{\partial f_i^{(0)}}{\partial x_j}\right|_{\mathbf{x}_0(t-t_0)}$$

Let us introduce the Melnikov distance $D(t, t_0) = \mathbf{n} \cdot \mathbf{d}$ which is the projection of $\mathbf{d}$ along a normal $\mathbf{n}$ to the unperturbed orbit $\mathbf{x}_0$ at time t, as $\mathbf{n}(t, t_0) = (-f_2^{(0)}, f_1^{(0)})$. One can write $D(t, t_0)$ in the form

$$D = D^s - D^u, \qquad \text{with} \qquad D^{s,u} = \mathbf{f}^{(0)} \times \mathbf{x}_1^{s,u} \qquad \text{(B.10)}$$

and $\mathbf{a} \times \mathbf{b} = a_1 b_2 - a_2 b_1$.

Taking the derivative of (B.10), using (B.8) and the conservative properties of the system, i.e. $\mathrm{Tr}\,\mathbf{M}(\mathbf{x}_0) = 0$, one obtains

$$\frac{dD^s}{dt} = \mathbf{f}^{(0)} \times \mathbf{f}^{(1)}, \tag{B.11a}$$

$$\frac{dD^u}{dt} = \mathbf{f}^{(0)} \times \mathbf{f}^{(1)}. \tag{B.11b}$$

By noting that $D^s(\infty, t_0) = \mathbf{f}^{(0)}\,(\mathbf{x}_0(\infty - t_0)) \times \mathbf{x}_1^s = 0$, the integration of (B.11a) from t_0 to ∞ gives:

$$D^s(t_0, t_0) = -\int_{t_0}^{\infty} \mathbf{f}^{(0)} \times \mathbf{f}^{(1)} dt.$$

In an analogous way, by integrating (B.11b) from $-\infty$ to t_0, one obtains

$$D^u(t_0, t_0) = \int_{-\infty}^{t_0} \mathbf{f}^{(0)} \times \mathbf{f}^{(1)} dt.$$

So finally one has for the Melnikov function:

$$D(t_0, t_0) = -\int_{-\infty}^{\infty} \mathbf{f}^{(0)} \times \mathbf{f}^{(1)} dt.$$

If $D(t_0, t_0)$, which is an explicitly computable function, changes sign at some t_0 then one has a homoclinic point and chaos appears around the separatrix.

Let us remark that, although the Melnikov method is based on a simple perturbation method, it is of considerable relevance.

The above analysis can be repeated for the case in which the unperturbed system has a heteroclinic orbit, which tends asymptotically to an unstable equilibrium point for $t \to -\infty$ and towards another one for $t \to +\infty$.

Appendix C

Characteristic and generalized Lyapunov exponents

C.1   Basic elements

The characteristic Lyapunov exponents (CLE) are a natural extension of the linear stability analysis to aperiodic motion in dynamical systems. Roughly speaking, they measure the typical rates of the exponential divergence of nearby trajectories i.e. they characterize in a quantitative way the sensitive dependence on initial conditions which is one of the main characteristics of deterministic chaos.

Consider a differentiable dynamical system with an evolution law given, in the case of continuous time t, by the differential equation

$$\frac{d\mathbf{x}}{dt} = \mathbf{f}[\mathbf{x}(t)] \tag{C.1}$$

or, in the case of discrete times t, by the map

$$\mathbf{x}(t+1) = \mathbf{g}[\mathbf{x}(t)]. \tag{C.2}$$

The variable $\mathbf{x}$ and the differentiable functions $\mathbf{f}$ and $\mathbf{g}$ vary in a phase space which can be R^d, or a compact manifold, or an infinite-dimensional space. Equations (C.1) and (C.2) generate a mapping of the phase space into itself

$$\mathbf{x}(0) \rightarrow \mathbf{x}(t) = \mathbf{T}^t \mathbf{x}(0) \tag{C.3}$$

where $\mathbf{T}$ is the nonlinear time-evolution operator.

To study the separation between two initially close points, one introduces the tangent vector $\mathbf{z}(t)$ which can be regarded as an infinitesimal perturbation $\delta\mathbf{x}(t)$

of the trajectory $\mathbf{x}(t)$. The time evolution of $\mathbf{z}(t)$ is described by a mapping of the tangent space into itself

$$\mathbf{z}(0) \to \mathbf{z}(t) = \mathbf{D}_{\mathbf{x}(0)}\mathbf{T}^t \, \mathbf{z}(0) \tag{C.4}$$

in terms of the linear operator $\mathbf{D}_{\mathbf{x}}\mathbf{T}^t$ (the Jacobian matrix).

The formal evolution law (C.4) is obtained from (C.2) and (C.3) as a linear differential equation

$$\frac{\mathrm{d}z_i(t)}{\mathrm{d}t} = \sum_{j=1}^{d} \left.\frac{\partial f_i}{\partial x_j}\right|_{\mathbf{x}(t)} z_j(t)$$

or a map

$$z_i(t+1) = \sum_{j=1}^{d} \left.\frac{\partial g_i}{\partial x_j}\right|_{\mathbf{x}(t)} z_j(t), \tag{C.5}$$

respectively.

For the sake of simplicity, we only discuss the case of maps. However all the results also hold for ordinary differential equations.

The study of the separation of nearby trajectories can then be reduced to the study of the properties of products of matrices. From the mathematical point of view, the most important result in this field is the ergodic multiplicative theorem of Oseledec [1968]. Let us consider an initial condition $\mathbf{x}(0)$ and the sequence of matrices $\mathbf{A}(1), \mathbf{A}(2), \ldots$ given by

$$A_{ij}(k) = \left.\frac{\partial g_i}{\partial x_j}\right|_{\mathbf{x}(k)},$$

and denote by $\mathbf{P}_N(\mathbf{x}(0))$ the product of the first N matrices of the sequence. The Oseledec theorem states:

Theorem Consider the transformation (C.2), $\mathbf{T}$ being a diffeomorphism of class C^1 of a compact connected Riemann manifold M onto itself, and μ an ergodic invariant measure of the system. Then, there exists a measurable subset $M_1 \subset M$, $\mu(M_1) = 1$, such that for almost all $\mathbf{x}(0) \in M_1$, with the exception of the set of initial conditions of zero μ probability measure,

$$\lim_{N \to \infty} \left(\mathbf{P}_N^\dagger(\mathbf{x}(0)) \, \mathbf{P}_N(\mathbf{x}(0))\right)^{1/(2N)} = \mathbf{V}(\mathbf{x}(0))$$

exists. The matrix $\mathbf{V}(\mathbf{x}(0))$ has d real positive eigenvalues $\exp[\lambda_i(\mathbf{x}(0))]$ repeated according to their multiplicity. The exponents $\lambda_1 \geq \lambda_2 \geq \ldots \geq \lambda_d$ are called *characteristic Lyapunov exponents*.

Moreover with $p \leq d$ 'generic' random vectors $[\mathbf{z}^{(1)}(0), \ldots, \mathbf{z}^{(p)}(0)]$ one has

$$\lambda_1 + \ldots + \lambda_p = \lim_{t \to \infty} \frac{1}{t} \ln\left(VOL_p \, [\mathbf{z}^{(1)}(t), \ldots, \mathbf{z}^{(p)}(t)]\right) \tag{C.6}$$

where VOL_p $[\mathbf{z}^{(1)}, \ldots, \mathbf{z}^{(p)}]$ is the volume of the open parallelepiped generated by the p vectors $\mathbf{z}^{(i)}$ and each $\mathbf{z}^{(i)}(k)$ obeys (C.5). As a consequence the sum of all the Lyapunov exponents is negative for a dissipative systems and zero for a conservative system. The theorem assures that the spectrum of Lyapunov exponents $\lambda_1 \geq \lambda_2 \geq \ldots \geq \lambda_d$ does not depend on the initial conditions if the dynamical system is ergodic, although it depends on the particular ergodic invariant measure μ considered. However, in real or in numerical experiments, the temporal averages are equal to the averages taken over a special probability measure, the so-called physical or 'natural' measure, that is selected by the effect of noise or round-off truncation [Eckmann and Ruelle 1985]. In the following we always implicitly refer to this probability measure. It is worth stressing that in low-dimensional symplectic systems, the value of the Lyapunov exponent can depend on initial conditions, since often there exist disconnected chaotic regions. An impressive example can be found in oval billiards [Benettin and Strelcyn 1978] where, for some values of the parameter, there are up to eight different stochastic regions, separated by invariant tori.

In the case of continuous flows, it is easy to see that, if the system does not approach a stable fixed point, at least one of the Lyapunov exponents has to be zero, since $\mathbf{z}(t)$ cannot grow exponentially in time in the direction tangent to the flow. In Hamiltonian systems, and more generally, in symplectic systems, one has that the Lyapunov exponents appear in couples of opposite sign, so that in the case of N degrees of freedom the spectrum is $\lambda_1 \geq \lambda_2 \geq \ldots \geq \lambda_N \geq -\lambda_N \geq -\lambda_{N-1} \geq \ldots \geq -\lambda_1$. Note that for symplectic flows, this implies that at least two Lyapunov exponents should vanish.

C.2   The Lyapunov dimension

The chaotic motion in dissipative systems evolves on foliated structures, the strange attractors, that are created by the competitive effect of stretching and folding. These structures are often fractals with a non-integer Hausdorff dimension. It is natural to expect that the value of this dimension is related to the Lyapunov exponents, since the positive ones correspond to the stretching along the corresponding eigendirections, while the negative ones are responsible for the contraction, and thus for the folding. It is easy to analytically derive such a relation for toy two-dimensional maps such as the generalized baker transformation, where one has that the information dimension, i.e. the Hausdorff dimension of the fractal probability measure, is

$$d_I = 1 + \frac{\lambda_1}{|\lambda_2|},$$

see, for example, Ott [1993]. This relation has been proved for hyperbolic maps of the plane [Ledrappier 1981, Young 1982]. In the case of higher-dimensional systems, Kaplan and Yorke [1979] conjectured that the fractal dimension can be estimated by a Lyapunov dimension

$$D_\lambda = J + \frac{\sum_{i=1}^{J} \lambda_i}{|\lambda_{J+1}|}$$

where J is the maximum integer such that $\sum_{i=1}^{J} \lambda_i \geq 0$. It can be shown that in many cases, D_λ coincides with the information dimension d_I. Let us briefly sketch the argument that leads to the Kaplan and Yorke formula. By definition of J, J-dimensional hypervolumes should increase or at least remain constant (since $\sum_{i=1}^{J} \lambda_i \geq 0$), while the $J+1$-dimensional hypervolumes should be contracted to zero (since $\sum_{i=1}^{J+1} \lambda_i < 0$). The dimension of an attractor thus is larger than J and smaller than $J+1$. Finally, a linear interpolation allows one to get the fractional part of the dimension as the intersection of the straight line connecting $\sum_{i}^{J} \lambda_i$ and $\sum_{i}^{J+1} \lambda_i$ with the n-axes in the diagram $\sum_{i}^{n} \lambda_i$ versus n. In the framework of high-dimensional systems, it is often sufficient to compute the number J since it provides the physically relevant information measuring the number of degrees of freedom necessary to reproduce the evolution.

C.3   A numerical algorithm to compute Lyapunov exponents

In the numerical computation of the CLE from equation (C.6) one faces two practical difficulties. Namely $z^{(i)}(n) = P_n z^{(i)}(0)$ diverges exponentially fast as $\exp(\lambda_1 n)$, for all generic $z^{(i)}(0)$, so one rapidly exceeds the overflow limit of the computer. In addition, the asymptotic direction of each $z^{(i)}(n)$ is given by a subspace corresponding to the first Lyapunov exponent. So if $\lambda_1 \neq \lambda_2$ the angle between $z^{(i)}(n)$ and $z^{(j)}(n)$, where $z^{(i)}(0) \neq z^{(j)}(0)$, becomes rapidly very small with obvious precision problems.

For the computation of λ_1 only the first problem exists. One can easily overcome it using the linearity of the evolution equation for the tangent vector and replacing after each m iterations the evolved vector z by another vector w with the same direction and a fixed norm. For convenience we introduce the matrices

$$P(n_1, n_2) = A(n_2 - 1) \cdot A(n_2 - 2) \cdot \ldots \cdot A(n_1 + 1) \cdot A(n_1)$$

and the vectors

$$w(k + 1) = P(mk, mk + k)w(k)/R_{k+1}(m)$$

where

$$\mathbf{w}(0) = \mathbf{z}(0)/|\mathbf{z}(0)|$$

and

$$R_{k+1}(m) = |\mathbf{P}(mk, mk + k)\mathbf{w}(k)|.$$

One thus has

$$|\mathbf{z}(mN)| = R_1(m) \cdot R_2(m) \cdot \ldots \cdot R_N(m)$$

so the first Lyapunov exponent is

$$\lambda_1 = \lim_{N \to \infty} \frac{1}{Nm} \sum_{i=1}^{N} \ln[R_i(m)].$$

The number of iterations m should be small enough to avoid overflow at each k.

The second problem arises when one deals with two or more vectors. Following the same approach one can replace after each m iterations all the evolved vectors $\mathbf{z}^{(i)}$, by a new set of vectors $\mathbf{w}^{(i)}$ ($i = 1, 2, \ldots, p$). If one has to compute the first p CLE one then proceeds as follows. A set of orthonormal p random vectors $\mathbf{z}^{(i)}$ is choosen and their evolution is computed by iterating equation (C.5). After m iterations, one applies the Gram–Schmidt method and introduces a new orthonormal set of vectors $\mathbf{w}^{(i)}$:

$$\mathbf{w}^{(i)}(k + 1) = \widetilde{\mathbf{w}}^{(i)}(k + 1)/R_{k+1}^{(i)}(m)$$

where $\widetilde{\mathbf{w}}^{(i)}(k)$ are obtained from the orthogonalization of the vectors $\mathbf{P}(mk, mk + k)\mathbf{w}^{(i)}(k)$ and

$$R_{k+1}^{(i)}(m) = |\widetilde{\mathbf{w}}^{(i)}(k + 1)|.$$

Of course if $p = 1$ then $R^{(1)} = R$. The first p Lyapunov exponents are obtained by the formula:

$$\lambda_1 + \ldots + \lambda_p = \lim_{N \to \infty} \frac{1}{Nm} \sum_{i=1}^{N} \ln[V_p(k, m)]$$

where $V_p(k, m)$ is the volume in the p-dimensional space of the parallelepiped generated by the vectors $\widetilde{\mathbf{w}}^{(1)}(k), \ldots, \widetilde{\mathbf{w}}^{(p)}(k)$.

C.4   Generalized Lyapunov exponents

Roughly speaking the Oseledec theorem states that for almost all perturbations, i.e. almost all $z(0)$, the distance between the trajectory and the perturbed one grows exponentially as

$$|z(t)| \sim |z(0)| e^{\lambda_1 t}.$$

Let us introduce the response R to an infinitesimal perturbation in $x(\tau)$ after a time t by the error growth rate

$$R_\tau(t) \equiv \frac{|z(\tau + t)|}{|z(\tau)|}.$$

The maximum Lyapunov exponent λ_1 can then be defined by averaging the logarithm of the response over the possible initial conditions along the trajectory

$$\lambda_1 = \lim_{t \to \infty} \frac{1}{t} \langle \ln R(t) \rangle \qquad\qquad (C.7)$$

where $\langle \cdot \rangle$ denotes the time average. The Oseledec theorem implies that for large time $\ln R(t)$ is a non-random quantity, in the sense that for almost all the initial conditions its value does not depend on the specific initial condition, so that the average in (C.7) can be neglected.

Since the typical growth of a generic tangent vector is given by the maximum Lyapunov exponent, it is clear the $R(t)$ alone cannot be used to extract the other Lyapunov exponents. To this end one introduces the n order response $R^{(n)}$ as

$$R^{(n)}(t) \equiv \frac{|z_1(t + \tau) \times z_2(t + \tau) \times \ldots \times z_n(t + \tau)|}{|z_1(\tau) \times z_2(\tau) \times \ldots \times z_n(\tau)|}$$

where z_i are n non-parallel, generic tangent vectors. It is possible to show that the sum of the first n Lyapunov exponents is

$$\sum_{i=1}^{n} \lambda_i = \lim_{t \to \infty} \frac{1}{t} \langle \ln R^{(n)}(t) \rangle.$$

In other words the sum of the first $n \leq d$ Lyapunov exponents gives the typical rate of exponential growth of a n-dimensional volume in the tangent space.

Due to its global nature, the maximum Lyapunov exponent cannot give any further information on intermittency properties, i.e. on the finite-time fluctuations of the growth rate. For these properties it is necessary to consider the full probability distribution of the response R.

A direct calculation of the probability distribution of R is a rather hard task. However, it can be reconstructed, under general conditions, from the knowledge of the moments $\langle R^q \rangle$. We then introduce the generalized Lyapunov exponent $L(q)$ of order q as [Benzi et al. 1985, Paladin and Vulpiani 1987a]

$$L(q) = \lim_{t \to \infty} \frac{1}{t} \ln \langle R(t)^q \rangle.$$

It is easy to see that

$$\lambda_1 = \left.\frac{dL(q)}{dq}\right|_{q=0}$$

and in the absence of fluctuations

$$L(q) = \lambda_1 q.$$

By general inequalities of probability theory, $L(q)$ is a concave function [Feller 1971].

These generalized Lyapunov experiments give an indication of the large fluctuations of $R_\tau(t)$ at finite time t. Define indeed an effective Lyapunov exponent γ as

$$R_\tau(t) \sim e^{\gamma(t)t} \qquad t \gg 1$$

and classify the trajectories $x(\tau), x(\tau+1), \ldots, x(\tau+t)$ of length t according to their γ-value.

In the limit $t \to \infty$ the probability of finding $\gamma \neq \lambda_1$ should vanish, as a consequence of the Oseledec theorem. Therefore for large t, the probability of having a trajectory of length t with a given γ is peaked about the most probable value λ_1. If $L(q)$ is finite for all finite q, one can assume the following ansatz

$$d\mathscr{P}_t(\gamma) = d\mu(\gamma)\, e^{-S(\gamma)t} \qquad \text{with} \qquad S(\gamma) \geq 0$$

where, for the Oseledec theorem, $S(\gamma) > 0$ for $\gamma \neq \lambda_1$ and $S(\lambda_1) = 0$. The function $\mu(\gamma)$ is a smooth function of γ.

The function $S(\gamma)$ is related to the generalized Lyapunov exponent $L(q)$. In fact, the moments $\langle R^q \rangle$ can be evaluated by averaging over the γ-distribution

$$\langle R(t)^q \rangle = \int d\mu(\gamma)\, e^{(q\gamma - S(\gamma))t} \sim e^{L(q)t}.$$

For large t the integral can be calculated by the saddle point method

$$L(q) = \max_\gamma \left[q\gamma - S(\gamma)\right]. \tag{C.8}$$

The Legendre transformation (C.8) shows that each value of q selects a particular $\hat{\gamma}$ given by

$$q = \left.\frac{dS(\gamma)}{d\gamma}\right|_{\hat{\gamma}}.$$

We can thus obtain $S(\gamma)$ from $L(q)$ by inverting the Legendre transformation (C.8).

The probability distribution of $R(t)$ for $t \gg 1$ is usually close to a log–normal distribution

$$\mathscr{P}[R(t)] \simeq \frac{1}{R(t)\sqrt{2\pi\mu t}} \exp\left[-\left(\ln R(t) - \lambda_1 t\right)^2/(2\mu t)\right] \tag{C.9}$$

where

$$\mu = \lim_{t \to \infty} \frac{1}{t} \langle (\ln R(t) - \lambda_1 t)^2 \rangle.$$

Indeed writing $t = \tilde{t}\tau_c$, where τ_c is the typical correlation decay time, one has

$$R_\tau(t) = \prod_{k=1}^{\tilde{t}} \tilde{R}(k) \qquad \text{with} \qquad \tilde{R}(k) = R_{\tau + (k-1)\tau_c}(\tau_c).$$

Therefore, since $\ln \tilde{R}(k)$ are practically uncorrelated variables, one can use the central limit theorem for $\ln R_m(n)$ and, after a change of variables, one obtains (C.9). Under the hypothesis that $R(t)$ is exactly a log–normal variable one has

$$L(q) = \lambda q + \frac{\mu}{2} q^2. \tag{C.10}$$

In general, even if the log–normal is a good approximation, (C.10) is correct only for small q. This is because the moments of the log–normal distribution grow very fast with q [Carleman 1922, Orszag 1970].

Appendix D

Convective instabilities and linear front propagation

Let us briefly review the methods to treat localized perturbations in unstable systems and to determine the convective exponents and the shape of the asymptotic disturbance. The same type of calculations have been independently performed in different fields and for various aspects we refer the reader to Criminale and Kovasznay [1962], Gaster [1962], Briggs [1964], Hall and Henckrotte [1968], Ben-Jakob et al. [1985], Crutchfield and Kaneko [1987], Deissler and Kaneko [1987], Bohr and Rand [1991], Conrado and Bohr [1995]. An interesting attempt to go beyond the linear approximation is given by van Saarloos [1988, 1989].

The starting point is a PDE of the form

$$\partial_t \psi(\mathbf{x}, t) = H[\psi(\mathbf{x}, t)]$$

where $H[\psi(\mathbf{x}, t)]$ contains some nonlinear terms and $\mathbf{x}$ lives in a d-dimensional vector space. Let us now consider the evolution of an infinitesimal field ϕ superimposed on ψ:

$$\partial_t \phi = \mathbf{L}\phi \tag{D.1}$$

where $\mathbf{L}$ is a linear differential operator related to the linearization of the functional H, i.e.

$$H[\psi + \phi] = H[\psi] + \mathbf{L}\phi + O(\phi^2).$$

In other words, it can be written as a functional derivative

$$\mathbf{L} = \frac{\delta H}{\delta \psi}$$

and in general it depends on ψ. Equation (D.1) is the analogue of the evolution of the tangent vector for differential equations. Assuming in the following that ψ is independent of x, one can Fourier transform (D.1) to get

$$\partial_t \phi_q = L(q)\, \phi_q \tag{D.2}$$

and thus the formal solution of the initial value problem is

$$\phi(x,t) = \int dq\, e^{(iq \cdot x + L(q)t)} \phi_q(0), \tag{D.3}$$

where the integral runs from $-\infty$ to ∞ in each of the d components of q. In addition, we assume that $\phi(x,0)$ is a localized disturbance, which implies that $\phi_q(0)$ is a slowly varying function of q.

If the system is unstable in the sense that the real part of $L(q)$ is positive for some range of q, it does not necessarily imply that $\phi(x,t)$ grows exponentially at fixed x. There might be terms in L, breaking the $x \to -x$ symmetry, which will convect the initial wave packet downstream so fast that the instability will not be felt in the rest frame. Such an instability is called *convective* as opposed to *absolute* instability. In the latter case the system is unstable in the rest frame.

To check these possibilities we must introduce a velocity v with which we move through the system. Thus, replace x by $x_0 + vt$, whereby

$$\phi(x,t) = e^{iq \cdot x_0} \int dq\, e^{[iq \cdot v + L(q)]t} \phi_q(0), \tag{D.4}$$

and, in the limit of large t we evaluate the integral in the saddle point approximation, see, for example, Bender and Orszag [1978]. We then have to look for saddle points in the complex q-plane of $L(q) + iq \cdot v$ and to choose only the contributions from those saddle points (which we shall call *allowed*), for which the contour can be deformed back to the real axis without encountering singularities. If the saddle point(s) occur(s) for $q = \tilde{q}(v)$, we can approximate the integral (D.4) – retaining only the part which is exponential in t – as

$$\phi(x,t) \sim e^{\lambda(v)t} \tag{D.5}$$

where the velocity-dependent stability exponent is

$$\lambda(v) = L(\tilde{q}) + i\tilde{q} \cdot v \tag{D.6}$$

and the allowed saddle point $\tilde{q}$ giving the largest real part of $\lambda(v)$ is chosen. Relation (D.6) gives λ as the Legendre transform of L [Bohr and Rand 1991] where $\tilde{q}$ and v are linked by the relations

$$\nabla_{\tilde{q}} L = -iv$$
$$\nabla_v \lambda = i\tilde{q}. \tag{D.7}$$

If $\lambda(v = 0)$ is positive the system is absolutely unstable: perturbations will grow even in a non-moving frame. In general, if there is a range of v with positive

$\lambda(\mathbf{v})$, the system is convectively stable for that range of velocities. Further, the shape of the growing 'spot' is given by the zeros of $\lambda(\mathbf{v})$: the zeros separate the exponentially growing region from the exponentially stable one and thus the curve $\mathbf{x}(s) = \mathbf{x_0} + \mathbf{v}_f(s)t$ (where $\mathbf{v}_f(s)$ parametrizes the zeros of $\lambda(\mathbf{v})$) is the boundary of the expanding spot. In dimensions higher than one, interesting growth shapes can occur from this procedure, depending on the terms in the operator $\mathbf{L}$ [Conrado and Bohr 1995].

D.1   The simplest example

Let us illustrate this technique in the one-dimensional PDE:

$$\partial_t \phi = a\phi - b\phi_x + \phi_{xx} \tag{D.8}$$

with a and b real. In the Fourier space, the linear operator $\mathbf{L}$ assumes the form $L(q) = a - ibq - q^2$ and the saddle point relation (D.7) gives

$$\left. \frac{dL}{dq} \right|_{q=\tilde{q}} = -iv \tag{D.9}$$

so that $\tilde{q} = i(v - b)/2$ is purely imaginary and one has

$$L(q) = -(q - \tilde{q})^2 + a - \frac{(v - b)^2}{4} - iqv. \tag{D.10}$$

Obviously the real line can be deformed into a contour passing this saddle point and the resulting stability exponent is

$$\lambda(v) = a - \frac{(v - b)^2}{4} \tag{D.11}$$

in agreement with (D.6). Thus $\lambda(v)$ is real for real v and the system is unstable in reference frames moving with speeds in the interval $[b - 2\sqrt{a}, b + 2\sqrt{a}]$. Absolute instability requires that the point $v = 0$ belongs to this interval, i.e. that $-2\sqrt{a} < b < 2\sqrt{a}$. The two edges of the disturbance move with velocities $v_f^{\pm} = -b \pm 2\sqrt{a}$.

Note that the relevant parameter is b^2/a, which corresponds to the Peclet number defined in section 9.1 as $Pe = vl/\chi$, where v is a characteristic advection velocity, l a characteristic length and χ the molecular diffusivity. From (D.8) we have the correspondence $v = b$ and $\chi = 1$. Moreover, the characteristic length should be the length of advection during the characteristic time $1/a$ of the instability: $L = b/a$. Consequently, $Pe = b^2/a$ and absolute instability requires a small Peclet number.

The example (D.8) is so simple, that we could have solved for $\lambda(v)$ directly, without Fourier transformation. First we transform to the frame moving with velocity b, i.e. we introduce $\xi = x - bt$, $\tau = t$. Thus $\partial_x = \partial\xi/\partial x\, \partial_\xi + \partial\tau/\partial x\, \partial_\tau = \partial_\xi$

whereas $\partial_t = \partial\xi/\partial t\,\partial_\xi + \partial\tau/\partial t\,\partial_\tau = \partial_\tau - b\partial_\xi$. This removes the term $-b\phi$. Further, we make the ansatz $\phi(\xi, \tau) = e^{a\tau}g(\xi, \tau)$. Now, g satisfies the diffusion equation $\partial_\tau g = g_{xx}$ with the solution $g(x, t) \approx e^{-x^2/4t}$ (disregarding non-exponential factors) from which (D.11) follows.

D.2   The next-simplest example

As a more complicated example, of relevance to the Kuramoto–Sivashinsky equation [Conrado and Bohr 1994], consider

$$\partial_t\phi = -b\phi_x - \phi_{xx} - c\phi_{xxx} - \phi_{xxxx}. \tag{D.12}$$

The linear operator in Fourier space is now

$$L(q) = -ibq + q^2 + icq^3 - q^4 \tag{D.13}$$

and the saddle points must satisfy (D.9) or the cubic equation

$$2q + 3icq^2 - 4q^3 = -i(v - b). \tag{D.14}$$

In terms of $p = -iq$ one has

$$4p^3 - 3cp^2 + 2p + v - b = 0. \tag{D.15}$$

This cubic polynomial in p with real coefficients has three roots, one real and one complex conjugate pair (as long as $c < \sqrt{8/3}$, which we shall assume in

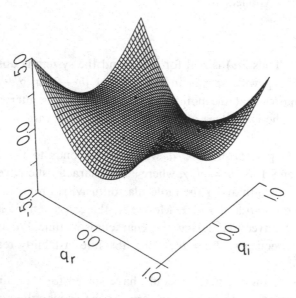

Figure D.1 Real part of $L(q)$ (D.13) (for $c = 0$) as a function of complex q. The saddle points are marked by dots.

the following). It is however easy to see that the path of integration can be deformed to pass over the complex roots in the direction of steepest descent, whereas it cannot for the real root, see Fig. D.1.

The complex conjugate pair is the 'right' saddle point and it can be explicitly found to be

$$p(v) = -\frac{1}{2}(s_+ + s_-) - \frac{c}{4} \pm i\frac{\sqrt{3}}{2}(s_+ - s_-), \qquad (D.16)$$

where

$$s_\pm = \frac{1}{2}\left((b-v) + \frac{c}{2} - \frac{c^3}{8}\right)$$

$$\pm \sqrt{v^2 + \frac{8}{27} + (b-v)c - \frac{c^2}{12} + \frac{(b-v)c^3}{4} + \frac{c^4}{4}}\Bigg)^{\frac{1}{3}}. \qquad (D.17)$$

From a knowledge of $p(v)$ one can calculate $\lambda(v)$ using (D.7). This quantity is complex in general. The real part of λ is related to the exponential growth/decay rate of the disturbance and it is positive in an interval $[v_f^-, v_f^+]$, where $v_f^\pm$ are the zeros of $\mathrm{Re}[\lambda(v)]$, i.e. the speeds of the back and front ends of the disturbance. For $c = 0$ the shape is symmetric and the front speed is $v_f^\pm \approx \pm 1.6$. For non-zero c the forward and the backward fronts move with different speeds. e.g. $v_f^- \approx -1.4$ and $v_f^+ \approx -1.8$ for $c = 0.1$. The imaginary part of λ gives rise to oscillations in ϕ whose wave vector is given by $\mathrm{Im}[\lambda]'(v)$.

D.3   Convective instability in coupled map lattices

In coupled map lattice systems with asymmetric coupling, used to mimic chaotic open-flows (section 4.5) one can have a convective instability which is quantitatively characterized in terms of velocity-dependent Lyapunov exponents [Deissler and Kaneko 1987]. As an example let us consider the map lattice:

$$x_i(n+1) = (1 - \epsilon)f(x_i(n)) + \epsilon[\alpha f(x_{i-1}(n)) + (1 - \alpha)f(x_{i+1}(n))] \qquad (D.18)$$

where $f(x) = 1 - ax^2$ and, if $\alpha \neq 1/2$, the coupling is asymmetric. It is possible to define the velocity-dependent Lyapunov exponents $\lambda_k(v)$ in the following way. Let us introduce the jacobian matrix $\mathbf{J}^{(v)}$ with elements

$$J_{i,j}^{(v)}(n) = \frac{\partial x_{i'}(n+1)}{\partial x_{j'}(n+1)} \qquad (D.19)$$

where

$$i' = i + [v\,n + 1] \qquad j' = i + [v\,n]$$

where $[v\,n]$ indicates the integer part of $v\,n$. It is easy to realize that the matrix (D.19) maps $\delta x_i(n)$ for $i_1 + [v\,n] \leq i \leq i_2 + [v\,n]$ into $\delta x_i(n+1)$ for

$i_1 + [v(n + 1)] \leq i \leq i_2 + [v(n + 1)]$, so that, when n increases, the region $i_1 + [vn] \leq i \leq i_2 + [vn]$ move 'downstream' with an average speed v. We can define the velocity-dependent Lyapunov exponents $\lambda_k(v)$ as the logarithms of the eigenvalues of the matrix

$$\left(\mathbf{B}(n)^\dagger \mathbf{B}(n) \right)^{1/(2n)}$$

where

$$\mathbf{B}(n) = \mathbf{J}^{(v)}(n)\mathbf{J}^{(v)}(n - 1)\ldots\mathbf{J}^{(v)}(1)$$

in the limit of large n.

In the limit case $v = 0$ one obtains the usual Lyapunov exponent. A typical situation for open flow, e.g. system (D.18) in a large range of values of ϵ and α, is that λ_1 is negative for $v = 0$. As v increases from zero, λ_1 increases until it reaches a maximum and then decreases [Deissler and Kaneko 1987]. This maximum provides a measure of the degree of chaoticity for open flows.

Appendix E

Generalized fractal dimensions and multifractals

Fractal structures appear in many physical phenomena such as turbulence, random walks and chaotic dynamical systems [Mandelbrot 1982]. The concept of dimension plays a central role in the characterization of fractals. Usually the dimension of an object is defined as the number of independent directions for a point moving on it. In this case, it is called the topological dimension d_T and is a positive integer number. It is equal to, or smaller than, the dimension d of the space in which the object is embedded. However a smooth line and a random walk trajectory have the same topological dimension $d_T = 1$ but very different characteristics since the latter densely fills a two-dimensional space. For this reason, it is necessary to introduce the fractal dimension d_F of a geometrical object, considering the scaling of the number $N(\epsilon)$ of hypercubes of size ϵ necessary to cover the object:

$$N(\epsilon) \sim \epsilon^{-d_F} \quad \text{for } \epsilon \to 0.$$

A more precise definition requires sophisticated mathematical methods and leads to the introduction of the Hausdorff dimension [Hausdorff 1919], which in some cases can be different from d_F. For a smooth geometrical object such as a line or a sphere, $d_T = d_F$ but, for instance, a random walk has $d_F = 2$.

However, the scaling laws appearing in nature cannot be characterized by just one geometrical parameter. One has to consider the scaling properties of an appropriate density $\mu(\mathbf{x})$ (in many cases a probability density) over the object. One then defines the coarse-grained measure

$$p_\epsilon(\mathbf{x}) = \int_{B(\mathbf{x},\epsilon)} \mu(\mathbf{y}) \, d\mathbf{y}$$

where $B(\mathbf{x}, \epsilon)$ is a hypercube of linear size ϵ centred in the point $\mathbf{x}$ of the object. In general $p_\epsilon(\mathbf{x})$ scales with an exponent α which depends on the particular point $\mathbf{x}$

$$p_\epsilon(\mathbf{x}) \sim \epsilon^\alpha \tag{E.1}$$

and, in general, $\alpha \neq d_F$. The object can be regarded as the superposition of different fractal sets

$$F(\alpha) = \{\mathbf{x} \text{ such that } p_\epsilon(\mathbf{x}) \sim \epsilon^\alpha \text{ for } \epsilon \to 0\} \tag{E.2}$$

each one characterized by a different exponent α. The object is called multifractal [Parisi and Frisch 1985, Benzi et al. 1984, Paladin and Vulpiani 1987a]. The fluctuations of the exponents α are ruled by a probability distribution which can be studied by analysing the scaling law of the moments

$$\langle p_\epsilon^q \rangle \equiv \sum_{k=1}^{N(\epsilon)} \left[p_\epsilon(\mathbf{x}(k)) \right]^{q+1} \sim \epsilon^{q\, d_{q+1}} \qquad \text{for } \epsilon \to 0, \tag{E.3}$$

where $\mathbf{x}(k)$ is centred in the kth hypercube and the average is a weighted sum over the hypercubes, i.e.

$$\langle (\cdots) \rangle \equiv \sum (\cdots) p_\epsilon(\mathbf{x}(k)).$$

The d_q are called generalized dimensions [Grassberger 1983, 1985, Hentschel and Procaccia 1983] and it can be shown that $d_F \equiv d_0$. In a homogeneous fractal $d_q = d_F$ for all q, and in general standard arguments of probability theory assure that d_q is a non-increasing function of q. The exponent $d_1 \leq d_0$ is the fractal dimension of the probability measure or *information dimension*.

The number of hypercubes of size ϵ necessary to cover a subset $F(\alpha)$ of the object should behave in the scaling hypothesis as

$$n(\alpha) \sim \epsilon^{-f(\alpha)}$$

where $f(\alpha) \leq d_F$ is the fractal dimension of the subset $F(\alpha)$ [Halsey et al. 1986]. Since the probability measure of a hypercube with centre $\mathbf{x} \in F(\alpha)$ scales as ϵ^α, the weight $\mathscr{P}_\epsilon(\alpha)$ of the corresponding subset should scale as

$$\mathscr{P}_\epsilon(\alpha) \sim \epsilon^{H(\alpha)} \qquad \text{with } H(\alpha) = \alpha - f(\alpha).$$

The function $H(\alpha \geq 0)$ is an entropy function, similar to the entropy function $S(\gamma)$ introduced for the finite-time fluctuations of the Lyapunov exponent in section C.4. For a more mathematical treatment of such entropy functions we refer to [Bohr and Rand 1991]. The sum over the hypercubes can be estimated as

$$\sum_{k=1}^{N(\epsilon)} \left[p_\epsilon(\mathbf{x}(k)) \right]^{q+1} \sim \int \mathrm{d}\alpha\, n(\alpha)\, \epsilon^{\alpha(q+1)}.$$

In the limit $\epsilon \to 0$, the integral is dominated by the saddle point value

$$\langle p_\epsilon^q \rangle \sim \epsilon^{\tilde{\alpha}q + H(\tilde{\alpha})}$$

where $\tilde{\alpha}$ is given by the minimum condition

$$\left. \frac{dH(\alpha)}{d\alpha} \right|_{\alpha = \tilde{\alpha}} = q.$$

The generalized dimensions are thus related to the $f(\alpha)$ function via the Legendre transformation

$$(q - 1)\, d_q = \min_\alpha \left[\alpha q - f(\alpha) \right]. \tag{E.4}$$

From this formula, it is evident that each order q moment selects a particular exponent α. The less probable the α, the larger the corresponding entropy function $H(\alpha)$. In particular the minimum $H(\alpha) = 0$ – corresponding to the relation $\alpha = f(\alpha)$ – is reached for $\alpha = d_1$, selected by $q = 1$ in (E.4) while the maximum of the $f(\alpha)$ curve is given by the fractal dimension

$$d_F \equiv d_0 = \max_\alpha f(\alpha).$$

selected by $q = 0$ in (E.4).

In Figs. E.1 and E.2 we show the typical shapes of $f(\alpha)$ as a function of α and of its Legendre transformation d_q as function of q. In the limit $\epsilon \to 0$, all the exponents different from d_1 cannot be detected, since their probabilities vanish. In this sense, d_1 is the most probable scaling exponent. This is a well-known result in the context of the information theory. Arguments borrowed from the Shannon–McMillan theorem [Khinchin 1957] show that the number

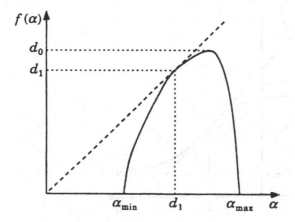

Figure E.1 Typical shape of $f(\alpha)$ as a function of α. The dashed line has slope 1.

$N_{\text{eff}}(\epsilon) \leq N(\epsilon)$ of the hypercubes which give the leading contribution to the information

$$I(\epsilon) = -\sum_{k=1}^{N(\epsilon)} p_\epsilon\big(\mathbf{x}(k)\big) \ln p_\epsilon\big(\mathbf{x}(k)\big) \simeq -d_1 \ln \epsilon$$

should scale as $N_{\text{eff}}(\epsilon) \sim \epsilon^{-d_1}$. In fact, the information dimension is the fractal dimension of the subset of the fractal that has full probability measure. In other words, the probability $p_\epsilon(\mathbf{x})$ defined in (E.1) scales as ϵ^{d_1} for all $\mathbf{x}$ on the fractal with the exception of a set that has zero measure when $\epsilon \to 0$.

The Legendre transformation becomes trivial in the limit $q \to \pm\infty$ where the minimum condition picks up the extreme values of the local exponents

$$\alpha_{\min} = \lim_{q\to\infty} d_q, \qquad \alpha_{\max} = \lim_{q\to-\infty} d_q.$$

If the α are random gaussian variables of mean d_1 and variance σ^2 – log–normal distribution of the coarse-grained measure p_ϵ – the entropy function has a parabolic shape

$$H(\alpha) = \frac{(\alpha - d_1)^2}{2\sigma^2}$$

with Legendre transformation $d_{q+1} = d_1 - \sigma^2 q/2$. In typical cases, this form is a good approximation only for small q, i.e. around the maximum of the probability distribution (bottom of the $H(\alpha)$ curve).

In practice appropriate probability measures p_ϵ are chosen according to the different physical phenomena. For instance, in fully developed turbulence p_ϵ is determined by the density of energy dissipation, in chaotic dynamical systems by the probability measure obtained from the frequency of visits on the invariant

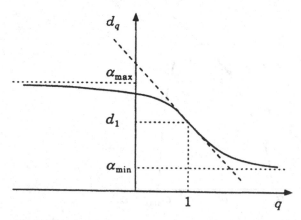

Figure E.2 Typical shape of d_q as a function of q. The dashed line has slope $-\sigma^2/2$.

set of the time evolution, in aggregates of particles by the growing probability and so on [Paladin and Vulpiani 1987a].

Let us briefly mention the fact that the relationship between $f(\alpha)$ and α (or $S(\gamma)$ and γ), on one hand, and d_q and q (or $L(q)$ and q), on the other hand, are suggestive of the relationship between the free energy and the entropy in thermodynamics: q, α and $f(\alpha)$ play the role of β (the inverse of the temperature), the internal energy and the entropy, respectively.

Appendix F

Multiaffine fields

Multifractal measures have an important role in many physical phenomena, and many simple probabilistic models, such as the random beta model [Benzi et al. 1984] or the two-scale Cantor set, have been proposed to generate them recursively.

The situation is much less simple for the scaling of field increments, such as the height of growing interfaces or the velocity in three-dimensional fully developed turbulence. In order to avoid a misleading terminology, we shall use the term self-affine for a field $\phi(x)$ such that

$$|\phi(x + r) - \phi(x)| \sim r^h, \tag{F.1a}$$

in all points x, while the term multiaffine for fields $\Phi(x)$ where the Hölder exponent h can depend on x so that the structure functions scale as

$$\langle |\Phi(x + r) - \Phi(x)|^q \rangle \sim r^{\zeta_q}, \tag{F.1b}$$

where the exponent ζ_q is a nonlinear function of q and $\langle \cdot \rangle$ indicates a spatial average. For a self-affine field $\zeta_p = hp$, of course. We saw in section 2.3 that the problem of the velocity difference is the playground where the multifractal formalism was originally proposed by Parisi and Frisch [1985] who considered the anomalous scaling of the velocity increments rather than the scaling of the energy dissipation (which is a positive defined measure [Mandelbrot 1974]).

Let us now briefly discuss a method to generate a d-dimensional multiaffine field, with a previously assigned set of exponents ζ_q. In one dimension, a different method has been introduced by Vicsek and Barabási [1991] and by Barabási et al. [1991] but its generalization to the multidimensional case seems to us rather difficult.

Let us stress, that it is important to have the possibility to generate a multi-affine signal with given scaling exponents, as a first step to understanding the dynamical mechanism explaining the intermittency in three-dimensional turbulence. Moreover, the existence of an algorithm for the construction of multiaffine fields is relevant for testing new methods for the treatment of experimental data. Finally, multiaffine fields are interesting not only in turbulence but also in various growth phenomena, like ballistic deposition, growth of thin films by vapour deposition, two phase viscous flow in porous media and sedimentation of granular material [Vicsek 1989].

The construction of a multiaffine function is a generalization of the recursive method used for obtaining a self-affine function. Indeed, it can be proved [Ausloos and Berman 1985] that the function

$$\Phi(x) = \lim_{N \to \infty} \sum_{n=-N}^{N} \gamma^{-nh} \left[1 - \exp(i2\pi\gamma^n x)\right] \exp(i\phi_n) \quad \text{with } \gamma > 1, \qquad \text{(F.2)}$$

is a self-affine function, i.e. the increments $F(\gamma\ell) = |\Phi(x + \gamma\ell) - \Phi(x)|$ have the same statistical properties as $\gamma^h F(\ell)$, if the phases ϕ_n are independent identically distributed random variables in the interval $[0, 2\pi]$. It follows

$$\langle F(\ell)^q \rangle \sim \ell^{\zeta_q} \qquad \text{with } \zeta_q = hq. \qquad \text{(F.3)}$$

If there are ultraviolet and an infrared cut-offs, i.e. if the index n in the sum runs from 0 to $N \gg 1$, the scaling properties (F.3) hold only for an appropriate range of scales $\gamma^{-N} \ll \ell \ll 1$.

Let us now describe the algorithm for the construction of multiaffine functions [Benzi et al. 1993b]. For this purpose, we shall consider $\Phi(x)$ given by the following wavelet decomposition:

$$\Phi(x) = \sum_{j=-\infty}^{+\infty} \sum_{k=-\infty}^{+\infty} a_{j,k} \psi_{j,k}(x), \qquad \text{(F.4)}$$

where

$$\psi_{j,k}(x) = 2^{j/2} \psi(2^j x - k),$$

and $\psi(x)$ is a wavelet function such that

$$\int_{-\infty}^{\infty} \psi(x) dx = 0.$$

In the discrete case, for $N = 2^n$ points x_i in the interval $[0, 1]$, the sums in (F.4) are restricted from zero to $n - 1$ for the index j and from zero to $2^j - 1$ for k. If the wavelet function fulfils certain properties, the functions $\phi_{j,k}$ will satisfy the following orthogonality conditions:

$$\int \psi_{j,k}(x)\psi_{j',k'}(x)\mathrm{d}x = \delta_{jj'}\delta_{kk'},$$

and form a complete set of functions in the Hilbert space L^2. Note that $|\psi(x)| \simeq 0$ for $|x| > l_c = O(1)$ which is the only property that we need from wavelet theory. A review and an introduction to the properties of orthogonal wavelet decomposition can be found in e.g. Meneveau [1991], Farge [1992].

The set of coefficients $a_{j,k}$ form a dyadic structure as shown Fig. F.1. We remark that in the discrete case the number of independent coefficients is $N-1$, as it should be, because the wavelet ψ has zero average. In the case of signals with non-zero average, a constant term must be included.

Consider now the random variable η with probability density distribution $P(\eta)$ and construct the following multiplicative process

$$\alpha_{1,0} = \epsilon_{1,0}\,\eta_{1,0}\,\alpha_{0,0}; \quad \alpha_{1,1} = \epsilon_{1,1}\,\eta_{1,1}\,\alpha_{0,0}$$

$$\alpha_{2,0} = \epsilon_{2,0}\,\eta_{2,0}\,\alpha_{1,0}; \quad \alpha_{2,1} = \epsilon_{2,1}\,\eta_{2,1}\,\alpha_{1,0}; \quad \alpha_{2,2} = \epsilon_{2,2}\,\eta_{2,2}\,\alpha_{1,1}; \quad \alpha_{2,3} = \epsilon_{2,3}\,\eta_{2,3}\,\alpha_{1,1}$$

and so on. The $\eta_{j,k}$ are independent random variables having the same distribution $P(\eta)$, the coefficient $\alpha_{0,0}$ is arbitrary and $\epsilon_{i,k} = \pm 1$ with equal probability. The general term is $\alpha_{j,k} = \epsilon_{j,k}\,\eta_{j,k}\alpha_{j-1,k'}$ with $k' = [k/2]$. It is easy to show that $|a_{j,k}|$ are random variables with moments:

$$\overline{|a_{j,k}|^p} = 2^{j\log_2(\overline{\eta^p})}\,\alpha_{0,0}. \tag{F.5}$$

The bar denotes the average over the ensemble of the realizations of the multiplicative process. Let us remark that the moments in (F.5) do not depend on k. In general, the scaling behaviour of the coefficients $\alpha_{j,k}$ does not imply that $\Phi(x)$ is multiaffine. However, from a heuristic point of view, one could guess that structure functions are power laws and calculate the exponents by

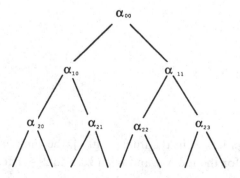

α_{00}

α_{10} α_{11}

α_{20} α_{21} α_{22} α_{23}

Figure F.1 The dyadic structure of the wavelet coefficients $\alpha_{j,k}$.

supposing that the scaling properties at scale $r \sim 2^{-j}$ are dominated by the jth term in the sum of equation (F.4). This leads to

$$\langle |\Phi(x+\ell) - \Phi(x)|^q \rangle \sim \ell^{\zeta_q} \quad \text{with } \zeta_q = -\log_2(\overline{\eta^q}) - q/2. \tag{F.6}$$

The above results can be also obtained in a rigorous way see Benzi et al. [1993b].

The generalization of the algorithm to more than one dimension is straightforward. Following Meneveau [1991], the three-dimensional field $\Phi(\mathbf{x})$ is decomposed as

$$\Phi(\mathbf{x}) = \sum_{j,k_1,k_2,k_3} \sum_{q=1}^{8} \alpha^{(q)}_{j,k_1,k_2,k_3} \psi^{(q)}_{j,k_1,k_2,k_3}(\mathbf{x}),$$

where the index j refers to the dilation factor, the indices k's to translations in the three possible directions and the index q is needed to take into account the internal degrees of freedom. The coefficients $\alpha^{(q)}_{j,k_1,k_2,k_3}$ are obtained by a multiplicative process in the same spirit as in the one-dimensional case. The condition of zero divergence is eventually imposed, as usual, in Fourier space.

Let us discuss a subtle point. It is possible to have

$$\langle \overline{|\Phi(x+r) - \Phi(x)|^q} \rangle \sim r^{\widetilde{\zeta}_q},$$

with a nonlinear $\widetilde{\zeta}_q$ even if the single realizations have no anomalous scaling. An example is provided by the following simple generalization of (F.2)

$$\Phi(x) = \lim_{N \to \infty} \sum_{n=-N}^{N} a_n[1 - \exp(i2\pi2^n x)] \exp(i\phi_n),$$

where the a_n are obtained by an uncorrelated multiplicative process $a_n = a_{n-1} b_n$ and b_n are independent identically distributed random variables. A rough heuristic argument, similar to that used to derive (F.6), confirmed by numerical computations, gives $\widetilde{\zeta}_q = -\log_2 \overline{b^q}$ while for a single realization there is no anomalous scaling and the single exponent is $-\log_2 \overline{b}$, i.e. one has:

$$\langle |\Phi(x+r) - \Phi(x)|^q \rangle \sim r^{\zeta_q}, \quad \text{with } \zeta_q = -\overline{\log_2 b}\, q.$$

The probability distribution function (PDF) of the variables $\delta\Phi_\ell(x) = \Phi(x+\ell) - \Phi(x)$ exhibits the typical shape of the PDF for the transverse velocity increments obtained in three-dimensional turbulent flows at high Reynolds numbers [Benzi et al 1991]. For large ℓ, the PDF is nearly gaussian, while on small scales the PDF becomes more and more peaked around zero with relatively high tails (corresponding to the presence of strong intermittency in the velocity gradients and hence in the energy dissipation).

Finally we stress that the multiaffine functions considered here do not satisfy any local scaling, that is to say for any point $\mathbf{x}$ the quantity $|\Phi(\mathbf{x}+\mathbf{r}) - \Phi(\mathbf{x})|$ has no scaling property in $\mathbf{r}$. The original approach by Parisi and Frisch [1985] should be considered only in a statistical sense. Recently there have been some attempts to extract exponents of local scaling from turbulent signals [Bacry et al. 1990, Vergassola et al. 1993]. For the multiaffine functions considered here this approach is meaningless because, by construction, the scaling exponents can be identified only in a statistical sense and not locally.

Appendix G

Reduction to a finite-dimensional dynamical system

The evolution law for a fluid is given by a PDE. Nevertheless in some cases, e.g. the onset of turbulence, it is possible to describe the system in terms of a finite-dimensional dynamical system. The basic idea of this approach, due to Ruelle and Takens [1971], and earlier to Arnold and Andronov (see Arnold [1983]) is the following. Let us consider an evolution equation of the form

$$\frac{\partial u}{\partial t} = \mathbf{L}u + \mathbf{B}(u, u), \tag{G.1}$$

where u represents the state of the system, e.g. in fluid dynamics the velocity field and the pressure field. $\mathbf{L}$ is a linear operator and $\mathbf{B}(\cdot, \cdot)$ a bilinear function, e.g. in the Navier–Stokes equations $\mathbf{L} = \nu\Delta$ and $\mathbf{B}(u, u)$ is the advection term. We look for the eigenfunctions ψ and eigenvalues λ given by the solution of the linearization of (G.1):

$$\mathbf{L}\psi_j = \lambda_j\psi_j.$$

In some hydrodynamical problems $\mathbf{L}$ is not self-adjoint so it is not possible to use the usual spectral theory, but nevertheless the completness of the set ψ_j is established including the generalized eigenfunctions (if there are any), i.e. vectors ψ such that

$$(\mathbf{L} - \lambda\mathbf{1})^{k+1}\psi = 0 , \qquad (\mathbf{L} - \lambda\mathbf{1})^{k}\psi \neq 0.$$

For moderate values of the Reynolds number only few eigenvalues λ_k ($k = 1, 2, \ldots, K$), have a positive real part. If we assume that at the initial time an arbitrary but very small perturbation is present, consisting of a linear combination of the eigenvalues:

$$\sum_{k=1}^{\infty} a_k \psi_k e^{\lambda_k t}$$

at large time only the first K terms will give a non-negligible contribution. So we can write

$$\sum_{k=1}^{K} a_k \psi_k e^{\lambda_k t},$$

where the a_k are constants. Taking into account the non-linearity one finds that the asymptotic motion takes place on an invariant K-dimensional surface or manifold M which is tangent at the origin to the linear manifold M_0 spanned by the eigenvalues $\psi_1, \ldots, \psi_K$. The manifold M is invariant under the flow determined by (G.1), i.e. any trajectory which starts from M remains on M. This manifold is locally attracting, there is a neighbourhood of the origin such that any trajectory starting from there approaches M at large time.

One of the possible ways to choose the coordinates $Q_1, \ldots, Q_K$ in M is a projection onto the linear manifold M_0 to which it is tangent. For any u in M, one considers the projector operator $\mathbf{P}$:

$$\mathbf{P}u = \sum_{k=1}^{K} Q_k \psi_k$$

where $Q_k = (\chi_k, u)$ and the χ_k are the eigenfunctions of the operator $\mathbf{L}^\dagger$. For the trajectories lying in M the equation of the motion (G.1) takes the form

$$\frac{dQ_k}{dt} = F_k(Q_1, \ldots, Q_K),$$

where

$$F_k(Q_1, \ldots, Q_K) = \lambda_k Q_k + \text{higher order terms}.$$

The computation of the functions $F_k(Q_1, \ldots, Q_K)$ is not a trivial task, for a detailed calculation of the invariant manifold for moderate values of the Reynolds number in the Taylor problem (fluid between two rotating cylinders) and Rayleigh–Bénard convection see Richtmyer [1981], Thual [1993].

G.1   Truncations of the Navier–Stokes equations

The use of invariant manifold techniques is not easy, in particular if the Reynolds number is much larger than the critical value R_c, corresponding to the onset of turbulence. In this case many degrees of freedom are involved and usually one uses a more practical approach, the so-called Galerkin method, in which the field is expanded in a suitable complete set of orthonormal functions, depending on the geometry and the boundary conditions. Inserting this expansion in the hydrodynamic equations and taking into account only a finite number of terms one obtains a set of ordinary differential equations.

We now briefly describe this standard procedure for the 2D Navier–Stokes equation [Lee 1987]. If we consider periodic boundary conditions on a square of length 2π and take into account the incompressibility condition, the expansion of $\mathbf{v}$ in Fourier series may be written:

$$\mathbf{v}(\mathbf{x}) = \sum_{\mathbf{k}} e^{i(\mathbf{k}\cdot\mathbf{x})} Q_{\mathbf{k}} \frac{\mathbf{k}^\perp}{|\mathbf{k}|},$$

where $\mathbf{k} = (k_1, k_2)$, $\mathbf{k}^\perp = (k_2, -k_1)$ and $Q_{\mathbf{k}} = -Q^*_{-\mathbf{k}}$ because $\mathbf{v}(\mathbf{x})$ is real. By expanding the pressure P and the external forcing of the Navier–Stokes equations $\mathbf{f}$ in a similar way, one has the evolution equations for a truncation of $Q_{\mathbf{k}}$

$$\frac{dQ_{\mathbf{k}}}{dt} = -i \sum_{\mathbf{k}'+\mathbf{k}''+\mathbf{k}=0} \frac{(\mathbf{k}')^\perp \cdot (\mathbf{k}'') \left(|\mathbf{k}''|^2 - |\mathbf{k}'|^2\right)}{2|\mathbf{k}'|\,|\mathbf{k}''|\,|\mathbf{k}|} Q^*_{\mathbf{k}'} Q^*_{\mathbf{k}''} - \nu|\mathbf{k}|^2 Q_{\mathbf{k}} + f_{\mathbf{k}},$$

where $\mathbf{k}', \mathbf{k}'', \mathbf{k}$ belong to a set $\mathscr{L}$ of wave vectors such that if $\mathbf{k} \in \mathscr{L}$ then $-\mathbf{k} \in \mathscr{L}$.

In an accurate simulation the number of modes $\mathscr{N}$ in the truncation $\mathscr{L}$ depends on the Reynolds number. A simple argument, similar to that for 3D discussed in section 2.3.5, shows that the number of degrees of freedom scales as

$$\mathscr{N} \sim Re.$$

Let us now derive the low-dimensional system which generates the Eulerian dynamic used in section 8.2. [Boldrighini and Franceschini 1979, Lee 1987]. We define the modes $1, 2, \ldots, 9$ corresponding to the following set:

$$\mathbf{k}_1 = (1,1), \quad \mathbf{k}_2 = (3,0), \quad \mathbf{k}_3 = (2,-1), \quad \mathbf{k}_4 = (1,2), \quad \mathbf{k}_5 = (0,1)$$

$$\mathbf{k}_6 = (1,0), \quad \mathbf{k}_7 = (1,-2), \quad \mathbf{k}_8 = (3,1), \quad \mathbf{k}_9 = (2,2),$$

and set

$$Q_{\mathbf{k}_1} = Q_1, \quad Q_{\mathbf{k}_2} = -iQ_2, \quad Q_{\mathbf{k}_3} = Q_3, \quad Q_{\mathbf{k}_4} = iQ_4, \quad Q_{\mathbf{k}_5} = Q_5$$

$$Q_{\mathbf{k}_6} = iQ_6, \quad Q_{\mathbf{k}_7} = iQ_7, \quad Q_{\mathbf{k}_8} = Q_8, \quad Q_{\mathbf{k}_9} = iQ_9.$$

After rescaling by a factor $\sqrt{10}$, letting $\nu = 1$ (that is equivalent to a change of time and length units) and assuming a forcing along the mode k_3, the equations for the amplitude Q_j $(j = 1, \ldots, 9)$ become:

$$\frac{dQ_1}{dt} = -2Q_1 + 4Q_2Q_3 + 4Q_4Q_5,$$

$$\frac{dQ_2}{dt} = -9Q_2 + 3Q_1Q_3 + 9Q_5Q_8 + 3Q_7Q_9,$$

$$\frac{dQ_3}{dt} = -5Q_3 - 7Q_1Q_2 + \frac{9}{\sqrt{5}}Q_1Q_7 - 5Q_4Q_8 + Re,$$

$$\frac{dQ_4}{dt} = -5Q_4 - Q_1Q_5 + 5Q_3Q_8 + 7Q_6Q_9,$$

$$\frac{dQ_5}{dt} = -Q_5 - 3Q_1Q_4 + \sqrt{5}Q_1Q_6 - Q_2Q_8, \qquad\qquad (G.2)$$

$$\frac{dQ_6}{dt} = -Q_6 - \sqrt{5}Q_1Q_5 - 3Q_4Q_9,$$

$$\frac{dQ_7}{dt} = -5Q_7 - \frac{9}{\sqrt{5}}Q_1Q_3 + Q_2Q_9,$$

$$\frac{dQ_8}{dt} = -10Q_8 - 8Q_2Q_5,$$

$$\frac{dQ_9}{dt} = -8Q_9 - 4Q_2Q_7 - 4Q_4Q_6.$$

Here $Re \propto f_3$ is the Reynolds number, the only control parameter for (G.2).

Many authors [Boldrighini and Franceschini 1979, Franceschini and Tebaldi 1979, Lee 1987] have performed detailed numerical analysis of the system (G.2) for Reynolds numbers up to around 10^3. Let us stress that this procedure is rather questionable from a physical point of view, since the system (G.2) is not an accurate approximation of the Navier–Stokes equation at large Re. However, this truncation of the Navier–Stokes equation was historically important since it represented the first bridge between hydrodynamics and the theory of chaotic systems with 'many' degrees of freedom.

The studies done on (G.2), considering 5, 6, 7, 8 and 9 modes [Boldrighini and Franceschini 1979, Lee 1987] are mainly interesting for the onset of turbulence and the following behaviour has been found for increasing Reynolds number:

(a) Stable fixed points.

(b) Hopf bifurcation to periodic cyclic orbits.

(c) Periodic/aperiodic/chaotic orbits.

For large Re (case (c)) the behaviour is strongly dependent on the truncation. For example, the 5-modes model becomes regular, while the 7-modes model remains chaotic at very large Reynolds number.

Appendix H

Directed percolation

Directed percolation is a special case of standard percolation (for a review on percolation theory, see Kinzel [1983], Stauffer and Aharony [1992]) in which only certain directions are allowed. To describe standard percolation, let us imagine a lattice where a given fraction p of the sites are 'occupied' and the rest are empty. Now define a *cluster* as a group of neighbouring occupied sites. All sites within a cluster are thus connected to each other by one unbroken chain of nearest-neighbour bonds connecting only occupied sites. Percolation theory deals with the number and properties of these clusters as the fraction p of occupied sites is varied. The occupied sites are assumed to be randomly distributed over the entire lattice and p is therefore the probability that a given site is occupied. (Note that the variant described here is called *site-percolation*: we started out with a lattice where the sites are either occupied or empty. One could as well introduce *bond-percolation*, where a fraction of the bonds are blocked.)

As the number p is increased, the clusters will, on average, grow in size and as p approaches a critical value, p_c, there will be a finite probability that one connected cluster will extend over the entire lattice. We call the value p_c the *percolation threshold* and to understand the statistical behaviour around this point it is necessary to take into account the strong critical fluctuations, which give rise to scaling laws with universal critical exponents. On the percolating cluster there will be *dangling bonds*, i.e. branches which do not return to the cluster. When all the dangling bonds are removed from the percolating cluster, what is left is the *backbone*. Both the percolating cluster and the backbone will be characterized by a correlation length ξ which diverges as $\xi \sim \mid p_c - p \mid^{-\nu}$ when the percolation threshold is approached.

Directed percolation is very similar to standard percolation, the only difference being that percolation (i.e. connection from one site to the next) can only occur in one fixed direction (i.e. downward) and never backwards. This is shown in Fig. H.1. On a tilted square lattice a certain fraction p of the sites are occupied, which is marked by black dots. The cluster is initiated in an occupied site indicated by the arrow, and it can only grow by connecting to other black dots along the allowed directions as marked on the figure. It can happen that the cluster will enclose an area of non-occupied sites (let us call it a 'hole'). Also, a branch can end without returning to the cluster: this is one of the dangling bonds. When the density of black sites approaches a certain critical value, the directed percolation threshold, p_c, the probability of having an infinite cluster becomes non-zero. It is quite obvious that the value of p_c for directed percolation should be larger than the corresponding threshold for normal percolation: it is simply more improbable to create an infinite cluster, when some connection directions are forbidden. Correspondingly the values of the critical exponent are also different. When $p \rightarrow p_c$ the size of the cluster diverges and so does the correlation length, but since this type of percolation has a chosen direction

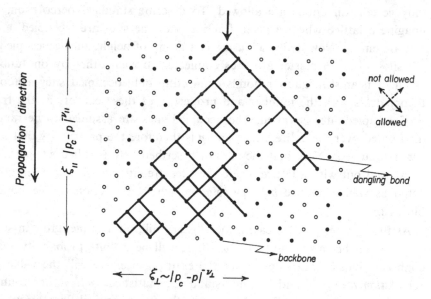

Figure H.1 Schematic drawing of directed percolation on a tilted square lattice. The black dots indicate occupied sites, the circles are empty sites. Starting at an occupied site (indicated by the arrow) and percolating along bonds connecting occupied sites, in the allowed directions, a cluster is generated. Note that there is a correlation length parallel to the propagating direction ($\xi_{\parallel}$) and one perpendicular to this direction ($\xi_{\perp}$).

there will be two correlation lengths which will scale differently: one along the chosen direction and one perpendicular to it:

$$\xi_{\parallel} \sim \mid p_c - p \mid^{-\nu_{\parallel}}, \tag{H.1a}$$

$$\xi_{\perp} \sim \mid p_c - p \mid^{-\nu_{\perp}} . \tag{H.1b}$$

Grassberger and de la Torre [1979] found that directed percolation is closely related to Reggeon field theory, described by a nonlinear Langevin equation with multiplicative noise. This field theory has an *absorbing state*: if the field vanishes at some instant, it has to remain so forever because the noise amplitude is proportional to the field itself. The absorbing state of directed percolation should be understood in the following way. Even if the probability p is above the critical value, $p > p_c$, the branching cluster in Fig. H.1 might suddenly end because a row might not contain any occupied site connecting to the cluster sites in the preceding row. In fact, in a finite system, this will always happen sooner or later, although, as the system size increases, the time before the absorbing state is reached also increases. This behaviour can be quantified using finite-size scaling. Defining an order parameter m as the average fraction of cluster sites at a given row (which corresponds to a given time) Grassberger and de la Torre [1979] and Kertesz and Wolf [1989] found that this order parameter depends on system size L, time t and the distance from criticality $\delta = \mid p - p_c \mid$ through the finite-size scaling form

$$m(L, t, \delta) \sim \xi^{-\beta/\nu} f_m(L/\xi, t/\xi^z) , \tag{H.2}$$

where we have now set $\xi \equiv \xi_{\perp}$, $\nu \equiv \nu_{\perp}$ and introduced the dynamical exponent $z = \nu_{\parallel}/\nu_{\perp}$, which leads to $\xi_{\parallel} = \xi^z$. The order parameter exponent β is defined by $m(\delta) \sim \delta^{\beta}$ for $p \rightarrow p_c^-$. It is interesting to note that the order parameter will always decrease in time. Thus, if the system size is fixed, one finds $m(L, t) \sim t^{-\beta/\nu z}$ which holds up to a cross-over scale determined by the system size.

The best current values for the critical exponents are in $d = 1+1$ [Jensen 1996] $\beta \simeq 0.276\,49$, $\nu_{\parallel} \simeq 1.733\,83$ and $\nu_{\perp} \simeq 1.096\,84$ and in $d = 2+1$ [Grassberger and Zhang 1996] $\beta \simeq 0.584$, $\nu_{\parallel} \simeq 1.295$ and $\nu_{\perp} \simeq 0.734$. For site DP in $d = 1+1$ on a tilted square lattice, as in the figure, one finds [Jensen 1996] $p_c \simeq 0.705\,485$.

References

Aggarwal, J. K. 1972, *Notes on Nonlinear Systems* (Van Nostrand, New York).

Alfaro, C., Benguria, R. and Depassier, M. 1992, *Physica D* **61**, 1.

Amar, J. G. and Family, F. 1990, *Phys. Rev. A* **41**, 3399.

Anselmet, F., Gagne, Y., Hopfinger, E. J. and Antonia, R. A. 1984, *J. Fluid Mech.* **140**, 63.

Antonia, R. A. and Sreenevasan, K. R. 1977, *Phys. Fluids* **20**, 1800.

Antonia, R. A. and Van Atta, C. 1978, *J. Fluid Mech.* **84**, 561.

Antonia, R. A., Hopfinger, E. J., Gagne, Y. and Anselmet, F. 1984, *Phys. Rev. A* **30**, 2704.

Aranson, I. S., Kramer, L. and Weber, A. 1991a, *Phys. Rev. Lett.* **67**, 404.

Aranson, I. S., Kramer, L. and Weber, A. 1991b, *Physica D* **53**, 376.

Aranson, I. S., Aranson, L., Kramer, L. and Weber, A. 1992a, *Phys. Rev. A* **46**, R2992.

Aranson, I. S., Golomb, D. and Sompolinsky, H. 1992b, *Phys. Rev. Lett.* **68**, 3495.

Aranson, I. S., Kramer, L. and Weber, A. 1993a, *Phys. Rev. E* **47**, 3231.

Aranson, I. S., Kramer, L. and Weber, A. 1993b, *Phys. Rev. E* **48**, R9.

Aranson, I. S., Kramer, L. and Weber, A. 1994, *Phys. Rev. Lett.* **72**, 2316.

Arecchi, F. T., Giacomelli, G., Ramazza, P. L. and Residori, S. 1990, *Phys. Rev. Lett.* **65**, 2531.

Arecchi, F. T., Giacomelli, G., Ramazza, P. L. and Residori, S. 1991, *Phys. Rev. Lett.* **67**, 3749.

Aref, H. 1983, *Ann. Rev. Fluid Mech.* **15**, 345.

Aref, H. 1984, *J. Fluid Mech.* **143**, 1.

Aref, H. and Balachandar, S. 1986, *Phys. Fluids* **29**, 3515.

Arnold, V. I. 1963, *Russ. Math. Survey* **18**, 9.

Arnold, V. I. 1965, *C. R. Acad. Sci. Paris A*, **261**, 17.

Arnold, V. I. 1976, *Methodes Mathematiques de la Mécanique Classique* (M.I.R. Publ., Moscow).

Arnold, V. I. 1983, *Geometrical Methods in the Theory of Ordinary Differential Equations* (Springer-Verlag, New York).

Arnold, V. I. and Avez, A. 1968, *Ergodic Problems of Classical Mechanics* (Benjamin, New York).

Aubry, N., Holmes, P., Lumley, J. L. and Stone, E. 1988, *J. Fluid Mech.* **192**, 115.

Aurell, E. and Gilbert, A. D. 1993, *Geophys. Astrophys. Fluid Dyn.* **73**, 5.

Aurell, E., Frick, P. and Shaidurov, V. 1994a, *Physica D* **72**, 95.

Aurell, E., Boffetta, G., Crisanti, A., Frick, P., Paladin, G. and Vulpiani, A. 1994b, *Phys. Rev. E* **50**, 4705.

Aurell, E., Boffetta, G., Crisanti, A., Paladin, G. and Vulpiani, A. 1996, *Phys. Rev. E* **53**, 2337.

Ausloos, M. and Berman, D. H. 1985, *Proc. Roy. Soc. A* **400**, 331.

Avellaneda, M. and Majda, A. 1989, *Phys. Rev. Lett.* **62**, 753.

Avellaneda, M. and Majda, A. 1991, *Commun. Math. Phys.* **138**, 339.

Babiano, A., Basdevant, C. and Sadourny, R. 1984 *C. R. Acad. Sci. Paris* **299**, 495.

Babiano, A., Basdevant, C., Legras, B. and Sadourny, R. 1987, *J. Fluid Mech.* **183**, 379.

Babiano, A., Boffetta, G., Provenzale, A. and Vulpiani, A. 1994, *Phys. Fluid A* **6**, 2465.

Bacry, E., Arnéodo, A., Frisch, U., Gagne, Y. and Hopfinger, E. J. 1990, in *Turbulence and Coherent Structures*, ed. Métais, O. and Lesieur, M. (Kluwer, Deventer).

Bagnoli, F., Rechtman, R. and Ruffo, S. 1992, *Phys. Lett. A* **172**, 34.

Bak, P., Tang, C. and Wiesenfeld, K. 1987, *Phys. Rev. Lett.* **59**, 381.

Bak, P. and Sneppen, K. 1993, *Phys. Rev. Lett.* **71**, 4083.

Balmforth, N. J., Cvitanović, P., Ierley, G. R., Spiegel, E. A. and Vattay, G. 1993, *Ann. N.Y. Acad. Sci.* **706**, 148.

Barabási, A. L. 1991, *J. Phys. A* **24**, L1013.

Barabási, A. L., Szépfalusy, P. and Vicsek, T. 1991, *Physica A* **178**, 17.

Barabási, A. L. and Vicsek, T. 1991, *Phys. Rev. A* **44**, 2730.

Barabási, A. L., Bourbonnais, R., Jensen, M. H., Kertész, J., Vicsek, T. and Zhang, Y.-C. 1992, *Phys. Rev. A* **45**, R6951.

Basdevant, C. and Couder, Y. 1986, *J. Fluid Mech.* **173**, 225.

Batchelor, G. K. 1952, *Proc. Roy. Soc. A* **213**, 349.

Batchelor, G. K. 1959, *J. Fluid Mech.* **5**, 113.

Batchelor, G. K. 1969, *Phys. Fluids* **12**, 233.

Bayly, B. J. 1986, *Phys. Rev. Lett.* **57**, 2800.

Bayly, B. J. and Childress, S. 1987, *Phys. Rev. Lett.* **59**, 1573.

Beck, C. 1994, *Phys. Rev. E* **49**, 3641.

Bekki, N. and Nosakki, K. 1985, *Phys. Lett. A* **110**, 133.

Bell, T. L. and Nelkin, M. 1977, *Phys. Fluids* **20**, 345.

Bender, C. M. and Orszag, S. A. 1978, *Advanced Mathematical Methods for Scientists and Engineers* (McGraw-Hill, New York).

Benettin, G., Galgani, L. and Strelcyn, J. M. 1976, *Phys. Rev. A* **14**, 2338.

Benettin, G. and Strelcyn, J. M. 1978, *Phys. Rev. A* **17**, 2157.

Benettin, G., Giorgilli, A., Galgani, L. and Strelcyn, J. M. 1980, *Meccanica* **15**, 9 and 21.

Ben-Jakob, E., Brand, H., Dee, G., Kramer, L. and Langer, J. 1985, *Physica D*, **14**, 348.

Benjamin, T. B. and Feir, J. E. 1967, *J. Fluid Mech.* **27**, 413.

Bensimon, D., Kadanoff, L. P., Liang, S., Shraiman, B. I. and Tang, C. 1986, *Rev. Mod. Phys.* **58**, 977.

Benzi, R., Sutera, A. and Vulpiani, A. 1981, *J. Phys. A* **14**, L453.

Benzi, R., Parisi, G., Sutera, A. and Vulpiani, A. 1982, *Tellus* **34**, 10.

Benzi, R., Paladin, G., Parisi, G. and Vulpiani, A. 1984, *J. Phys. A* **17**, 3521.

Benzi, R., Paladin, G., Parisi, G. and Vulpiani, A. 1985, *J. Phys. A* **18**, 2157.

Benzi, R., Paladin, G., Patarnello, S., Santangelo, P. and Vulpiani, A. 1986, *J. Phys. A* **19**, 3771.

Benzi, R., Patarnello, S. and Santangelo, P. 1987, *Europhys. Lett.* **3**, 811.

Benzi, R., Paladin, G. and Vulpiani, A. 1990, *Phys. Rev. A* **42**, 3654.

Benzi, R., Biferale, L., Paladin, G., Vulpiani, A. and Vergassola, M. 1991, *Phys. Rev. Lett.* **67**, 2299.

Benzi, R., Biferale, L. and Parisi, G. 1993a, *Physica D* **65**, 163.

Benzi, R., Biferale, L., Crisanti, A., Paladin,

G., Vergassola, M. and Vulpiani, A. 1993b, *Physica D* **65**, 352.

Benzi, R., Biferale, L., Kerr, R. M. and Trovatore, E. 1996, *Phys. Rev. E* **53**, 3541.

Berkooz, G., Holmes, P. and Lumley, J. L. 1993, *Ann. Rev. Fluid Mech.* **25**, 539.

Bertini, L., Cancrini, N. and Jona Lasinio, G. 1994, *Commun. Math. Phys.* **165**, 211.

Bessis, D., Paladin, G., Turchetti, G. and Vaienti, S. 1988, *J. Stat. Phys.* **51**, 109.

Bhagavatula, R., Grinstein, G., He, Y. and Jayaprakash, C. 1992, *Phys. Rev. Lett.* **69**, 3482.

Biferale, L., Jensen, M. H., Paladin, G. and Vulpiani, A. 1992, *Physica A* **185**, 19.

Biferale, L., Crisanti, A., Falcioni, M. and Vulpiani, A. 1993, *J. Phys. A* **26**, L923.

Biferale, L., Blank, M. and Frisch, U. 1994, *J. Stat. Phys.* **75**, 781.

Biferale, L. and Kerr, R. 1995, *Phys. Rev. E* **52**, 6113.

Biferale, L., Lambert, A., Lima, R. and Paladin, G. 1995a, *Physica D* **80**, 105.

Biferale, L., Crisanti, A., Vergassola, M. and Vulpiani, A. 1995b, *Phys. Fluids A* **7**, 2725.

Biktashev, V. N. 1989, in *Nonlinear Waves 2*, ed. Gaponov-Grekhov, A. V. and Rabinovich, M. I. (Springer-Verlag, Heidelberg) p. 87.

Birkhoff, G. D. 1927, *Dynamical Systems* (AMS Publications, New York).

Bodenschatz, E., Weber, A. and Kramer, L. 1989, in *Nonlinear Processes in Excitable Media*, ed. A. V. Holden, M. Marcus and H. G. Othmer (Plenum Press, New York).

Boffetta, G., Paladin, G. and Vulpiani, A. 1996, *J. Phys. A* **29**, 2291.

Boffetta, G., Celani, A., Crisanti, A. and Vulpiani, A. 1997, *Phys. Fluids* **A9**, 724.

Bohr, T. 1990, in *Applications of Statistical Mechanics and Field Theory to Condensed Matter*, p. 157 eds. A.R. Bishop, D. Baeriswyl and J. Carmelo (Plenum Press, New York).

Bohr, T., and Rand, D., 1987, *Physica D* **26**, 387.

Bohr, T., Grinstein, G., Jayaprakash, C. and He, Y. 1987, *Phys. Rev. Lett.* **58**, 2155.

Bohr, T. and Christensen, O. B. 1989, *Phys. Rev. Lett.* **63**, 2161.

Bohr, T., Jensen, M. H., Pedersen, A. W. and Rand, D. A. 1989, in *New Trends in Nonlinear Dynamics and Pattern Forming Phenomena*, ed. Coullet, P. and Huerre, P. (Plenum Press, New York) p. 185.

Bohr, T., Pedersen, A. W. and Jensen, M. H. 1990a, *Phys. Rev. A* **42**, 3626.

Bohr, T., Pedersen, A. W., Jensen, M. H. and Rand, D. A. 1990b, in *Nonlinear Evolution of Spatio-Temporal Structures in Dissipative Continuous Systems*, eds. Busse, F. and Kramer, L. (Plenum Press, New York) p. 425.

Bohr, T. and Rand, D. 1991, *Physica D* **52**, 532.

Bohr, T., Grinstein, G., Jayaprakash, C., Jensen, M. H., Krug, J. and Mukamel, D. 1992, *Physica D* **59**, 177.

Bohr, T. and Pikovsky, A., 1993, *Phys. Rev. Lett.* **70**, 2892.

Bohr, T., Bosch, E. and van de Water, W. 1994, *Nature* **372**, 48.

Bohr, T., Grinstein, G. and Jayaprakash, C. 1995, *Chaos* **5**, 1995.

Bohr, T., Huber, G. and Ott, E. 1996, *Europhysics Letters* **33**, 589.

Bohr, T. and Lundbek Hansen, J. 1996, *Chaos* **6**, 554.

Bohr, T., Huber, G. and Ott, E. 1997, *Physica D* **106**, 95.

Boldrighini, C. and Franceschini, V. 1979, *Commun. Math. Phys.* **64**, 159.

Boots, B., Okabe, A. and Sugihara, K. 1992, *Spatial Tessellations: Concepts and Applications of Voronoi Diagrams* (Wiley and Sons, New York).

Borue, V. 1994, *Phys. Rev. Lett.* **72**, 1475.

Bosch, E. G. T. 1995, *An Experimental Investigation of Faraday Waves and Spatio-Temporal Chaos*, Ph.D. Thesis, Eindhoven Technical University.

Bouchaud, J. P., Mezard, M. and Parisi, G. 1995, *Phys. Rev. E* **52**, 3656.

Bourzutschky, M. S. and Cross, M. C. 1992, *J. Chaos* **2**, 173.

Brandenburg, A. 1992, *Phys. Rev. Lett.* **69**, 605.

Briggs, R. G. 1964, *Electron Stream*

Interaction with Plasmas (MIT, Cambridge, MA).

Buldyrev, S. V., Barabási, A. L., Caserta, F., Havlin, S., Stanley, H. E. and Vicsek, T. 1992, *Phys. Rev. A* **46**, R8313.

Bulsara, A. R., Jacobs, E. W. and Schieve, W. C. 1990, *Phys. Rev. A* **42**, 4614.

Bunimovich, L. A. and Sinai, Y. G. 1988 *Nonlinearity* **1**, 491.

Burgers, J. M. 1940, *Proc. Roy. Neth. Acad. Soc.* **43** (1), 2.

Burgers, J. M. 1974, *The Nonlinear Diffusion Equation* (D. Reidel Publishing Company, Dordrecht).

Caglioti, E. and Loreto, V. 1996, *Phys. Rev. E* **53**, 2953.

Cantwell, B. J., Coles, D. and Dimotakis, P. 1978, *J. Fluid Mech.* **87**, 641.

Carbone, V. 1994, *Europhys. Lett.* **27**, 581.

Carbone, V. 1995, *Europhys. Lett.* **29**, 377.

Cardoso, O., and Tabeling, P. 1988, *Europhys. Lett.* **7**, 225.

Carleman, T. 1922, *C. R. Acad. Sci. Paris* **174**, 1680.

Castaing, B., Gunaratne, G. H., Heslot, F., Kadanoff, L. P., Libchaber, A., Thomae, S., Wu, X.-Z., Zaleski, S. and Zanetti, G. 1989, *J. Fluid. Mech.* **204**, 1.

Castaing, B., Gagne, Y. and Hopfinger, E. J. 1990, *Physica D* **46**, 177.

Chaiken, J., Chu, C. K., Tabor, M. and Tan, Q. M. 1987, *Phys. Fluids* **30**, 687.

Chaikin, P. and Lubensky, T. 1995, *Principles of Condensed Matter Physics* (Cambrigde University Press, Cambridge).

Chandrasekhar, S. 1943, *Rev. Mod. Phys.* **15**, 1.

Chang, H.-C. 1994, *Ann. Rev. Fluid Mech.* **26**, 103.

Chang, H.-C., Demekhin, E. and Kopelevich, D. 1995, *Phys. Rev. Lett.* **75**, 1747.

Chaté, H. and Manneville, P. 1988, *Phys. Rev. A* **38**, 4351.

Chaté, H. and Manneville, P. 1989, *Physica D* **32**, 409.

Chaté, H. and Manneville, P. 1995, *Physica D* **224**, 348.

Chen, K., Bak, P. and Jensen, M. H. 1990, *Phys. Lett. A* **149**, 207.

Chen, Z. Y. 1990, *Phys. Rev. A* **42**, 5837.

Childress, S. 1970, *J. Math. Phys.* **11**, 3063.

Childress, S. and Gilbert, A. D. 1995, *Stretch, Twist, Fold: The Fast Dynamo* (Springer-Verlag, Berlin).

Chow, C. C. and Hwa, T. 1995, *Physica D* **84**, 494.

Cieplak, M. and Robbins, M. O. 1988, *Phys. Rev. Lett.* **60**, 2042.

Cieplak, M. and Robbins, M. O. 1990, *Phys. Rev. B* **41**, 11508.

Ciliberto, S. and Bigazzi, P. 1988, *Phys. Rev. Lett.* **60**, 286.

Cocke, W. J. 1969, *Phys. Fluids* **12**, 2488.

Cocke, W. J. 1971, *Phys. Fluids* **14**, 1624.

Cole, J. D. 1951, *Quart. Appl. Math.* **9**, 225.

Collet, P. 1989, *Workshop on Small Diffusivity Dynamos and Dynamical Systems*, (Nice).

Collet, P. and Eckmann, J.-P. 1980, *Iterated Maps on the Interval as Dynamical Systems* (Birkhäuser, Boston).

Coniglio, A. 1981, *Phys. Rev. Lett.* **46**, 250.

Conrado, C. and Bohr, T. 1994, *Phys. Rev. Lett.* **72**, 3522.

Conrado, C. and Bohr, T. 1995, *Phys. Rev. E* **51**, 4485.

Coppersmith, S. N. 1993, in *Phase Transitions and Relaxation in Systems with Competing Energy Scales*, ed. Riste, T. and Sherrington, D. (Kluwer, Boston), p. 317.

Corrsin, S. 1951, *J. Appl. Phys.* **22**, 469.

Corrsin, S. 1962, in *Mécanique de la Turbulence* p. 27; ed. A. Favre (CNRS, Paris).

Coullet, P., Gil, L. and Lega, J. 1989, *Phys. Rev. Lett.* **62**, 1619.

Criminale, W. and Kovasznay 1962, *J. Fluid Mech.* **14**, 59.

Crisanti, A., Falcioni, M., Paladin, G. and Vulpiani, A. 1990a, *Physica A* **166**, 305.

Crisanti, A., Falcioni, M., Paladin, G. and Vulpiani, A. 1990b, *J. Phys. A: Math. Gen.* **23**, 3307.

Crisanti, A., Falcioni, M., Paladin, G. and

Vulpiani, A. 1991, *La Rivista del Nuovo Cimento* **14**, (12), 1.

Crisanti, A., Jensen, M. H., Vulpiani, A. and Paladin, G. 1992, *Phys. Rev. A* **46**, R7363.

Crisanti, A., Jensen, M. H., Vulpiani, A. and Paladin, G. 1993a, *Phys. Rev. Lett.* **70**, 166.

Crisanti, A., Jensen, M. H., Vulpiani, A. and Paladin, G. 1993b, *J. Phys. A* **26**, 6943.

Cross, M. C. and Hohenberg, P. C. 1993, *Rev. Mod. Phys.* **65**, 851.

Crutchfield, J. P., Farmer, D. J. and Huberman, B. A. 1982, *Phys. Rep.* **92**, 45.

Crutchfield, J. P. and Kaneko, K. 1987, in *Directions in Chaos* vol. 1, ed. B.-L. Hao (World Scientific, Singapore).

Cvitanović, P. 1989, *Universality in Chaos* (Adam Hilger, Bristol).

Daviaud, F., Dubois, M. and Bergé, P. 1989, *Europhys. Lett.* **9**, 441.

Davidenko, J. M., Pertsov, A. M., Salomonsz, R., Baxter, W. and Jalife, J. 1991, *Nature* **353**, 349.

Deane, A. E. and Sirovich, L. 1991a, *J. Fluid Mech.* **222**, 231.

Deane, A. E. and Sirovich, L. 1991b, *J. Fluid Mech.* **222**, 251.

Deane, A. E., Kevrekidis, I. G., Karniadakis, G. E. and Orszag, S. A. 1991, *Phys. Fluids A* **3**, 2337.

Decker, W., Pesch, W., and Weber, A., 1994 *Phys. Rev. Lett.* **73**, 648.

Dee, G. and Langer, J. S. 1982, *Phys. Rev. Lett.* **50**, 383.

Deissler, R. and Kaneko, K. 1987, *Phys. Lett. A* **119**, 397.

Desnyansky, V. N. and Novikov, E. A. 1974a, *Prinkl. Mat. Mekh.* **38**, 507.

Desnyansky, V. N. and Novikov, E. A. 1974b, *Atmos. Oceanic Phys.* **10**, 127.

Ditlevsen, P. D. 1996, *Phys. Rev. E* **54**, 985.

Ditlevsen, P. D. and Mogensen, I. A. 1996, *Phys. Rev. E* **53**, 4785.

Dombre, T., Frisch, U., Greene, J. M., Hénon, M., Mehr, A. and Soward, A. M. 1986, *J. Fluid Mech.* **167**, 353.

Drummond, I. T. and Münch, W. 1990, *J. Fluid Mech.* **215**, 45.

Drummond, I. T. and Münch, W. 1991, *J. Fluid Mech.* **225**, 529.

Du, Y., Tél, T. and Ott, E. 1994, *Physica D* **76**, 168.

Dubrulle, B. 1994, *Phys. Rev. Lett.* **73**, 959.

Eckmann, J.-P. and Ruelle, D. 1985, *Rev. Mod. Phys.* **57**, 617.

Eckmann, J.-P. and Wayne, C. E. 1988, *J. Stat. Phys.* **50**, 853.

Eckmann, J.-P. and Wayne, C. E. 1989, *Commun. Math. Phys.* **121**, 147.

Eckmann, J.-P. and Ruelle, D. 1992, *Physica D* **56**, 185.

Edwards, S. F. and Wilkinson, D. R. 1982, *Proc. Roy. Soc. London A* **381**, 17.

Eggers, J. 1992, *Phys. Rev. A* **46**, 1951.

Eggers, J. and Grossmann, S. 1991, *Phys. Lett. A* **156**, 444.

Egolf, D. A. and Greenside, H. S. 1994, *Nature* **369** 129.

Egolf, D. A. and Greenside, H. S. 1995, *Phys. Rev. Lett.* **74** 1751.

Elhmaidi, D., Provenzale, A. and Babiano, A. 1993, *J. Fluid Mech.* **257**, 533.

Elphick, C. and Meron, E. 1991, *Physica D*, **53**, 385.

Elphick, C., Ierley, G., Regev, O. and Spiegel, E. 1991, *Phys. Rev. A* **44**, 1110.

Erzan, A., Pietronero, L. and Vespignani, A. 1995, *Rev. Mod. Phys.* **67**, 545.

Eyink, G. and Goldenfeld, N. 1994, *Phys. Rev. A* **50**, 4679.

Falcioni, M., Paladin, G. and Vulpiani, A. 1988, *J. Phys. A: Math. Gen.* **21**, 3451.

Falcioni, M. and Vulpiani, A. 1989, *J. Phys. A* **22**, 1201.

Falcioni, M., Paladin, G. and Vulpiani, A. 1989, *Europhys. Lett.* **10**, 201.

Falk, J., Jensen, M. H. and Sneppen, K. 1994, *Phys. Rev. E* **49**, 2804.

Family, F. 1986, *J. Phys. A* **19**, L441.

Family, F. and Vicsek, T. 1985, *J. Phys. A* **18**, L75.

Farge, M. 1992, *Ann. Rev. Fluid Mech.* **24**, 395.

Feigenbaum, M. J. 1978, *J. Stat. Phys.* **19**, 25.

Feigenbaum, M. J. 1979, *J. Stat. Phys.* **21**, 669.

Feingold, M., Kadanoff, L. P. and Piro, O. 1988, *J. Stat. Phys.* **50**, 529.

Feller, W. 1971, *An Introduction to Probability Theory and Its Applications* (Wiley and Sons, New York).

Finn, J. M. and Ott, E. 1990, *Phys. Fluids B* **2**, 916.

Ford, J. 1983, *Physics Today* **36**, April, 40.

Forster, D., Nelson, D. and Stephen, M. 1977, *Phys. Rev. A* **16**, 732.

Fournier, J. D. and Frisch, U. 1983, *J. de Méc. Théor. et Appl.* **2**, 699.

Franceschini, V. and Tebaldi, C. 1979, *J. Stat. Phys.* **21**, 707.

Fraser, A. M. and Swinney, H. L. 1986, *Phys. Rev. A* **33**, 1134.

Frick, P. and Aurell, E. 1993, *Europhys. Lett.* **24**, 725.

Frisch, U. 1995, *Turbulence: the Legacy of A. N. Kolmogorov* (Cambridge University Press, Cambridge).

Frisch, U., Sulem, P. L. and Nelkin, M. 1978, *J. Fluid Mech.* **87**, 719.

Frisch, U. and Sulem, P. L. 1984, *Phys. Fluids* **27**, 1921.

Frisch, U., Hasslacher, B. and Pomeau, Y. 1986, *Phys. Rev. Lett.* **56**, 1505.

Frisch, U., She, Z. S. and Thual, O. 1986, *J. Fluid Mech.* **168**, 221.

Frisch, U. and She, Z.-S. 1991, *Fluid Dyn. Res.* **8**, 139.

Frisch, U. and Vergassola, M. 1991, *Europhys. Lett.* **14** , 439.

Fujisaka, H. 1983, *Prog. Theor. Phys.* **70**, 1264.

Fujisaka, H. and Yamada, T. 1977, *Prog. Theor. Phys.* **57**, 734.

Gagne, Y. and Castaing, B. 1991, *C. R. Acad. Sci. Paris* **312**, 441.

Gallavotti, G. 1983, *The Elements of Mechanics* (Springer-Verlag, Berlin).

Galloway, D. J. and Frisch, U. 1987, *J. Fluid Mech.* **180**, 557.

Galluccio, S. and Vulpiani, A. 1994, *Physica A* **212**, 75.

Gaster, M. 1962, *J. Fluid Mech.* **14**, 222.

Gat, O., Procaccia, I. and Zeitak, R. 1995, *Phys. Rev. E* **51**, 1148.

Gerisch, G. and Hess, B. 1974, *Proc. Natl. Acad. Sci. USA* **71**, 2118.

Ginzburg, V. and Landau, L. 1950, *JETP* **20**, 1064.

Gledzer, E. B. 1973, *Sov. Phys. Dokl.* **18**, 216.

Gollub, J. P. 1991, *Physica D* **51**, 501.

Grappin, R., Leorat, J. and Pouquet, A. 1986, *J. Physique* **47**, 1127.

Grassberger, P. 1983, *Phys. Lett. A* **97**, 227.

Grassberger, P. 1985, *Phys. Lett. A* **107**, 101.

Grassberger, P. and Sundermeyer, K. 1978, *Phys. Lett. B* **77**, 220.

Grassberger, P. and de la Torre, A. 1979, *Ann. Phys.* **122**, 373.

Grassberger, P. and Procaccia, I. 1983, *Phys. Rev. A* **28**, 2591.

Grassberger, P. and Zhang, Y.-C. 1996, *Physica A* **224**, 16.

Grauer, R. and Sideris, T. C. 1991, *Phys. Rev. Lett.* **67**, 3511.

Grinstein, G., Jayaprakash. C. and Bolker. B. 1991, *Phys. Rev. A* **44**, 4923.

Grinstein, G., Jayaprakash, C. and Pandit, R. 1996, *Physica D* **90**, 96.

Grossmann, S. and Lohse, D. 1992, *Z. Phys. B.* **89**, 11.

Guckenheimer, J. and Holmes, P. 1983, *Nonlinear Oscillations, Dynamical Systems and Bifurcations of Vector Fields*, Applied Mathematical Sciences Vol. 42 (Springer-Verlag, Berlin).

Gundlach, V. M. and Rand, D. 1993, *Nonlinearity* **6**, 165; *ibid*, 201; *ibid*, 215.

Gunton, J. D., San Miguel, M. and Sahni, P. S. 1983, in *Phase Transitions and Critical Phenomena*, Vol. 8, ed. Domb, C. and Lebowitz, J. L. (Academic Press, New York, 1983).

Gurbatov, S. N., Saichev, A. I. and Shandarin, S. F. 1989, *Mon. Not. Roy. Astro. Soc.* **236**, 385.

Gurbatov, S. N., Malachov, A. and Saichev, A. I. 1991, *Nonlinear Random Waves and Turbulence in Nondispersive Media: waves, rays and particles* (Manchester University Press, Manchester).

Hagan, P. 1982, *SIAM J. Appl. Math.* **42**, 762.

Hall, L. and Henckrotte, W. 1968, *Phys. Rev.* **166**, 120.

Halpin-Healey, T. and Zhang, Y. C. 1995, *Phys. Rep.* **254**, 215.

Halsey, T. C., Jensen, M. H., Kadanoff, L. P., Procaccia, I. and Shraiman, B. 1986, *Phys. Rev. A* **33**, 1141.

Hansel, D. and Sompolinsky, H, 1993, *Phys. Rev. Lett.* **71**, 2710.

Hausdorff, F. 1919, *Mathematische Annalen* **79**, 157.

Havlin, S., Barabási, A. L., Buldyrev, S. V., Peng, C. K., Schwartz, M., Stanley, H. E. and Vicsek, T. 1993, in Proc. of Granada Conference on *Growth Patterns in Physical Sciences and Biology*, ed. Garcia-Ruiz, J. M., Louis, E., Meakin. P. and Sander, L. (Plenum Press, New York) p. 85.

He, S., Kahanda, G. and Wong, P.-Z. 1992, *Phys. Rev. Lett.* **69**, 3731.

Heisenberg, W. 1948a, *Z. Physik* **124**, 628.

Heisenberg, W. 1948b, *Proc. Roy. Soc. A* **195**, 402.

Hele-Shaw, H. S. 1898, *Nature* **58**, 34.

Hénon, M. 1966, *C. R. Acad. Sci. Paris A* **262** 312.

Hénon, M. and Heiles, C. 1964, *Astron. J.* **69**, 73.

Hentschel, H. G. E. and Procaccia, I. 1983, *Physica D* **8**, 435.

Herweijer, J. and van de Water, W. 1995, *Phys. Rev. Lett.* **74**, 4653.

Herzel, H. and Pompe, B. 1987, *Phys. Lett. A* **122**, 121.

Heslot, F., Castaing, B. and Libchaber, A. 1987, *Phys. Rev. A* **36**, 5870.

Hohenberg, P., and Shraiman, P., 1989, *Physica D* **37**, 109.

Holmes, P. 1990, *Phys. Rep.* **193**, 137.

Hopf, E. 1950, *Comm. Pure Appl. Mech.* **3**, 201.

Horita, T., Hata, H., Ishizaki, R. and Mori, H. 1990, *Prog. Theor. Phys.* **83**, 1065.

Horvárth, V. K., Family, F. and Vicsek, T. 1990, *Phys. Rev. Lett.* **65**, 1388.

Horvárth, V. K., Family, F. and Vicsek, T. 1991, *J. Phys. A* **24**, L25.

Houlrik, J. M., Webman, I. and Jensen, M. H. 1990, *Phys. Rev. A* **41**, 4210.

Houlrik, J. M. and Jensen, M. H. 1992, *Phys. Lett. A* **163**, 275.

Huber, G. 1993, *The Onset of Vortex Turbulence*. PhD Thesis, UMI# 9309755 (Boston University).

Huber, G. 1994, *Spatio-Temporal Patterns* (SFI Proceedings **XXI**), ed. Palffy-Muhoray, P. and Cladis, P. (Addison-Wesley).

Huber, G., Alstrøm, P. and Bohr, T. 1990, *Phys. Rev. Lett.* **69**, 2380.

Huber, G., Alstrøm, P. and Bohr, T. 1992, *Phys. Rev. Lett.* **69**, 2380.

Huber, G., Jensen, M. H. and Sneppen, K. 1995, *Phys. Rev. E* **52**, R2133.

Huberman, B. A. and Rudnick, J. 1980, *Phys. Rev. Lett.* **45**, 154.

Hyman, J. M., Nicoalenko, B. and Zaleski, S. 1986, *Physica D* **23D**, 265.

Hynne, F. and Graae Sørensen, P. 1990, *J. Chem. Phys.* **92**, 1747.

Hynne, F. and Graae Sørensen, P. 1993, *Phys. Rev. E* **92**, 1747.

Iooss, G. and Joseph, D. D. 1990, *Elementary Stability and Bifurcation Theory* (Springer-Verlag, New York).

Isichenko, M. B. 1992, *Rev. Mod. Phys.* **64**, 961.

Isola, S., Politi, A., Ruffo, S. and Torcini, A. 1990, *Phys Lett. A* **143**, 365.

Jakubith, S., Rotermund, H. H., Engel, W., von Oertzen and Ertl, G. 1990, *Phys. Rev. Lett.* **65**, 3013.

Jayaprakash, C., Hayot, F. and Pandit, R. 1993, *Phys. Rev. Lett.* **71**, 12.

Jayaprakash, C., Hayot, F. and Pandit, R. 1994, *Phys. Rev. Lett.* **72**, 308 (Reply).

Jensen, H. J. 1993, in *Phase Transitions and Relaxation in Systems with Competing Energy Scales*, ed. Riste, T. and Sherrington, D. (Kluwer, Boston), p. 137.

Jensen, I. 1996, *J. Phys. A* **29**, 7013.

Jensen, M. H. 1989a, *Phys. Rev. Lett.* **62**, 1361.

Jensen, M. H. 1989b, *Physica D* **38**, 203.

Jensen, M. H. and Procaccia I. 1991, *J. de Physique II* **1**, 1139.

Jensen, M. H., Paladin, G. and Vulpiani, A. 1991a, *Phys. Rev. A* **43**, 798.

Jensen, M. H., Paladin, G. and Vulpiani, A. 1991b, *Phys. Rev. Lett.* **67**, 208.

Jensen, M. H., Paladin, G. and Vulpiani, A. 1992, *Phys. Rev. A* **45**, 7214.

Jullien, R. and Botet, R. 1985, *J. Phys. A* **18**, 2279.

Kadanoff, L. P., Nagel, S. R., Wu, L. and Zhou, S.-M. 1989, *Phys. Rev. A* **39**, 6524.

Kadanoff, L. P., Lohse, D., Wang, J. and Benzi, R. 1995, *Phys. Fluids* **7**, 617.

Kadanoff, L. P., Lohse, D. and Schörghofer, N. 1997, *Physica D* **100**, 165.

Kahanda, G., Zou, X., Farrell, R. and Wong, P.-Z. 1992, *Phys. Rev. Lett.* **68**, 3741.

Kahanda, G. and Wong, P.-Z. 1993, *Phys. Rev. Lett.* **71**, 806.

Kaneko, K. 1984, *Prog. Theor. Phys.* **72**, 480.

Kaneko, K. 1985, *Prog. Theor. Phys.* **74**, 1033.

Kaneko, K. 1989a, *Phys. Lett. A* **139**, 47.

Kaneko, K. 1989b, *Physica D* **34**, 1.

Kaneko, K. 1993, *Theory and Applications of Coupled Map Lattices* (Wiley and Sons, New York).

Kaplan, J. L. and Yorke, J. A. 1979, *Lect. Notes in Math.* **730**, 204.

Kapral, R., Livi, R., Oppo, G. L. and Politi, A. 1994, *Phys. Rev. E* **49**, 2009.

Kardar, M. and Zhang, Y.-C. 1987, *Phys. Rev. lett.* **58**, 2087.

Kardar, M., Parisi, G. and Zhang, Y. C. 1986, *Phys. Rev. Lett.* **56**, 889.

Kaspar, F. and Schuster, H. 1986, *Phys. Lett. A* **113**, 451.

Kawara, T. 1983, *Phys. Rev. Lett.* **51**, 381.

Keener, J. and Tyson, J. 1986, *Physica D* **21**, 307.

Kerr, R. M. and Siggia, E. D. 1978, *J. Stat. Phys.* **19**, 543.

Kertész, J. and Wolf, D. E. 1988, *J. Phys. A* **21**, 747.

Kertész, J. and Wolf, D. E. 1989, *Phys. Rev. Lett.* **62**, 2571.

Kessler, D. A., Levine, H. and Tu, Y. 1991, *Phys. Rev. A* **43**, 4551.

Khinchin, A. I. 1957, *Mathematical Foundation of Information Theory* (Dover Publ., New York).

Kida, S. 1979, *J. Fluid Mech.* **93**, 337.

Kim, J. M. and Kosterlitz, M. 1989, *Phys. Rev. Lett.* **62**, 2289.

Kinzel, W. 1983, in *Percolation Structures and Processes* ed. Deutcher, G., Zallen, R. and Adler, J., *Ann. Isr. Phys. Soc.* **5**, 425.

Kolmogorov, A. N. 1941, *C. R. Acad. Sci. USSR* **30**, 301 and **32**, 16.

Kolmogorov, A. N. 1954, *Dokl. Akad. Nauk. SSSR* **98**, 527.

Kolmogorov, A. V. 1962, *J. Fluid Mech.* **12**, 82.

Kook, H., Ling, F. H. and Schmidt, G. 1991, *Phys. Rev. A* **43**, 2700.

Koplik, J. and Levine, H. 1985, *Phys. Rev. B* **32**, 280.

Kraichnan, R. H. 1967, *Phys. Fluids* **10**, 1417.

Kraichnan, R. H. 1974, *J. Fluid Mech.* **64**, 737.

Kraichnan, R. H. and Montgomery, D. 1980, *Rep. Prog. Phy.* **43**, 547.

Krug, J., 1994, *Phys. Rev. Lett.* **72**, 2907.

Krug, J. 1995, in *Proceedings on the International Colloquium on Modern Field Theory II* p. 141. (Tata Institute, India) ed. Das, S., Mandal, G., Mukhi, S. and Wadia, S. (World Scientific, Singapore).

Krug, J. and Spohn, H. 1991, in *Solids far from Equilibrium*, ed. Godrèche, C. (Cambridge University Press, Cambridge).

Krug, J., Meakin, P. and Halpin-Healy, T. 1992, *Phys. Rev. A* **45**, 638.

Kubo, R. 1957 *J. Phys. Soc. Japan* **12**, 570.

Kuramoto, Y. 1984, *Chemical Oscillations, Waves and Turbulence* (Springer-Verlag, Berlin).

Kuznetsov, S. P., 1993, in *Theory and Applications of Coupled Map Lattices*, ed. Kaneko, K. (Wiley and Sons, New York).

Kuznetsov, S. P. and Pikovsky, A. S. 1986, *Physica D* **19**, 384.

Kuznetsova, L. M., Rabinovich, M. I. and Sushchik, M. M. 1983, *Izv. Atmospheric and Oceanic Physics* **19**, 36.

Lamb, H. 1945, *Hydrodynamics* (Dover Publ., New York).

Landau, L. 1937, *JETP* **11**, 545.

Landau, L. D. and Lifshitz, L. 1959, *Statistical Mechanics* (Pergamon Press, New York).

Landau, L. D. and Lifshitz, L. 1987, *Fluid Mechanics* (Pergamon Press, New York).

Laplace, S. 1814, *Essai philosophique sur les probabilités* (Courcier, Paris).

Lechleiter, J., Girard, S., Peralta, E. and Clapham, D. 1991, *Science* **252**, 123.

Ledrappier, F. 1981, *Commun. Math. Phys.* **81**, 229.

Lee, J. 1987, *Physica D* **24**, 54.

Leith, C. E. and Kraichnan, R. H. 1972, *J. Atmos. Sci.* **29**, 1041.

Lenormand, R. and Zarcone, C. 1985, *Phys. Rev. Lett.* **54**, 2226.

Leschhorn, H. and Tang, L.-H. 1994, *Phys. Rev. E.* **49**, 1238.

Lesieur, M. 1990, *Turbulence in Fluids* (Kluwer, Dordrecht).

Lesieur, M. and Herring, J. 1985, *J. Fluid Mech.* **161**, 77.

Leslie, D. C. 1973, *Developments in the Theory of Turbulence* (Clarendon Press, Oxford).

Lichtenberg, A. J. and Liebermann, M. A. 1983, *Regular and Stochastic Motions* (Springer Verlag, Berlin).

Lilly, D. K. 1973, in *Dynamic Meteorology*, ed. Morel, P. (Reidel Publishing Company, Boston) p. 353.

Livi, R., Politi, A. and Ruffo, S. 1986, *J. Phys. A* **19**, 2033.

Livi, R., Politi, A., Ruffo, S. and Vulpiani, A. 1987, *J. Stat. Phys.* **46**, 147.

Loomis, W. F. 1982, *The Development of Dictyostelium Discoideum* (Academic Press, London).

Lorenz, E. N. 1963, *J. Atm. Sci.* **20**, 130.

Lorenz, E. N. 1969, *Tellus* **21**, 3.

Loreto, V., Paladin, G. and Vulpiani, A. 1996, *Phys. Rev. E* **53**, 2087.

Lumley, J. L. 1962, in *Mécanique de la Turbulence* ed. Favre, A. (CNRS, Paris) p. 17.

Lundbek Hansen, J. and Bohr, T. 1996, *Phys. Rev. Lett.* **77**, 5441.

Lundbek Hansen, J. and Bohr, T. 1998, *Physica D*, to appear.

Lupini, R. and Siboni, S. 1989, *Il Nuovo Cimento B* **103**, 237.

L'vov, V. S., Lebedev, V. V., Paton, M. and Procaccia, I. 1993, *Nonlinearity* **6**, 25.

L'vov, V. S. and Procaccia, I. 1994, *Phys. Rev. Lett.* **72**, 307 (Comment).

L'vov, V. S. and Procaccia, I. 1995, *Phys. Rev. E* **52**, 3840; *ibid* 3852.

Lynov, J. P., Nielsen, A. H., Pécseli, H. L. and Rasmussen, J. J. 1991, *J. Fluid Mech.* **224**, 485.

Malagoli, A., Paladin, G. and Vulpiani, A. 1986, *Phys. Rev. A* **34**, 1550.

Mandelbrot, B. B. 1974, *J. Fluid Mech.* **62**, 331.

Mandelbrot, B. B. 1975, *J. Fluid Mech.* **72**, 401.

Mandelbrot, B. B. 1982, *The Fractal Geometry of Nature* (Freeman, New York).

Manneville, P. 1981, *Phys. Lett. A* **84**, 129.

Manneville, P. 1988, in *Propagation in Systems far from Equilibrium*, ed. Wesfreid, J. E. et al. (Springer Series in Synergetics, Vol. 41) (Springer-Verlag, Berlin) p. 265.

Marcq, P., Chaté, H. and Manneville, P. 1996, *Phys. Rev. Lett.* **77**, 4003.

Markstein, G. H. 1951, *J. Aeronaut. Sci.* **18**, 199.

Marsden, J. E. and McCracken, M. 1975, *The Hopf Bifurcation and its Applications* (MIT Press, Cambridge, Mass.).

Maslov, S. and Paczuski, M. 1994, *Phys. Rev. E* **50**, R643.

Maslov, S., Paczuski, M. and Bak, P. 1994, *Phys. Rev. Lett.* **73**, 2162.

Mathews, J. and Walker, R. 1964 *Mathematical Methods of Physics* (Addison-Wesley, Redwood City).

Matsumoto, K. and Tsuda, I. 1983, *J. Stat. Phys.* **31**, 87.

Maurer, J. and Libchaber, A. 1979, *J. Phys. Lett.* **40**, 419.

Maxey, M. R. and Riley, J. J. 1983, *Phys. Fluids* **26**, 883

Maxwell, J. C. 1890, *The Scientific Papers*, vol. 2 (Cambridge University Press, Cambridge).

McComb, W. D. 1991, *The Physics of Fluid Turbulence* (Oxford Science Publications, Oxford).

Meakin, P., Ramanlal, P., Sander, L. M. and Ball, R. C. 1986, *Phys. Rev. A* **34**, 5091.

Medina, E., Hwa, T., Kardar, M. and Zhang, Y.-C. 1989, *Phys. Rev. A* **39**, 3053.

Melnikov, V. K. 1963, *Trans. Moscow Math. Soc.* **12**, 1.

Meneveau, C. 1991, *J. Fluid Mech.* **232**, 469.

Meneveau, C. and Sreenivasan, K. R. 1987, *Phys. Rev. Lett.* **59**, 1424.

Metheron, G. and De Marsily, G. 1980, *Water Resources Res.* **16**, 901.

Miller, J., Weichman, P. B. and Cross, M. C. 1992, *Phys. Rev. A* **45**, 2328.

Millionshchikov, M. D. 1941a, *Dokl. Akad. Nauk SSSR* **32**, 611.

Millionshchikov, M. D. 1941b, *Izv. Akad. Nauk SSSR Ser. Geogr. Geofiz.* **5**, 433.

Misguich, J. M., Balescu, R., Pécseli, H. L., Mikkelsen, T., Larsen, S. E. and Qiu Xiaoming 1987, *Plasma Phys. Contr. Fusion* **29**, 825.

Moffatt, H. K. 1983, *Rep. Prog. Phys.* **46**, 621.

Molchanov, S. A., Ruzmaikin, A. A. and Sokoloff, D. D. 1984, *Geophys. Astrophys. Fluid Dyn.* **30**, 241.

Monin, A. S. and Yaglom, A. M. 1971, *Statistical Fluid Mechanics* Vol. 1 (MIT Press, Cambridge Mass.).

Monin, A. S. and Yaglom, A. M. 1975, *Statistical Fluid Mechanics* Vol. 2 (MIT Press, Cambridge Mass.).

Morris, S. W., Bodenschatz, E., Canell, D. S. and Ahlers, G. 1993, *Phys. Rev. Lett.* **71**, 2026.

Moser, J. 1962, *Nachr. Akad. Wiss. Göttingen Math-Phys KC 2* **1**, 15.

Müller, S. C., Plesser, T. and Hess, B. 1987, *Biophys. Chem.* **36**, 357.

Musha, T., Kosugi, Y., Matsumoto, G. and Suzuki, M. 1981, *IEEE Trans. Biomed. Eng.* **28**, 616.

Nekhoroshev, N. N. 1977, *Russ. Math. Survey* **32**, 1.

Nepomnyashchy, A. A. 1974, *Fluid Dynamics*, **9**, 354 (Plenum 1975) (Translation of *Izv. Akad. Nauk SSSR, Mekhanika Zhidkosti i Gaza*, No. 3. p. 28, May–June 1974 (in Russian)). See also earlier work in *Proc. of the Perm. State Univ.* **316**, 91 and 105 (1974).

Newell, A. C. 1974, *Lect. Appl. Math* **15**, 157 (Am. Math. Soc.).

Newell, A. and Whitehead, J. 1969, *J. Fluid Mech.* **38** 279.

Newell, A. C., Passot, T. and Lega, J. 1993, *Annual Rev. Fluid Mech.* **25**, 399.

Nicolis, G. and Progogine, I. 1977, *Self-organization in Nonequilibrium Systems* (Wiley and Sons, New York).

Novikov, E. A 1994, *Phys. Rev. E* **50**, R3303.

Novikov, E. A. and Steward, R. W. 1964, *Izv. Akad. Nauk. SSSR. Ser. Geofiz.* **3**, 408.

Novikov, E. A. and Sedov, Yu. B. 1978, *Sov. Phys. JETP* **48**, 440.

Nozakki, K. and Bekki, N. 1983, *Phys. Rev. Lett.* **51**, 2171.

Obukhov, A. M. 1949, *Izv. Akad. Nauk. SSSR. Ser. Geogr. Geofiz.* **13**, 58.

Obukhov, A. M. 1962, *J. Fluid. Mech.* **13**, 77.

Obukhov, A. M. 1971, *Atmos. Oceanic. Phys.* **7**, 41.

Ogura, Y. 1963, *J. Fluid. Mech.* **16**, 38.

Ohkitani, K. and Yamada, M. 1988, *Phys. Rev. Lett.* **60**, 983.

Ohkitani, K. and Yamada, M. 1989, *Prog. Theor. Phys.* **81**, 329.

Olami, Z., Procaccia, I. and Zeitak, R. 1994, *Phys. Rev. E.* **49**, 1232.

Oono, Y. and Kohmoto, M. 1985, *Phys. Rev. Lett.* **55**, 2927.

Oono, Y. and Yeung, C. 1987, *J. Stat. Phys.* **48**, 593.

Orszag, S. A. 1970, *Phys. Fluids* **13**, 2211.

Orszag, S. A. 1977, in *Fluid Dynamics*, ed. Balian R. and Peube J. L. (Gordon and Breach, London) p. 235.

Orszag, S. A., Sulem, P. L. and Goldirsch, I. 1987, *Physica D* **27**, 311.

Osborne, A. R., Kirwan, A. D., Provenzale, A. and Bergamasco, L. 1986, *Physica D* **23**, 75.

Oseledec, V. I. 1968, *Trans. Moscow Math. Soc.* **19**, 197.

Oseledec, V. I. 1993, *Geophys. Astrophys. Fluid Dyn.* **73**, 133.

Ott, E. 1993, *Chaos in Dynamical Systems* (Cambridge University Press, Cambridge).

Ott, E. and Antonsen, T. M., Jr. 1988, *Phys. Rev. Lett.* **61**, 2839.

Ott, E. and Antonsen, T. M., Jr. 1989, *Phys. Rev. A* **39**, 3660.

Ott, E. and Antonsen, T. M., Jr. 1990, *Phys. Rev. Lett.* **64**, 699.

Ottino, J. M. 1989, *The Kinematics of Mixing: Stretching, Chaos and Transport* (Cambridge University Press, Cambridge).

Ottino, J. M. 1990, *Ann. Rev. Fluid Mech.* **22**, 207.

Ouyang, Q. and Swinney, H. 1991, *Chaos* **1**, 411.

Ouyang, Q. and Flesselles, J.-M. 1996, *Nature* **379**, 143.

Packard, N. H., Crutchfield, J. P., Farmer, J. D. and Shaw, R. S. 1980, *Phys. Rev. Lett.* **45**, 712.

Paczuski, M., Maslov, S. and Bak, P. 1994, *Europhys. Lett.* **27**, 97.

Paczuski, M., Bak, P. and Maslov, S. 1995, *Phys. Rev. Lett.* **74**, 4253.

Paladin, G. and Vulpiani, A. 1986, *J. Phys. A* **19**, 1881.

Paladin, G., Peliti, L. and Vulpiani, A. 1986, *J. Phys. A* **19**, L991.

Paladin, G. and Vulpiani, A. 1987a, *Phys. Rep.* **156**, 147.

Paladin, G. and Vulpiani, A. 1987b, *Phys. Rev. A* **35**, 1971.

Paladin, G. and Vulpiani, A. 1994, *J. Phys. A* **27**, 4911.

Paladin, G., Serva, M. and Vulpiani, A. 1995, *Phys. Rev. Lett.* **74**, 66.

Parisi, G. 1979, *Phys. Lett. A* **73**, 203.

Parisi, G. 1990, preprint of the University of Rome II, unpublished.

Parisi, G. 1992, *Euro. Phys. Lett.* **17**, 637.

Parisi, G. and Frisch, U. 1985, in *Turbulence and Predictability of Geophysical Flows and Climatic Dynamics* ed. Ghil, N.,

Benzi, R. and Parisi, G. (North Holland, Amsterdam) p. 84.

Pasmanter, R. A. 1988, *Fluid Dyn. Res.* **3**, 320.

Pasmanter, R. A. 1994, *Phys. Fluids* **6**, 1236.

Pietronero, L. 1990, *Physica A* **163**, 316.

Pietronero, L. and Tosatti, E. (eds) 1986, *Fractal in Physics* (North Holland, Amsterdam).

Pike, R. and Stanley, H. E. 1981, *J. Phys. A* **14**, L169.

Pikovsky, A. 1991, *Phys. Lett. A* **156**, 223.

Pikovsky, A. 1992, *Phys. Lett. A* **168**, 276.

Pikovsky, A. 1993, *Chaos* **3**, 225.

Pisarenko, D., Biferale, L., Courvasier, D., Frisch, U. and Vergassola, M. 1993, *Phys. Fluids A* **65**, 2533.

Pismen, L. and Nepomnyashchy, A. A. 1991, *Phys. Rev. A* **44**, R2243.

Platzman, G. F. 1964, *Tellus* **XVI**, 422.

Poincaré, H. 1899, *Les Methodes Nouvelles de la Mécanique Celeste*, Vol. 1–3 (Gauthier Villars, Paris).

Pomeau, Y. 1985, *C. R. Acad. Sci. Paris* **301**, 1323.

Pomeau, Y. and Manneville, P. 1980, *Commun. Math. Phys.* **74**, 149.

Pomeau, Y., Pumir, A. and Pelce, P. 1984, *J. Stat. Phys.* **37**, 39.

Pomeau, Y., Pumir, A., and Young, W. R. 1989, *Phys. Fluids A* **1**, 462.

Procaccia, I. 1984, *J. Stat. Phys.* **36**, 649.

Procaccia, I., Jensen, M. H., L'vov, V. S., Sneppen, K. and Zeitak, R. 1992, *Phys. Rev. A* **46**, 3220.

Pumir, A., Manneville, P. and Pomeau, Y. 1983, *J. Fluid Mech.* **135**, 27.

Redner, S. 1989, *Physica D* **38**, 287.

Résibois, P. and De Leener, M. 1977, *Classical Kinetic Theory of Fluids* (Wiley and Sons, New York).

Reynolds, O. 1883, *Phil. Trans. R. Soc. A* **175**, 935.

Rica, S. and Tirapegui, E. 1990, *Phys. Rev. Lett.* **64**, 878.

Richardson, L. F. 1922, *Weather Prediction by Numerical Process* (Cambridge University Press, Cambridge).

Richardson, L. F. 1926, *Proc. Roy. Soc. A* **110**, 706.

Richtmyer, R. D. 1981, *Principles of Advanced Mathematical Physics* Vols. I and II (Springer-Verlag, Berlin).

Robert, R. and Sommeria, J. 1991, *J. Fluid Mech.* **229**, 291.

Rolf, J., Bohr, T. and Jensen, M. H. 1998, *Phys. Rev. E*, to appear.

Rose, H. A. and Sulem, P. L. 1978, *J. Physique* **39**, 441.

Rost, M. and Krug, J. 1995a, *Physica D* **88**, 1.

Rost, M. and Krug, J. 1995b, *Phys. Rev. Lett.* **75**, 3894.

Roux, J. C., Simoyi, R. H. and Swinney, H. L. 1983, *Physica D* **8**, 257.

Rubio, M. A., Edwards, C. A., Dougherty, A. and Gollub, J. P. 1989, *Phys. Rev. Lett.* **63**, 1685.

Rubio, M. A., Edwards, C. A., Dougherty, A. and Gollub, J. P. 1990, *Phys. Rev. Lett.* **65**, 1339.

Ruelle, D. 1979, *Phys. Lett. A* **72**, 81.

Ruelle, D. 1982, *Commun. Math. Phys.* **87**, 287.

Ruelle, D. and Takens, F. 1971, *Commun. Math. Phys.* **20**, 167 and **23**, 343.

Ruiz, R. and Nelson, D. R. 1981, *Phys. Rev. A* **23**, 3224.

van Saarloos, W. 1988, *Phys. Rev. A* **37**, 211.

van Saarloos, W. 1989, *Phys. Rev. A* **39**, 6367.

van Saarloos, W. and Hohenberg, P. C. 1992, *Physica D* **56**, 303.

Sakagutchi, H. 1990a, *Prog. Theor. Phys.* **83**, 169.

Sakagutchi, H. 1990b, *Prog. Theor. Phys.* **84**, 792.

Sauer, T., Yorke, J. A. and Casdagli, M. 1991, *J. Stat. Phys.* **65**, 579.

Schögl, F. 1972, *Z. Phys.* **253**, 147.

Schörghofer, N., Kadanoff, L. P. and Lohse, D. 1995, *Physica D* **88**, 40.

Schwarz, M. and Edwards, S. F. 1992, *Europhys. Lett.* **20**, 301.

Segel, L. 1969, *J. Fluid Mech.* **38**, 203.

She, Z. S., Aurell, E. and Frisch, U. 1992, *Commun. Math. Phys.* **148**, 623.

She, Z. S. and Levêque, E. 1994, *Phys. Rev. Lett.* **72**, 336.

Sherratt, J. 1994, *Physica D* **70**, 370.

Shraiman, B. I. 1986, *Phys. Rev. Lett.* **57**, 325.

Shraiman, B. I. 1987, *Phys. Rev A* **36**, 261.

Shraiman, B. I., Pumir, A., van Saarloos, W., Hohenberg, P. C., Chaté, H. and Holen, H. 1992, *Physica D* **57**, 241.

Siggia, E. D. 1977, *Phys. Rev. A* **15**, 1730.

Siggia, E. D. 1978, *Phys. Rev. A* **17**, 1166.

Siggia, E. D. 1981, *J. Fluid Mech.* **107**, 375.

Sinai, Ya. G. 1992, *Commun. Math. Phys.* **148**, 601.

Sirovich, L. 1990, *Empirical Eigenfunctions and Low-Dimensional Systems*, Cent. Fluid Mech. Rep. 90–202, Brown University.

Sivashinsky, G. I. 1977, *Acta Astron.* **4**, 1177.

Sivashinsky, G. I. and Michelson, D. M. 1980, *Prog. Theor. Phys.* **63**, 2112.

Smith, L. 1988, *Phys. Lett. A* **133**, 283.

Smith, M. and Yakhot, V. 1993, *Phys. Rev. Lett.* **71**, 352.

Sneppen, K. 1992, *Phys. Rev. Lett.* **69**, 3539.

Sneppen, K., Krug, J., Jensen, M. H., Jayaprakash, C. and Bohr, T. 1992, *Phys. Rev. A* **46**, R7351.

Sneppen K. and Jensen, M. H. 1993, *Phys. Rev. Lett.* **71**, 101.

Sneppen, K. and Jensen, M. H. 1994, *Phys. Rev. E* **49**, 919.

Solomon, T. H. and Gollub, J. P. 1988a, *Phys. Rev. A* **38**, 6280.

Solomon, T. H. and Gollub, J. P. 1988b, *Phys. Fluids* **31**, 1372.

Sørensen, P. G. and Hynne, F. 1989, *J. Phys. Chem.* **93**, 5467.

Sparrow, E. M., Husar, R. B. and Goldstein, R. J. 1970, *J. Fluid Mech.* **41**, 793.

Sreenivasan, K. R. and Meneveau, C. 1988, *Phys. Rev. A* **38**, 6287.

Stanley, H. E. 1977, *J. Phys. A* **10**, L211.

Stassinpoulos, D. and Alstrøm, P. 1992, *Phys. Rev. A* **45**, 675.

Stauffer, D. and Aharony, A. 1992, *Introduction to Percolation Theory* (Taylor and Francis, London).

Stewartson, K. and Stuart, J. 1971, *J. Fluid Mech.* **48**, 529.

Stiller, O., Popp, S., Aranson, I. and Kramer, L. 1995, *Physica D* **87**, 361.

Sun, Y. S. 1983, *Celest. Mech.* **30**, 7.

Sun, Y. S. 1984, *Celest. Mech.* **33**, 111.

Suzuki, E. and Toh, S. 1995, *Phys. Rev. E*, **51**, 5628.

Takens, F. 1981, *Lect. Notes in Math.* **898**, 366.

Tang, C. and Bak, P. 1988, *Phys. Rev. Lett.* **60**, 2347.

Tang, L.-H., Kertész, J. and Wolf, D. E. 1991, *J. Phys. A* **24**, L1193.

Tang, L.-H. and Leschhorn, H. 1992, *Phys. Rev. A* **45**, R8309.

Tang, L.-H. and Leschhorn, H. 1993, *Phys. Rev. Lett.* **70**, 3832.

Tang, L.-H., Kardar, M. and Dhar, D. 1995, *Phys. Rev. Lett.* **74**, 920.

Taylor, G. I. 1921, *Proc. London Math. Soc.* **20**, 196.

Taylor, G. I. 1938, *Proc. Roy. Soc. A* **164**, 476.

Taylor, G. I. 1953, *Proc. Roy. Soc. A* **219**, 186.

Taylor, G. I. 1954, *Proc. Roy. Soc. A* **225**, 473.

Thual, O. 1993, in *Astrophysical Fluid Dynamics*, ed. Zahn, J. P. and Zinn-Justin, S. (North-Holland, Amsterdam) p. 39.

Tritton, D. J. 1988, *Physical Fluid Dynamics* (Oxford University Press, Oxford).

Tyson, J. and Keener, J. 1988, *Physica D* **32** 327.

Van Atta, C. W. and Park, G. 1972, in *Statistical Models and Turbulence*, ed. Rosenblatt, M. and Van Atta, C. W. (Lect. Notes in Phys. **12**, Springer-Verlag, Berlin).

Vergassola, M., Benzi, R., Biferale, L. and Pisarenko, D. 1993, *J. Phys. A* **26**, 6093.

Vergassola, M., Dubrulle, B., Frisch, U. and Noullez, A. 1994, *Astron. Astrophys.* **289**, 325.

Vicsek, T. 1989, *Fractal Growth Phenomena* (World Scientific, Singapore).

Vicsek, T., Cserzö, M. and Horváth, V. 1990, *Physica A* **167**, 315.

Vicsek, T. and Barabási, A. L. 1991, *J. Phys. A* **24**, L845.

Vincent, A. and Meneguzzi, M. 1991, *J. Fluid Mech.* **225**, 1.

Voronoi, G. 1908, *J. Reine Ang. Math.* **134**, 198.

Waleffe, F. 1992, *Phys. Fluids A* **4**, 350.

Wang, H.-X. and Wang, K.-L. 1994, *Phys. Rev. E* **49**, 5853.

van de Water, W. and Bohr, T. 1993, *Chaos* **3**, 747.

van de Water, W. and Herweijer, J. 1996, *Physica Scripta* **T67**, 136.

Weaire, D. and Rivier, N. 1984, *Contemp. Phys.* **25**, 59.

Weiss, J. B. and Knoblock, E. 1989, *Phys. Rev. A* **40**, 2579.

Weiss, J. B. and McWilliams, J. C. 1991, *Phys. Fluids A* **3**, 835.

Weiss, N. O. 1985, in *Solar system magnetic fields* ed. Priest, E. R. (D. Reidel P.C., Boston) p. 156.

Wiggins, S. 1988, *Global Bifurcations and Chaos: Analytical Methods* (Springer-Verlag, Berlin).

Wilkinson, D. and Willemsen, J. F. 1983, *J. Phys. A* **16**, 3365.

Winfree, A. T. 1972, *Science* **175**, 634.

Winfree, A. T. 1987, *When Time Breaks Down* (Princeton University Press, Princeton).

Winfree, A. T. 1989, *J. Theor. Biol.* **138**, 353.

Winfree, A. T., Winfree, E. M. and Seifert, H. 1985, *Physica D* **17**, 109.

Wolf, A., Swift, J. B., Swinney, H. L. and Vastano, J. A. 1985, *Physica D* **16**, 285.

Wolf, D. E. and Kertész, J. 1987, *Euro. Phys. Lett.* **4**, 651.

Wolfram, C. 1986, *Theory and Applications of Cellular Automata* (World Scientific, Singapore).

Wu, X.-Z., Kadanoff, L. P., Libchaber, A. and Sano, M. 1990, *Phys. Rev. Lett.* **64**, 2140.

Wyld, H. W. 1961, *Ann. Phys.* **14**, 143.

Yakhot, V. 1981, *Phys. Rev. A* **24**, 642.

Yamada, M. and Ohkitani, K. 1987, *J. Phys. Soc. of Japan* **56**, 4210.

Yamada, M. and Ohkitani, K. 1988a, *Progr. Theo. Phys.* **79**, 1265.

Yamada, M. and Ohkitani, K. 1988b, *Phys. Lett. A* **134**, 165.

Yanagita, T. and Kaneko, K. 1993, *Phys. Lett. A* **175**, 415.

Young, L.-S. 1982, *Ergodic Theory and Dyn. Syst.* **2**, 109.

Young, W. R. and Jones, S. 1991, *Phys. Fluids A* **3**, 1087.

Zaikin, A. N. and Zhabotinsky, A. M. 1970, *Nature* **225**, 535.

Zaitsev, S. I. 1992, *Physica A* **189**, 411.

Zaleski, S. 1989, *Physica D* **34**, 427.

Zaleski, S. and Lallemand, P. 1985, *J. Phys. (Paris) Lett.* **46**, L793.

Zaslavsky, G. M. 1994, *Physica D* **76**, 110.

Zaslavsky, G. M., Stevens, D. and Weitzner, H. 1993, *Phys. Rev. E* **48**, 1683.

Zel'dovich, Ya. B. 1957, *Sov. Phys. JEPT* **4**, 460.

Zel'dovich, Ya. B. 1970, *Astron. Astrophys.* **5**, 84.

Zel'dovich, Ya. B., Molchanov, S. A., Ruzmaikin, A. A. and Sokoloff, D. D. 1984, *J. Fluid Mech.* **144**, 1.

Zel'dovich, Ya. B. and Ruzmaikin, A. A. 1986, in *Nonlinear Phenomena in Plasma Physics and Hydrodynamics*, ed. Sagdeev R. Z. (M.I.R. Publ., Moscow) p. 119.

Zhang, J., Zhang, Y.-C., Alstroem, P. and Levinsen, M. T. 1992, *Physica A* **189**, 383.

Zhang, Y.-C. 1990, *J. de Physique* **31**, 2129.

Zhang, Y.-C. 1992, *J. de Physique I* **2**, 2175.

Zheng, X. and Glauser, M. N. 1990, *ASME Comput. Eng.* **2**, 121.

Zumofen, G., Kafter, J. K. and Blumen, A. 1990, *Phys. Rev. A* **42**, 4601.

Index